D0847357

SPECIAL FUNCTIONS

SPECIAL FUNCTIONS

Earl D. Rainville, Ph.D.
PROFESSOR OF MATHEMATICS
IN THE UNIVERSITY OF MICHIGAN

CHELSEA PUBLISHING COMPANY
Bronx, New York

The present work is an unabridged reprint, with various minor changes, of a work first published in 1960. This edition is printed by arrangement with The Macmillan Company. It is published, 1971, at Bronx, New York and is printed on special long-life alkaline paper.

International Standard Book Number: 0-8284-0258-2

Library of Congress Catalog Card Number: 70-172380

Library of Congress Classification Number: QA 351

Dewey Decimal Classification Number: 515'.5

Printed in the United States of America

Preface

I have attempted to write this book in such a way that it can be read not only by professional mathematicians, physicists, engineers, and chemists, but also by well-trained graduate students in those and closely allied fields. Even the research worker in special functions may notice, however, some results or techniques with which he is not already familiar.

Many of the standard concepts and methods which are useful in the detailed study of special functions are included. The reader will also find here other tools, such as the Sheffer classification of polynomial sets and Sister Celine's technique for obtaining recurrence relations, which deserve to become more widely used.

Those who know me will not be surprised to find a certain emphasis on generating functions and their usefulness. That functions of hypergeometric character pervade the bulk of the book is but a reflection of their frequent occurrence in the subject itself.

More than fifty special functions appear in this work, some of them treated extensively, others barely mentioned. There are dozens of topics, numerous methods, and hundreds of special functions which could well have been included but which have been omitted. The temptation to approach the subject on the encyclopedic level intended by the late Harry Bateman was great. To me it seems that such an approach would have resulted in less, rather than more, usefulness; the work would never have reached the stage of publication.

The short bibliography at the end of the book should give the reader ample material with which to start on a more thorough study of the field.

This book is based upon the lectures on Special Functions which I have been giving at The University of Michigan since 1946. The enthusiastic reception accorded the course here has encouraged me to present the material in a form which may facilitate the teaching of similar courses elsewhere.

I wish to acknowledge the assistance given me in the way of both corrections and comments on the manuscript by Professor Phillip E. Bedient of Franklin and Marshall College, Professor Jack R. Britton of The University of Colorado, and Professor Ralph L. Shively of Western Reserve University. I was also aided and encouraged by the late Fred Brafman who was Associate Professor of Mathematics at The University of Oklahoma at the time of his death. Professor Brafman read the first ten chapters critically and discussed some of the later material with me. Several students now taking the course have been helpful in catching errors and pointing out rough spots in the presentation.

Professor Bedient has also aided me by an independent reading of the proof sheets.

EARL D. RAINVILLE

Ann Arbor, Michigan

Contents

Chapter 8: GENERATING FUNCTIONS

Chapter 9: ORTHOGONAL POLYNOMIALS

Chapter 10: LEGENDRE POLYNOMIALS

Chapter 11: HERMITE POLYNOMIALS

Chapter 17: ULTRASPHERICAL AND GEGENBAUER POLYNOMIALS

Chapter 18: OTHER POLYNOMIAL SETS

Chapter 19: ELLIPTIC FUNCTIONS

Chapter 20: THETA FUNCTIONS

Chapter 21: JACOBIAN ELLIPTIC FUNCTIONS

CHAPTER 1

Infinite Products

1. Introduction. Two topics, infinite products and asymptotic series, which are seldom included in standard courses are treated to some extent in short preliminary chapters.

The variables and parameters encountered are to be considered complex except where it is specifically stipulated that they are real.

Exercises are included not only to present the reader with an opportunity to increase his skill but also to make available a few results for which there seemed to be insufficient space in the text.

A short bibliography is included at the end of the book. All references are given in a form such as Fasenmyer [2], meaning item number two under the listing of references to the work of Sister M. Celine Fasenmyer, or Brafman [1;944], meaning page 944 of item number one under the listing of references to the work of Fred Brafman. In general, specific reference to material a century or more old is omitted. The work of the giants in the field, Euler, Gauss, Legendre, etc., is easily located either in standard treatises or in the collected works of the pertinent mathematician.

2. Definition of an infinite product. The elementary theory of infinite products closely parallels that of infinite series. Given a sequence a_k defined for all positive integral k, consider the finite product

$$(1) \qquad P_n = \prod_{k=1}^{n}(1 + a_k) = (1 + a_1)(1 + a_2)\cdots(1 + a_n).$$

If $\lim_{n\to\infty} P_n$ exists and is equal to $P \neq 0$, we say that the infinite product

$$(2) \qquad \prod_{n=1}^{\infty} (1 + a_n)$$

converges to the value P. If at least one of the factors of the product (2) is zero, if only a finite number of the factors of (2) are zero, and if the infinite product with the zero factors deleted converges to a value $P \neq 0$, we say that the infinite product *converges to zero.*

If the infinite product is not convergent, it is said to be *divergent.* If that divergence is due not to the failure of $\lim_{n\to\infty} P_n$ to exist but to the fact that the limit is zero, the product is said to *diverge to zero.* We make no attempt to treat products with an infinity of zero factors.

The peculiar role which zero plays in multiplication is the reason for the slight difference between the definition of convergence of an infinite product and the analogous definition of convergence of an infinite series.

3. A necessary condition for convergence. The general term of a convergent infinite series must approach zero as the index of summation approaches infinity. A similar result will now be obtained for infinite products.

THEOREM 1. *If* $\prod_{n=1}^{\infty} (1 + a_n)$ *converges,*

$$\lim_{n\to\infty} a_n = 0.$$

Proof: If the product converges to $P \neq 0$,

$$1 = \frac{P}{P} = \frac{\lim_{n\to\infty} \prod_{k=1}^{n} (1 + a_k)}{\lim_{n\to\infty} \prod_{k=1}^{n-1} (1 + a_k)} = \lim_{n\to\infty} (1 + a_n).$$

Hence $\lim_{n\to\infty} a_n = 0$, as desired. If the product converges to zero, remove the zero factors and repeat the argument.

4. The associated series of logarithms. Any product without zero factors has associated with it the series of principal values of the logarithms of the separate factors in the following sense.

THEOREM 2. *If no* $a_n = -1$, $\prod_{n=1}^{\infty} (1 + a_n)$ *and* $\sum_{n=1}^{\infty} \operatorname{Log}(1 + a_n)$ *converge or diverge together.*

Proof: Let the partial product and partial sum be indicated as follows:

$$P_n = \prod_{k=1}^{n} (1 + a_k), \qquad S_n = \sum_{k=1}^{n} \operatorname{Log}(1 + a_k).$$

Then* $\exp S_n = P_n$. We know from the theory of complex variables that $\lim_{n\to\infty} \exp S_n = \exp \lim_{n\to\infty} S_n$. Therefore P_n approaches a limit if and only if S_n approaches a limit, and P_n cannot approach zero because the exponential function cannot take on the value zero.

5. Absolute convergence. Assume that the product $\prod_{n=1}^{\infty} (1 + a_n)$ has had its zero factors, if any, deleted. We define *absolute convergence* of the product by utilizing the associated series of logarithms.

The product $\prod_{n=1}^{\infty} (1 + a_n)$, with zero factors deleted, is said to be absolutely convergent if and only if the series $\sum_{n=1}^{\infty} \operatorname{Log}(1 + a_n)$ is absolutely convergent.

THEOREM 3. *The product* $\prod_{n=1}^{\infty} (1 + a_n)$, *with zero factors deleted, is absolutely convergent if and only if* $\sum_{n=1}^{\infty} a_n$ *is absolutely convergent.*

Proof: First throw out any a_n's which are zero; they contribute only unit factors in the product and zero terms in the sum and thus have no bearing on convergence.

We know that if either the series or the product in the theorem converges, $\lim_{n\to\infty} a_n = 0$. Let us then consider n large enough, $n > n_0$, so that $|a_n| < \frac{1}{2}$ for all $n > n_0$. We may now write

$$\frac{\operatorname{Log}(1 + a_n)}{a_n} = \sum_{k=0}^{\infty} \frac{(-1)^k a_n^k}{k + 1}, \tag{1}$$

from which it follows that

$$\left| \frac{\operatorname{Log}(1 + a_n)}{a_n} - 1 \right| \leqq \sum_{k=1}^{\infty} \frac{|a_n|^k}{k+1} < \sum_{k=1}^{\infty} \frac{1}{2^{k+1}} = \frac{1}{2}.$$

*We make frequent use of the common notation $\exp u = e^u$.

Thus we have

$$\frac{1}{2} < \left|\frac{\text{Log}\,(1 + a_n)}{a_n}\right| < \frac{3}{2},$$

from which

$$\left|\frac{\text{Log}\,(1 + a_n)}{a_n}\right| < \frac{3}{2} \quad \text{and} \quad \left|\frac{a_n}{\text{Log}\,(1 + a_n)}\right| < 2.$$

By the comparison test it follows that the absolute convergence of either of $\sum_{n=1}^{\infty} \text{Log}\,(1 + a_n)$ or $\sum_{n=1}^{\infty} a_n$ implies the absolute convergence of the other. We then use the definition of absolute convergence of the product to complete the proof of Theorem 3.

Because of Theorem 2 it follows at once that an infinite product which is absolutely convergent is also convergent.

Example (*a*): Show that the following product converges and find its value:

$$\prod_{n=1}^{\infty} \left[1 + \frac{1}{(n + 1)(n + 3)}\right].$$

The series of positive numbers

$$\sum_{n=1}^{\infty} \frac{1}{(n + 1)(n + 3)}$$

is known to be convergent. It can easily be tested by the polynomial test or by comparison with the series $\sum_{n=1}^{\infty} \frac{1}{n^2}$. Hence our product is absolutely convergent by Theorem 3.

The partial products are often useful in evaluating an infinite product. When the following method is employed, there is no need for the separate testing for convergence made in the preceding paragraph. Consider the partial products

$$P_n = \prod_{k=1}^{n} \left[1 + \frac{1}{(k + 1)(k + 3)}\right] = \prod_{k=1}^{n} \frac{(k + 2)^2}{(k + 1)(k + 3)}$$

$$= \frac{[3 \cdot 4 \cdot 5 \cdots (n + 2)]^2}{[2 \cdot 3 \cdot 4 \cdots (n + 1)][4 \cdot 5 \cdot 6 \cdots (n + 3)]} = \frac{n + 2}{2} \cdot \frac{3}{n + 3}.$$

At once $\lim_{n \to \infty} P_n = \frac{3}{2}$, from which we conclude both that the infinite product converges and that its value is $\frac{3}{2}$.

Example (*b*): Show that if z is not a negative integer,

$$\lim_{n\to\infty} \frac{(n-1)!\ n^z}{(z+1)(z+2)(z+3)\cdots(z+n-1)}$$

exists.

We shall form an infinite product for which the expression

$$P_n = \frac{(n-1)!\ n^z}{(z+1)(z+2)(z+3)\cdots(z+n-1)}$$

is a partial product, prove that the infinite product converges, and thus conclude that $\lim_{n\to\infty} P_n$ exists.

Write

$$P_{n+1} = \frac{n!\ (n+1)^z}{(z+1)(z+2)\cdots(z+n)}$$

$$= \frac{n!}{(z+1)(z+2)\cdots(z+n)} \cdot \frac{2^z}{1^z} \cdot \frac{3^z}{2^z} \cdot \frac{4^z}{3^z} \cdots \frac{(n+1)^z}{n^z}$$

$$= \prod_{k=1}^{n} \left[\frac{k}{z+k} \cdot \frac{(k+1)^z}{k^z}\right] = \prod_{k=1}^{n} \left[\left(1+\frac{z}{k}\right)^{-1}\left(1+\frac{1}{k}\right)^z\right].$$

Consider now the product*

(2) $$\prod_{n=1}^{\infty} \left[\left(1+\frac{z}{n}\right)^{-1}\left(1+\frac{1}{n}\right)^z\right].$$

Since

$$\lim_{n\to\infty} n^2\left[\left(1+\frac{z}{n}\right)^{-1}\left(1+\frac{1}{n}\right)^z - 1\right]$$

$$= \lim_{\beta\to 0} \frac{(1+z\beta)^{-1}(1+\beta)^z - 1}{\beta^2} = \lim_{\beta\to 0} \frac{(1+\beta)^z - 1 - z\beta}{\beta^2}$$

$$= \lim_{\beta\to 0} \frac{z[(1+\beta)^{z-1} - 1]}{2\beta} = \lim_{\beta\to 0} \frac{z(z-1)(1+\beta)^{z-2}}{2} = \tfrac{1}{2}z(z-1),$$

we conclude with the aid of the comparison test and the convergence of $\sum_{n=1}^{\infty} \frac{1}{n^2}$ that the product (2) converges. Therefore $\lim_{n\to\infty} P_n$ exists.

6. Uniform convergence. Let the factors in the product $\prod_{n=1}^{\infty} [1 + a_n(z)]$ be dependent upon a complex variable z. Let R

*We shall find in Chapter 2 that this product has the value $z\Gamma(z)$.

be a closed region in the z-plane. If the product converges in such a way that, given any $\epsilon > 0$, there exists an n_0 independent of z for all z in R such that

$$\left| \prod_{k=1}^{n_0+p} [1 + a_k(z)] - \prod_{k=1}^{n_0} [1 + a_k(z)] \right| < \epsilon$$

for all positive integral p, we say that the product $\prod_{n=1}^{\infty} [1 + a_n(z)]$ is *uniformly convergent* in the region R.

Again the convergence properties parallel those of infinite series. We need a Weierstrass M-test.

THEOREM 4. *If there exist positive constants M_n such that $\sum_{n=1}^{\infty} M_n$ is convergent and $|a_n(z)| < M_n$ for all z in the closed region R, the product $\prod_{n=1}^{\infty} [1 + a_n(z)]$ is uniformly convergent in R.*

Proof: Since $\sum_{n=1}^{\infty} M_n$ is convergent and $M_n > 0$, $\prod_{n=1}^{\infty} (1 + M_n)$ is convergent and $\lim_{n \to \infty} \prod_{k=1}^{n} (1 + M_k)$ exists. Therefore, given any $\epsilon > 0$, there exists an n_0 such that

$$\prod_{k=1}^{n_0+p} (1 + M_k) - \prod_{k=1}^{n_0} (1 + M_k) < \epsilon$$

for all positive integers p. For all z in R, each $a_k(z)$ is such that $|a_k(z)| < M_k$. Hence

$$\begin{aligned} &\left| \prod_{k=1}^{n_0+p} [1 + a_k(z)] - \prod_{k=1}^{n_0} [1 + a_k(z)] \right| \\ &\qquad = \left| \prod_{k=1}^{n_0} [1 + a_k(z)] \right| \cdot \left| \prod_{k=n_0+1}^{n_0+p} [1 + a_k(z)] - 1 \right| \\ &\qquad < \prod_{k=1}^{n_0} (1 + M_k) \left[\prod_{k=n_0+1}^{n_0+p} (1 + M_k) - 1 \right] \\ &\qquad < \prod_{k=1}^{n_0+p} (1 + M_k) - \prod_{k=1}^{n_0} (1 + M_k) < \epsilon, \end{aligned}$$

which was to be proved.

EXERCISES

1. Show that the following product converges, and find its value:

$$\prod_{n=1}^{\infty}\left[1+\frac{6}{(n+1)(2n+9)}\right]. \qquad \textit{Ans. } \frac{21}{8}.$$

2. Show that $\prod_{n=2}^{\infty}\left(1-\frac{1}{n^2}\right)=\frac{1}{2}$.

3. Show that $\prod_{n=2}^{\infty}\left(1-\frac{1}{n}\right)$ diverges to zero.

4. Investigate the product $\prod_{n=0}^{\infty}(1+z^{2^n})$ in $|z|<1$.

Ans. Abs. conv. to $\frac{1}{1-z}$.

5. Show that $\prod_{n=1}^{\infty}\exp\left(\frac{1}{n}\right)$ diverges.

6. Show that $\prod_{n=1}^{\infty}\exp\left(-\frac{1}{n}\right)$ diverges to zero.

7. Test $\prod_{n=1}^{\infty}\left(1-\frac{z^2}{n^2}\right)$. *Ans.* Abs. conv. for all finite z.

8. Show that $\prod_{n=1}^{\infty}\left[1+\frac{(-1)^{n+1}}{n}\right]$ converges to unity.

9. Test for convergence: $\prod_{n=2}^{\infty}\left(1-\frac{1}{n^p}\right)$ for real $p\neq 0$.

Ans. Conv. for $p>1$· div. for $p\leqq 1$.

10. Show that $\prod_{n=1}^{\infty}\frac{\sin(z/n)}{z/n}$ is absolutely convergent for all finite z with the usual convention at $z=0$. *Hint:* Show first that

$$\lim_{n\to\infty} n^2\left[\frac{\sin(z/n)}{z/n}-1\right]=-\frac{z^2}{6}.$$

11. Show that if c is not a negative integer,

$$\prod_{n=1}^{\infty}\left[\left(1-\frac{z}{c+n}\right)\exp\left(\frac{z}{n}\right)\right]$$

is absolutely convergent for all finite z. *Hint*: Show first that

$$\lim_{n\to\infty} n^2\left[\left(1-\frac{z}{c+n}\right)\exp\left(\frac{z}{n}\right)-1\right]=z\left(c-\frac{1}{2}z\right).$$

CHAPTER 2

The Gamma and Beta Functions

7. The Euler or Mascheroni constant γ. At times we need to use the constant γ, defined by

$$\gamma = \lim_{n\to\infty} (H_n - \operatorname{Log} n), \tag{1}$$

in which, as usual,

$$H_n = \sum_{k=1}^{n} \frac{1}{k}. \tag{2}$$

We shall prove that γ exists and that $0 \leqq \gamma < 1$. Actually $\gamma = 0.5772$, approximately.

Let $A_n = H_n - \operatorname{Log} n$. Then the A_n form a decreasing sequence because

$$\begin{aligned} A_{n+1} - A_n &= H_{n+1} - H_n - \operatorname{Log}(n+1) + \operatorname{Log} n \\ &= \frac{1}{n+1} + \operatorname{Log}\frac{n}{n+1} = \frac{1}{n+1} + \operatorname{Log}\left(1 - \frac{1}{n+1}\right) \\ &= -\sum_{k=1}^{\infty} \frac{1}{(k+1)(n+1)^{k+1}} < 0. \end{aligned}$$

Furthermore, since $1/t$ decreases steadily as t increases,

$$\frac{1}{k} < \int_{k-1}^{k} \frac{dt}{t} < \frac{1}{k-1}, \qquad k \geqq 2. \tag{3}$$

We sum the inequalities (3) from $k = 2$ to $k = n$ and thus obtain

$$H_n - 1 < \int_1^2 \frac{dt}{t} + \int_2^3 \frac{dt}{t} + \cdots + \int_{n-1}^n \frac{dt}{t} < H_{n-1},$$

or

$$H_n - 1 < \operatorname{Log} n < H_{n-1},$$

from which it follows that

$$-1 < -H_n + \operatorname{Log} n < -\frac{1}{n},$$

or

$$1 > A_n > \frac{1}{n}.$$

Thus we see that the A_n decrease steadily, are all positive, and are less than unity. It follows that γ exists and is non-negative and less than unity.

8. The Gamma function. We follow Weierstrass in defining the function $\Gamma(z)$ by

$$(1) \qquad \frac{1}{\Gamma(z)} = ze^{\gamma z} \prod_{n=1}^{\infty} \left[\left(1 + \frac{z}{n}\right)\exp\left(-\frac{z}{n}\right)\right],$$

in which γ is the Euler constant of Section 7. The product in (1) is absolutely convergent for all finite z as was seen in Ex. 11, page 7, the special case $c = 0$ and z replaced by $(-z)$. That the product is also uniformly convergent in any closed region in the z-plane is easily shown by employing the associated series of logarithms.

We shall see in Section 15 that the function $\Gamma(z)$ defined by (1) is identical with that defined by Euler's integral; that is,

$$\Gamma(z) = \int_0^\infty e^{-t}t^{z-1}\,dt, \qquad \operatorname{Re}(z) > 0.$$

The right member of (1) is analytic for all finite z. Its only zeros are simple ones at $z = 0$ and at each negative integer. We may therefore conclude that

(a) $\Gamma(z)$ is analytic except at $z =$ nonpositive integers and $z = \infty$;

(b) $\Gamma(z)$ has a simple pole at $z =$ each nonpositive integer, $z = 0, -1, -2, -3, \cdots$;

(c) $\Gamma(z)$ has an essential singularity at $z = \infty$, a point of condensation of poles;

(d) $\Gamma(z)$ is never zero [because $1/\Gamma(z)$ has no poles].

9. A series for $\Gamma'(z)/\Gamma(z)$. By taking logarithms of each member of equation (1) of Section 8, we obtain

$$\log \Gamma(z) = -\text{Log}\, z - \gamma z - \sum_{n=1}^{\infty}\left[\text{Log}\left(1 + \frac{z}{n}\right) - \frac{z}{n}\right].$$

Term-by-term differentiation of the members of the foregoing equation yields

$$\frac{\Gamma'(z)}{\Gamma(z)} = -\frac{1}{z} - \gamma - \sum_{n=1}^{\infty}\left(\frac{1}{z+n} - \frac{1}{n}\right),$$

or

$$(1)\qquad \frac{\Gamma'(z)}{\Gamma(z)} = -\gamma - \frac{1}{z} + \sum_{n=1}^{\infty}\frac{z}{n(z+n)},$$

the series on the right being absolutely and uniformly convergent in any closed region excluding the singular points of $\Gamma(z)$, a result easily deduced by using the Weierstrass M-test and the convergence of $\sum_{n=1}^{\infty}\frac{1}{n^2}$.

10. Evaluation of $\Gamma(1)$ and $\Gamma'(1)$. In the Weierstrass definition of $\Gamma(z)$ put $z = 1$ to get

$$\begin{aligned}\frac{1}{\Gamma(1)} &= e^{\gamma}\prod_{n=1}^{\infty}\left[\left(1+\frac{1}{n}\right)\exp\left(-\frac{1}{n}\right)\right]\\ &= e^{\gamma}\lim_{n\to\infty}\prod_{k=1}^{n}\left[\frac{k+1}{k}\exp\left(-\frac{1}{k}\right)\right]\\ &= e^{\gamma}\lim_{n\to\infty}(n+1)\exp(-H_n)\\ &= e^{\gamma}\lim_{n\to\infty}(n+1)\exp(-\gamma - \text{Log}\, n - \epsilon_n),\end{aligned}$$

in which $\epsilon_n \to 0$ as $n \to \infty$. It follows that

$$\frac{1}{\Gamma(1)} = e^{\gamma}\lim_{n\to\infty}\frac{n+1}{n}e^{-\gamma} = 1,$$

so that $\Gamma(1) = 1$.

We know from the series for $\Gamma'(z)/\Gamma(z)$ obtained in Section 9 that

$$\frac{\Gamma'(1)}{\Gamma(1)} = -\gamma - 1 + \sum_{n=1}^{\infty}\frac{1}{n(n+1)},$$

so that

$$\Gamma'(1) = -\gamma - 1 + \sum_{n=1}^{\infty}\left(\frac{1}{n} - \frac{1}{n+1}\right)$$
$$= -\gamma - 1 + \lim_{n\to\infty}\left(1 - \frac{1}{n+1}\right),$$

since the series involved telescopes. Thus we find that $\Gamma'(1) = -\gamma$.

11. The Euler product for $\Gamma(z)$. From the Weierstrass product definition of $\Gamma(z)$ we obtain

$$z\Gamma(z) = \frac{\exp(-\gamma z)}{\prod_{n=1}^{\infty}\left[\left(1 + \frac{z}{n}\right)\exp\left(-\frac{z}{n}\right)\right]},$$

so that

$$\text{(1)} \qquad z\Gamma(z) = \exp(-\gamma z) \lim_{n\to\infty} \prod_{k=1}^{n}\left[\left(1 + \frac{z}{k}\right)^{-1}\exp\left(\frac{z}{k}\right)\right].$$

But

$$\gamma = \lim_{n\to\infty}(H_n - \operatorname{Log} n) = \lim_{n\to\infty}[H_n - \operatorname{Log}(n+1)]$$
$$= \lim_{n\to\infty}\left[H_n - \sum_{k=1}^{n}\operatorname{Log}\frac{k+1}{k}\right].$$

Hence

$$\exp(-\gamma z) = \lim_{n\to\infty}\exp\left[-zH_n + z\sum_{k=1}^{n}\operatorname{Log}\frac{k+1}{k}\right]$$
$$= \lim_{n\to\infty}\prod_{k=1}^{n}\left[\left(\frac{k+1}{k}\right)^{z}\exp\left(-\frac{z}{k}\right)\right].$$

Therefore (1) can be written

$$z\Gamma(z) = \lim_{n\to\infty}\prod_{k=1}^{n}\left[\left(1 + \frac{1}{k}\right)^{z}\exp\left(-\frac{z}{k}\right)\left(1 + \frac{z}{k}\right)^{-1}\exp\left(\frac{z}{k}\right)\right],$$

from which it follows that

$$\text{(2)} \qquad \Gamma(z) = \frac{1}{z}\prod_{n=1}^{\infty}\left[\left(1 + \frac{1}{n}\right)^{z}\left(1 + \frac{z}{n}\right)^{-1}\right],$$

which is Euler's product for $\Gamma(z)$. Note that for real $x > 0$, $\Gamma(x) > 0$.

Refer now to Example (*b*), page 5, to conclude that

$$\text{(3)} \qquad \Gamma(z) = \lim_{n\to\infty}\frac{(n-1)!\, n^{z}}{z(z+1)(z+2)\cdots(z+n-1)}.$$

It will be of value to us later to note that, since

$$\lim_{n\to\infty} \frac{(n+1)^z}{n^z} = 1,$$

we can equally well write the result (3) in the form

$$(4) \qquad \Gamma(z) = \lim_{n\to\infty} \frac{n!\, n^z}{z(z+1)(z+2)\cdots(z+n)}.$$

12. The difference equation $\Gamma(z+1) = z\Gamma(z)$. From Euler's product for $\Gamma(z)$ we obtain

$$\frac{\Gamma(z+1)}{\Gamma(z)} = \frac{z}{z+1} \frac{\prod_{n=1}^{\infty}\left[\left(1+\frac{1}{n}\right)^{z+1}\left(1+\frac{z+1}{n}\right)^{-1}\right]}{\prod_{n=1}^{\infty}\left[\left(1+\frac{1}{n}\right)^{z}\left(1+\frac{z}{n}\right)^{-1}\right]}$$

$$= \frac{z}{z+1}\prod_{n=1}^{\infty}\left[\left(1+\frac{1}{n}\right)\left(1+\frac{z}{n}\right)\left(1+\frac{z+1}{n}\right)^{-1}\right]$$

$$= \frac{z}{z+1}\lim_{n\to\infty}\prod_{k=1}^{n}\left(\frac{k+1}{k}\cdot\frac{k+z}{k+z+1}\right)$$

$$= \frac{z}{z+1}\lim_{n\to\infty}\frac{n+1}{1}\cdot\frac{1+z}{n+z+1} = z.$$

Therefore

$$(1) \qquad \Gamma(z+1) = z\Gamma(z)$$

for all finite z except for the poles of $\Gamma(z)$.

If $z = m$, a positive integer, iterated use of the equation (1) yields $\Gamma(m+1) = m!$. Since $\Gamma(1) = 1$, this is another of the many reasons we define $0! = 1$.

13. The order symbols o and O. Let R be a region in the complex z-plane. If and only if

$$\lim_{z\to c \text{ in } R} \frac{f(z)}{g(z)} = 0,$$

we write

$$f(z) = o[g(z)], \qquad \text{as } z \to c \text{ in } R.$$

If and only if $\left|\frac{f(z)}{g(z)}\right|$ is bounded as $z \to c$ in R, we write

$$f(z) = O[g(z)], \qquad \text{as } z \to c \text{ in } R.$$

It is common practice to omit the qualifying expressions such as "$z \to c$ in R" whenever the surrounding text is deemed to make such qualification unnecessary to a trained reader. The point $z = c$ may on occasion be the point at infinity. Also, the symbols o and O are sometimes used when the variable z is real, the approach is along the real axis, and even when z takes on only integral values.

EXAMPLE (*a*): Since $\lim\limits_{z\to 0} \dfrac{\sin^2 z}{z} = 0$, we may write

$$\sin^2 z = o(z), \qquad \text{as } z \to 0,$$

noting that in this instance the manner of approach is immaterial.

EXAMPLE (*b*): For real x, $|\cos x| \leqq 1$, from which it is easy to conclude that

$$\cos x - 4x = O(x), \qquad \text{as } x \to \infty,\ x \text{ real}.$$

EXAMPLE (*c*): In Chapter 3 we shall show that if

$$s_n(x) = \sum_{k=0}^{n} k!\, x^k,$$

$$\left| \int_0^\infty \frac{e^{-t}\, dt}{1 - xt} - s_n(x) \right| \leqq (n+1)!\, |x|^{n+1}, \qquad \text{for } \operatorname{Re}(x) \leqq 0.$$

From the preceding inequality we may conclude that, for fixed n,

$$\int_0^\infty \frac{e^{-t}\, dt}{1 - xt} - s_n(x) = o(x^n), \qquad \text{as } x \to 0 \text{ in } \operatorname{Re}(x) \leqq 0.$$

14. Evaluation of certain infinite products. The Weierstrass infinite product for $\Gamma(z)$ yields a simple evaluation of all infinite products whose factors are rational functions of the index n. The most general such product must take the form

$$(1) \qquad P = \prod_{n=1}^{\infty} \frac{(n + a_1)(n + a_2) \cdots (n + a_s)}{(n + b_1)(n + b_2) \cdots (n + b_s)}$$

$$= \prod_{n=1}^{\infty} \frac{\prod\limits_{k=1}^{s} \left(1 + \dfrac{a_k}{n}\right)}{\prod\limits_{k=1}^{s} \left(1 + \dfrac{b_k}{n}\right)}$$

because convergence requires that the nth factor approach unity as $n \to \infty$, which in turn forces the numerator and denominator polynomials to be of the same degree and to have equal leading coefficients. Now the nth factor in the right member of (1) may be put in the form

$$1 + \frac{1}{n}\left(\sum_{k=1}^{s} a_k - \sum_{k=1}^{s} b_k\right) + O\left(\frac{1}{n^2}\right),$$

so that we must also insist, to obtain convergence, that

$$\sum_{k=1}^{s} a_k = \sum_{k=1}^{s} b_k. \tag{2}$$

If (2) is not satisfied, the product in (1) diverges; we get absolute convergence or no convergence.

We now have an absolutely convergent product (1) in which the a's and b's satisfy the condition (2).

Since

$$\exp\left(\frac{1}{n}\sum_{k=1}^{s} a_k\right) = \exp\left(\frac{1}{n}\sum_{k=1}^{s} b_k\right),$$

we may, without changing the value of the product (1), insert the appropriate exponential factors to write

$$P = \prod_{n=1}^{\infty} \frac{\prod_{k=1}^{s}\left[\left(1 + \frac{a_k}{n}\right)\exp\left(-\frac{a_k}{n}\right)\right]}{\prod_{k=1}^{s}\left[\left(1 + \frac{b_k}{n}\right)\exp\left(-\frac{b_k}{n}\right)\right]}. \tag{3}$$

The Weierstrass product, page 9, for $1/\Gamma(z)$ yields

$$\prod_{n=1}^{\infty}\left[\left(1 + \frac{z}{n}\right)\exp\left(-\frac{z}{n}\right)\right] = \frac{1}{z\exp(\gamma z)\Gamma(z)} = \frac{1}{\Gamma(z+1)\exp(\gamma z)}.$$

Thus we obtain from (3) the result

$$\begin{aligned} P &= \prod_{k=1}^{s} \frac{\Gamma(1+b_k)\exp(\gamma b_k)}{\Gamma(1+a_k)\exp(\gamma a_k)} \\ &= \exp\left[\gamma\left(\sum_{k=1}^{s} b_k - \sum_{k=1}^{s} a_k\right)\right]\prod_{k=1}^{s}\frac{\Gamma(1+b_k)}{\Gamma(1+a_k)} \\ &= \prod_{k=1}^{s}\frac{\Gamma(1+b_k)}{\Gamma(1+a_k)}. \end{aligned}$$

Theorem 5. *If* $\sum_{k=1}^{s} a_k = \sum_{k=1}^{s} b_k$, *and if no* a_k *or* b_k *is a negative integer,*

$$\prod_{n=1}^{\infty} \frac{(n+a_1)(n+a_2)\cdots(n+a_s)}{(n+b_1)(n+b_2)\cdots(n+b_s)} = \frac{\Gamma(1+b_1)\Gamma(1+b_2)\cdots\Gamma(1+b_s)}{\Gamma(1+a_1)\Gamma(1+a_2)\cdots\Gamma(1+a_s)}.$$

If one or more of the a_k is a negative integer, the product on the left is zero, which agrees with the existence of one or more poles of the denominator factors on the right.

Example: Evaluate

$$\prod_{n=1}^{\infty} \frac{(c-a+n-1)(c-b+n-1)}{(c+n-1)(c-a-b+n-1)}.$$

Since $(c-a-1)+(c-b-1)=(c-1)+(c-a-b-1)$, we may employ Theorem 5 if no one of the quantities c, $c-a$, $c-b$, $c-a-b$ is either zero or a negative integer. With those restrictions we obtain

$$(4) \qquad \prod_{n=1}^{\infty} \frac{(c-a+n-1)(c-b+n-1)}{(c+n-1)(c-a-b+n-1)} = \frac{\Gamma(c)\Gamma(c-a-b)}{\Gamma(c-a)\Gamma(c-b)}.$$

15. Euler's integral for $\Gamma(z)$. Elementary treatments of the Gamma function are usually based on an integral definition. Theorem 6 connects the function $\Gamma(z)$ defined by the Weierstrass product with that defined by Euler's integral.

Theorem 6. *If* $\text{Re}(z) > 0$,

$$(1) \qquad \Gamma(z) = \int_0^\infty e^{-t}t^{z-1}\,dt.$$

We shall establish four lemmas intended to break the proof of Theorem 6 into simple parts.

Lemma 1. If $0 \leqq \alpha < 1$, $1+\alpha \leqq \exp(\alpha) \leqq (1-\alpha)^{-1}$.

Proof: Compare the three series

$$1+\alpha=1+\alpha, \qquad \exp(\alpha)=1+\alpha+\sum_{n=2}^{\infty}\frac{\alpha^n}{n!}, \qquad (1-\alpha)^{-1}=1+\alpha+\sum_{n=2}^{\infty}\alpha^n.$$

Lemma 2. If $0 \leqq \alpha < 1$, $(1-\alpha)^n \geqq 1-n\alpha$, *for* n *a positive integer.*

Proof: For $n=1$, $1-\alpha = 1-1\cdot\alpha$, as desired. Next assume that

$$(1-\alpha)^k \geqq 1-k\alpha,$$

and multiply each member by $(1-\alpha)$ to obtain

$$(1-\alpha)^{k+1} \geqq (1-\alpha)(1-k\alpha) = 1-(k+1)\alpha + k\alpha^2,$$

so that

$$(1 - \alpha)^{k+1} \geqq 1 - (k + 1)\alpha.$$

Lemma 2 now follows by induction.

Lemma 3. If $0 \leqq t < n$, *n a positive integer,*

$$0 \leqq e^{-t} - \left(1 - \frac{t}{n}\right)^n \leqq \frac{t^2e^{-t}}{n}.$$

Proof: Use $\alpha = t/n$ in Lemma 1 to get

$$1 + \frac{t}{n} \leqq \exp\left(\frac{t}{n}\right) \leqq \left(1 - \frac{t}{n}\right)^{-1}$$

from which

$$\left(1 + \frac{t}{n}\right)^n \leqq e^t \leqq \left(1 - \frac{t}{n}\right)^{-n} \tag{2}$$

or

$$\left(1 + \frac{t}{n}\right)^{-n} \geqq e^{-t} \geqq \left(1 - \frac{t}{n}\right)^n,$$

so that

$$e^{-t} - \left(1 - \frac{t}{n}\right)^n \geqq 0. \tag{3}$$

But also

$$e^{-t} - \left(1 - \frac{t}{n}\right)^n = e^{-t}\left[1 - e^t\left(1 - \frac{t}{n}\right)^n\right]$$

and, by (2), $e^t \geqq \left(1 + \frac{t}{n}\right)^n$. Hence

$$e^{-t} - \left(1 - \frac{t}{n}\right)^n \leqq e^{-t}\left[1 - \left(1 - \frac{t^2}{n^2}\right)^n\right]. \tag{4}$$

Now Lemma 2 with $\alpha = t^2/n^2$ yields

$$\left(1 - \frac{t^2}{n^2}\right)^n \geqq 1 - \frac{t^2}{n}$$

which may be used in (4) to obtain

$$e^{-t} - \left(1 - \frac{t}{n}\right)^n \leqq e^{-t}\left[1 - 1 + \frac{t^2}{n}\right] = \frac{t^2e^{-t}}{n}. \tag{5}$$

The inequalities (3) and (5) constitute the result stated in Lemma 3.

Lemma 4. If n is integral and $\text{Re}(z) > 0$,

$$\Gamma(z) = \lim_{n\to\infty} \int_0^n \left(1 - \frac{t}{n}\right)^n t^{z-1}\, dt. \tag{6}$$

Proof: In the integral on the right in (6) put $t = n\beta$ and thus obtain

$$\int_0^n \left(1 - \frac{t}{n}\right)^n t^{z-1}\, dt = n^z \int_0^1 (1 - \beta)^n \beta^{z-1}\, d\beta. \tag{7}$$

An integration by parts gives us the reduction formula

$$\int_0^1 (1 - \beta)^n \beta^{z-1}\, d\beta = \frac{n}{z} \int_0^1 (1 - \beta)^{n-1} \beta^z\, d\beta,$$

iteration of which yields

$$\int_0^1 (1 - \beta)^n \beta^{z-1}\, d\beta = \frac{n(n-1)(n-2)\cdots 1}{z(z+1)(z+2)\cdots(z+n-1)} \int_0^1 \beta^{z+n-1}\, d\beta$$

$$= \frac{n!}{z(z+1)(z+2)\cdots(z+n)}.$$

Now (7) becomes

$$\int_0^n \left(1 - \frac{t}{n}\right)^n t^{z-1}\, dt = \frac{n!n^z}{z(z+1)(z+2)\cdots(z+n)}$$

so that

$$\lim_{n\to\infty} \int_0^n \left(1 - \frac{t}{n}\right)^n t^{z-1}\, dt = \lim_{n\to\infty} \frac{n!n^z}{z(z+1)\cdots(z+n)} = \Gamma(z)$$

by equation (4), page 12.

We are now in a position to prove Theorem 6, which states that

$$\Gamma(z) = \int_0^\infty e^{-t} t^{z-1}\, dt, \qquad \text{Re}(z) > 0. \tag{8}$$

The integral on the right in (8) converges for $\text{Re}(z) > 0$. With the aid of Lemma 4, write

$$\int_0^\infty e^{-t} t^{z-1}\, dt - \Gamma(z) = \lim_{n\to\infty} \left[\int_0^\infty e^{-t} t^{z-1}\, dt - \int_0^n \left(1 - \frac{t}{n}\right)^n t^{z-1}\, dt\right]$$

$$= \lim_{n\to\infty} \left[\int_0^n \left\{e^{-t} - \left(1 - \frac{t}{n}\right)^n\right\} t^{z-1}\, dt + \int_n^\infty e^{-t} t^{z-1}\, dt\right].$$

From the convergence of the integral on the right in (8) it follows that

$$\lim_{n\to\infty} \int_n^\infty e^{-t}t^{z-1}\,dt = 0.$$

Hence

$$(9)\quad \int_0^\infty e^{-t}t^{z-1}\,dt - \Gamma(z) = \lim_{n\to\infty} \int_0^n \left[e^{-t} - \left(1 - \frac{t}{n}\right)^n\right] t^{z-1}\,dt.$$

But, by Lemma 3 and the fact that $|t^z| = t^{\mathrm{Re}(z)}$,

$$\left|\int_0^n \left[e^{-t} - \left(1 - \frac{t}{n}\right)^n\right] t^{z-1}\,dt\right| \leqq \int_0^n \frac{t^2e^{-t}}{n} \cdot t^{\mathrm{Re}(z)-1}\,dt$$

$$\leqq \frac{1}{n}\int_0^n e^{-t}t^{\mathrm{Re}(z)+1}\,dt.$$

Now $\int_0^\infty e^{-t}t^{\mathrm{Re}(z)+1}\,dt$ converges, so $\int_0^n e^{-t}t^{\mathrm{Re}(z)+1}\,dt$ is bounded. Therefore

$$\lim_{n\to\infty} \int_0^n \left[e^{-t} - \left(1 - \frac{t}{n}\right)^n\right] t^{z-1}\,dt = 0,$$

and we may conclude from equation (9) that (8) is valid.

16. The Beta function. We define the Beta function $B(p, q)$ by

$$(1)\quad B(p,q) = \int_0^1 t^{p-1}(1-t)^{q-1}\,dt, \qquad \mathrm{Re}(p) > 0,\ \mathrm{Re}(q) > 0.$$

Another useful form for this function can be obtained by putting $t = \sin^2 \varphi$, thus arriving at

$$(2)\quad B(p,q) = 2\int_0^{\frac{1}{2}\pi} \sin^{2p-1}\varphi \cos^{2q-1}\varphi\,d\varphi, \qquad \mathrm{Re}(p) > 0,\ \mathrm{Re}(q) > 0.$$

The Beta function is intimately related to the Gamma function. Consider the product

$$(3)\quad \Gamma(p)\Gamma(q) = \int_0^\infty e^{-t}t^{p-1}\,dt \cdot \int_0^\infty e^{-v}v^{q-1}\,dv.$$

In (3) use $t = x^2$ and $v = y^2$ to obtain

$$\Gamma(p)\Gamma(q) = 4\int_0^\infty \exp(-x^2)x^{2p-1}\,dx \cdot \int_0^\infty \exp(-y^2)y^{2q-1}\,dy,$$

$$\Gamma(p)\Gamma(q) = 4\int_0^\infty\int_0^\infty \exp(-x^2 - y^2)x^{2p-1}y^{2q-1}\,dx\,dy.$$

Next turn to polar coordinates for the iterated integration over the first quadrant in the xy-plane. Using $x = r\cos\theta$, $y = r\sin\theta$, we may write

$$\Gamma(p)\Gamma(q) = 4\int_0^\infty\int_0^{\frac{1}{2}\pi} \exp(-r^2)r^{2p+2q-2}\cos^{2p-1}\theta\sin^{2q-1}\theta\, r\,d\theta\,dr$$

$$= 2\int_0^\infty \exp(-r^2)r^{2p+2q-1}dr \cdot 2\int_0^{\frac{1}{2}\pi} \cos^{2p-1}\theta\sin^{2q-1}\theta\,d\theta.$$

Now put $r = \sqrt{t}$ and $\theta = \frac{1}{2}\pi - \varphi$ to obtain

$$\Gamma(p)\Gamma(q) = \int_0^\infty e^{-t}t^{p+q-1}\,dt \cdot 2\int_0^{\frac{1}{2}\pi} \sin^{2p-1}\varphi\cos^{2q-1}\varphi\,d\varphi,$$

from which it follows that

$$\Gamma(p)\Gamma(q) = \Gamma(p+q)B(p,q).$$

Theorem 7. *If* $\mathrm{Re}(p) > 0$ *and* $\mathrm{Re}(q) > 0$,

$$B(p,q) = \frac{\Gamma(p)\Gamma(q)}{\Gamma(p+q)}. \tag{4}$$

By (4), $B(p,q) = B(q,p)$, a result just as easily obtained directly from (1) or (2).

Equations (2) and (4) yield a generalization of Wallis' formula of elementary calculus. In (2) put $2p - 1 = m$, $2q - 1 = n$, and use (4) to write

$$\int_0^{\frac{1}{2}\pi} \sin^m\varphi\cos^n\varphi\,d\varphi = \frac{\Gamma\left(\frac{m+1}{2}\right)\Gamma\left(\frac{n+1}{2}\right)}{2\Gamma\left(\frac{m+n+2}{2}\right)}, \tag{5}$$

valid for $\mathrm{Re}(m) > -1$, $\mathrm{Re}(n) > -1$.

17. The value of $\Gamma(z)\Gamma(1-z)$. The important relation (4) of Section 16 suggests that the product of two Gamma functions whose arguments have the sum unity may possess some pleasant property, since if $p + q = 1$, $\Gamma(p+q) = \Gamma(1) = 1$.

If z is such that $0 < \mathrm{Re}(z) < 1$, both z and $(1-z)$ have real part positive, and we may use (4) of Section 16 to write

$$\Gamma(z)\Gamma(1-z) = B(z, 1-z) = \int_0^1 t^{z-1}(1-t)^{-z}\,dt$$

$$= \int_0^1 \left(\frac{t}{1-t}\right)^z \frac{dt}{t}.$$

Now put $t/(1-t) = y$ to arrive at

$$(1) \qquad \Gamma(z)\Gamma(1-z) = \int_0^\infty \frac{y^{z-1}\,dy}{1+y}, \qquad 0 < \text{Re}(z) < 1.$$

The integral on the right in (1) can be evaluated with the aid of contour integration in an α-plane where $\text{Re}(\alpha) = y$. The contour

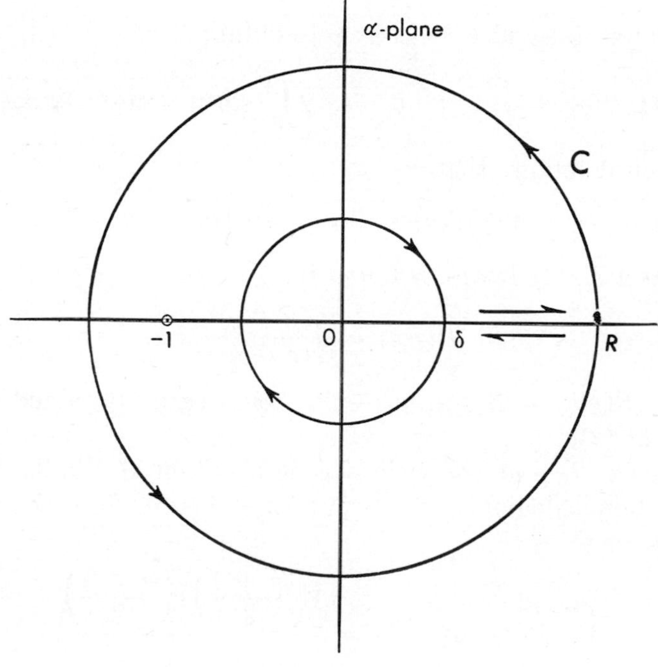

Figure 1

C in Figure 1 encircles a single simple pole $\alpha = -1$ of the integrand in

$$\int_C \frac{\alpha^{z-1}\,d\alpha}{1+\alpha}$$

so that the residue theory at once yields

$$(2) \qquad \int_C \frac{\alpha^{z-1}\,d\alpha}{1+\alpha} = 2\pi i(-1)^{z-1} = 2\pi i \exp[\pi i(z-1)].$$

The integral on the left in (2) may be split into four parts, as indicated in the figure. In detail we use

(a) $\alpha = Re^{i\theta}$, θ from 0 to 2π;
(b) $\alpha = ye^{2\pi i}$, y from R to δ;
(c) $\alpha = \delta e^{i\theta}$, θ from 2π to 0;
(d) $\alpha = ye^{0i}$, y from δ to R.

Thus (2) can be written in the form

$$(3)\quad \int_0^{2\pi} \frac{iR^z \exp(iz\theta)\, d\theta}{1 + R\exp(i\theta)} + \int_R^{\delta} \frac{y^{z-1}\exp(2\pi iz)\, dy}{1+y}$$
$$+ \int_{2\pi}^{0} \frac{i\delta^z \exp(iz\theta)\, d\theta}{1 + \delta\exp(i\theta)} + \int_{\delta}^{R} \frac{y^{z-1}\exp(0iz)\, dy}{1+y} = 2\pi i \exp[\pi i(z-1)].$$

Now let $\delta \to 0$ and $R \to \infty$ and use $0 < \text{Re}(z) < 1$ to conclude that the first and third integrals on the left in (3) approach zero. Then the limiting form of (3) is

$$\exp(2\pi iz)\int_{\infty}^{0} \frac{y^{z-1}\, dy}{1+y} + \int_0^{\infty} \frac{y^{z-1}\, dy}{1+y} = -\, 2\pi i \exp(\pi iz),$$

from which we obtain

$$\int_0^{\infty} \frac{y^{z-1}\, dy}{1+y} = \frac{2\pi i \exp(\pi iz)}{\exp(2\pi iz) - 1} = \frac{2\pi i}{\exp(\pi iz) - \exp(-\pi iz)}.$$

We have thus shown that, for $0 < \text{Re}(z) < 1$,

$$(4)\qquad \Gamma(z)\Gamma(1-z) = \int_0^{\infty} \frac{y^{z-1}\, dy}{1+y} = \frac{\pi}{\sin \pi z}.$$

But each member of (4) is analytic for all nonintegral z, and the theory of analytic continuation permits us to come to the useful conclusion of Theorem 8.

THEOREM 8. *If z is nonintegral,*

$$\Gamma(z)\Gamma(1-z) = \frac{\pi}{\sin \pi z}.$$

Our first, and extremely simple, application of Theorem 8 is the evaluation of $\Gamma(\tfrac{1}{2})$. Use $z = \tfrac{1}{2}$ to get

$$\Gamma(\tfrac{1}{2})\Gamma(\tfrac{1}{2}) = \pi,$$

which, since $\Gamma(\tfrac{1}{2}) > 0$, yields

$$(5)\qquad \Gamma(\tfrac{1}{2}) = \sqrt{\pi}.$$

18. The factorial function. Throughout this book we make frequent use of the common notation

$$(1)\qquad \begin{aligned}(\alpha)_n &= \prod_{k=1}^{n} (\alpha + k - 1) \\ &= \alpha(\alpha + 1)(\alpha + 2)\cdots(\alpha + n - 1), \qquad n \geqq 1, \\ (\alpha)_0 &= 1, \qquad \alpha \neq 0.\end{aligned}$$

The function $(\alpha)_n$ is called the *factorial function.* It is an immediate generalization of the elementary factorial, since $n! = (1)_n$.

In manipulations with $(\alpha)_n$ it is important to keep in mind that $(\alpha)_n$ is a product of n factors, starting with α and with each factor larger by unity than the preceding factor.

Lemma 5. $(\alpha)_{2n} = 2^{2n}\left(\frac{\alpha}{2}\right)_n\left(\frac{\alpha+1}{2}\right)_n.$

Proof: In the product

$$(\alpha)_{2n} = \alpha(\alpha + 1)(\alpha + 2)(\alpha + 3)\cdots(\alpha + 2n - 1),$$

group alternate factors, factor 2 out of each factor on the right, and thus conclude that

$$(\alpha)_{2n} = 2^{2n}\left(\frac{\alpha}{2}\right)_n\left(\frac{\alpha+1}{2}\right)_n.$$

Lemma 6. If k is a positive integer and n a non-negative integer,

$$(2)\qquad (\alpha)_{kn} = k^{nk}\left(\frac{\alpha}{k}\right)_n\left(\frac{\alpha+1}{k}\right)_n\cdots\left(\frac{\alpha+k-1}{k}\right)_n.$$

The proof of Lemma 6 is like that of Lemma 5 except that the factors of $(\alpha)_{kn}$ are grouped into k sets of n factors each, and then k is factored out of each factor to obtain (2).

Other properties of $(\alpha)_n$ will be introduced when needed, particularly in series manipulations involving functions of hypergeometric character. At present we are concerned only with the relation of $(\alpha)_n$ to the Gamma function.

We know that $\Gamma(1 + z) = z\Gamma(z)$. It follows that, for n a positive integer,

$$\begin{aligned}\Gamma(\alpha + n) &= (\alpha + n - 1)\Gamma(\alpha + n - 1) \\ &= (\alpha + n - 1)(\alpha + n - 2)\Gamma(\alpha + n - 2) \\ &= \cdots \\ &= (\alpha + n - 1)(\alpha + n - 2)\cdots\alpha\Gamma(\alpha).\end{aligned}$$

THEOREM 9. *If α is neither zero nor a negative integer,*

$$(\alpha)_n = \frac{\Gamma(\alpha + n)}{\Gamma(\alpha)}. \tag{3}$$

We have already had, in equation (3), page 11, the result

$$\Gamma(z) = \lim_{n\to\infty} \frac{(n-1)!\, n^z}{z(z+1)(z+2)\cdots(z+n-1)},$$

which can now be written in the form

$$\Gamma(z) = \lim_{n\to\infty} \frac{(n-1)!\, n^z}{(z)_n}. \tag{4}$$

Equation (4), reinterpreted in the light of Theorem 9, yields a result of value to us in the subsequent two sections.

Lemma 7. If n is integral and z is not a negative integer,

$$\lim_{n\to\infty} \frac{(n-1)!\, n^z}{\Gamma(z+n)} = 1. \tag{5}$$

19. Legendre's duplication formula. Let us turn to Lemma 5, page 22, and use $\alpha = 2z$. We thus obtain

$$(2z)_{2n} = 2^{2n}(z)_n(z + \tfrac{1}{2})_n.$$

In view of Theorem 9 we may rewrite the above as

$$\frac{\Gamma(2z+2n)}{\Gamma(2z)} = \frac{2^{2n}\Gamma(z+n)\Gamma(z+\frac{1}{2}+n)}{\Gamma(z)\Gamma(z+\frac{1}{2})},$$

or

$$\frac{\Gamma(2z)}{\Gamma(z)\Gamma(z+\frac{1}{2})} = \frac{\Gamma(2z+2n)}{2^{2n}\Gamma(z+n)\Gamma(z+\frac{1}{2}+n)},$$

which, since the left member is independent of n, also implies

$$\frac{\Gamma(2z)}{\Gamma(z)\Gamma(z+\frac{1}{2})} = \lim_{n\to\infty} \frac{\Gamma(2z+2n)}{2^{2n}\Gamma(z+n)\Gamma(z+\frac{1}{2}+n)}. \tag{1}$$

We next insert in the right member of (1) the appropriate factors to permit us to make use of the result in Lemma 7. From (1) we write

$$\frac{\Gamma(2z)}{\Gamma(z)\Gamma(z+\frac{1}{2})}$$

$$= \lim_{n\to\infty} \frac{\Gamma(2z+2n)}{(2n-1)!(2n)^{2z}} \cdot \frac{(n-1)!n^z}{\Gamma(z+n)} \cdot \frac{(n-1)!n^{z+\frac{1}{2}}}{\Gamma(z+\frac{1}{2}+n)} \cdot \frac{2^{2z}(2n-1)!}{2^{2n}n^{\frac{1}{2}}[(n-1)!]^2},$$

which, because of Lemma 7, becomes

$$\frac{\Gamma(2z)}{\Gamma(z)\Gamma(z+\frac{1}{2})} = \lim_{n\to\infty} \frac{2^{2z}(2n-1)!}{2^{2n}n^{\frac{1}{2}}[(n-1)!]^2}.$$

It follows that

$$\frac{\Gamma(2z)}{2^{2z}\Gamma(z)\Gamma(z+\frac{1}{2})} = c,$$

in which c is independent of z. To evaluate c we use $z = \frac{1}{2}$ and find that

$$c = \frac{\Gamma(1)}{2\Gamma(\frac{1}{2})\Gamma(1)} = \frac{1}{2\sqrt{\pi}}.$$

We have thus discovered an expression for $\Gamma(2z)$ in terms of $\Gamma(z)$ and $\Gamma(z+\frac{1}{2})$. It is *Legendre's duplication formula,*

(2) $$\sqrt{\pi}\,\Gamma(2z) = 2^{2z-1}\Gamma(z)\Gamma(z+\tfrac{1}{2}).$$

20. Gauss' multiplication theorem. Following the technique used to discover and prove Legendre's duplication formula, we readily move on to a theorem of Gauss involving the product of k Gamma functions.

Lemma 6, page 22, can be written

$$(\alpha)_{nk} = k^{nk}\prod_{s=1}^{k}\left(\frac{\alpha+s-1}{k}\right)_n$$

and by Theorem 9, page 23, $(\alpha)_n = \Gamma(\alpha+n)/\Gamma(\alpha)$. We thus obtain

(1) $$\frac{\Gamma(\alpha+nk)}{\Gamma(\alpha)} = k^{nk}\prod_{s=1}^{k}\frac{\Gamma\left(\frac{\alpha+s-1}{k}+n\right)}{\Gamma\left(\frac{\alpha+s-1}{k}\right)}.$$

In (1) put $\alpha = kz$ and rearrange the members of the equation to arrive at

(2) $$\frac{\Gamma(kz)}{\prod_{s=1}^{k}\Gamma\left(z+\frac{s-1}{k}\right)} = \frac{\Gamma(kz+kn)}{k^{nk}\prod_{s=1}^{k}\Gamma\left(z+n+\frac{s-1}{k}\right)}$$

$$= \lim_{n\to\infty}\frac{\Gamma(kz+kn)}{k^{nk}\prod_{s=1}^{k}\Gamma\left(z+n+\frac{s-1}{k}\right)}.$$

By Lemma 7, page 23, we know that

$$\lim_{n\to\infty} \frac{(n-1)!\, n^{z+\frac{s-1}{k}}}{\Gamma\left(z+n+\frac{s-1}{k}\right)} = 1,$$

and

$$\lim_{n\to\infty} \frac{(nk-1)!\,(nk)^{kz}}{\Gamma(kz+kn)} = 1.$$

We now use the foregoing two limits in conjunction with (2) to obtain

$$\frac{\Gamma(kz)}{\prod_{s=1}^{k} \Gamma\left(z+\frac{s-1}{k}\right)}$$

$$= \lim_{n\to\infty} \frac{\Gamma(kz+kn)}{(nk-1)!(nk)^{kz}} \cdot \frac{(nk)^{kz}(nk-1)!}{k^{nk}} \prod_{s=1}^{k} \frac{(n-1)!n^{z+\frac{s-1}{k}}}{\Gamma\left(z+n+\frac{s-1}{k}\right)} \cdot \frac{1}{(n-1)!n^{z+\frac{s-1}{k}}}$$

$$= \lim_{n\to\infty} \frac{(nk)^{kz}(nk-1)!}{k^{nk}} \prod_{s=1}^{k} \frac{1}{(n-1)!\, n^{z+\frac{s-1}{k}}}$$

$$= \lim_{n\to\infty} \frac{(nk)^{kz}(nk-1)!}{k^{nk}[(n-1)!]^{k} n^{kz+\frac{1}{2}(k-1)}}.$$

Therefore,

$$\frac{\Gamma(kz)}{k^{kz} \prod_{s=1}^{k} \Gamma\left(z+\frac{s-1}{k}\right)} = c,$$

in which c is independent of z. To determine c, we put $z = 1/k$, use the fact that $\Gamma(1) = 1$, and obtain

$$\frac{1}{kc} = \prod_{s=1}^{k-1} \Gamma\left(\frac{s}{k}\right) = \prod_{s=1}^{k-1} \Gamma\left(\frac{k-s}{k}\right).$$

Then

$$\frac{1}{k^2c^2} = \prod_{s=1}^{k-1} \Gamma\left(\frac{s}{k}\right)\Gamma\left(\frac{k-s}{k}\right) = \prod_{s=1}^{k-1} \frac{\pi}{\sin \frac{\pi s}{k}},$$

or

$$k^2c^2\pi^{k-1} = \prod_{s=1}^{k-1} \sin \frac{\pi s}{k}. \tag{3}$$

We can obtain c once we know the value of the product on the right in (3).

Lemma 8. If $k \geqq 2$, $\prod_{s=1}^{k-1} \sin \frac{\pi s}{k} = \frac{k}{2^{k-1}}$.

Proof: Let $\alpha = \exp(2\pi i/k)$ be a primitive kth root of unity. Then for all x,

$$x^k - 1 = (x - 1) \prod_{s=1}^{k-1} (x - \alpha^s),$$

from which, by differentiation of both members, we get

$$kx^{k-1} = \prod_{s=1}^{k-1} (x - \alpha^s) + (x - 1)g(x), \tag{4}$$

in which $g(x)$ is a polynomial in x. Put $x = 1$ in (4) to obtain

$$k = \prod_{s=1}^{k-1} (1 - \alpha^s).$$

But

$$1 - \alpha^s = 1 - \exp\left(\frac{2\pi i s}{k}\right) = -\exp\left(\frac{\pi i s}{k}\right)\left[\exp\left(\frac{\pi i s}{k}\right) - \exp\left(\frac{-\pi i s}{k}\right)\right]$$

$$= -2i \exp\left(\frac{\pi i s}{k}\right) \sin \frac{\pi s}{k}.$$

Hence

$$k = (-2i)^{k-1} \exp[\tfrac{1}{2}\pi i(k-1)] \prod_{s=1}^{k-1} \sin \frac{\pi s}{k} = 2^{k-1} \prod_{s=1}^{k-1} \sin \frac{\pi s}{k},$$

which yields the desired result.

With the aid of Lemma 8, equation (3) can be written

$$k^2 c^2 \pi^{k-1} = \frac{k}{2^{k-1}}.$$

The constant c is positive because the Gamma function is positive for positive argument. Hence

$$c = (2\pi)^{-\frac{1}{2}(k-1)} k^{-\frac{1}{2}}.$$

This completes the proof of the Gauss multiplication theorem.

THEOREM 10 $\prod_{s=1}^{k} \Gamma\left(z + \frac{s-1}{k}\right) = (2\pi)^{\frac{1}{2}(k-1)} k^{\frac{1}{2} - kz} \Gamma(kz)$.

21. A summation formula due to Euler. Let

$$P(x) = x - [x] - \tfrac{1}{2},$$

in which $[x]$ means the greatest integer $\leqq x$, a notation also used frequently in later chapters. The function $P(x)$ is periodic,

$$P(x + 1) = P(x),$$

and is represented graphically in Figure 2.

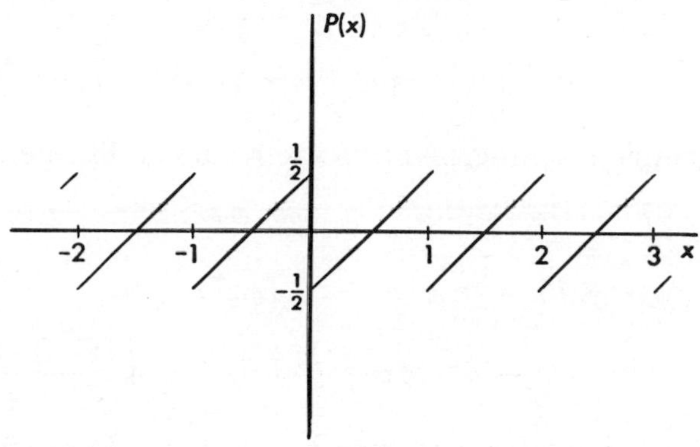

Figure 2

Euler employed $P(x)$ in obtaining some useful summation formulas, of which we use only that in Theorem 11.

THEOREM 11. *If $f'(x)$ is continuous for $x \geqq 0$,*

$$\sum_{k=0}^{n} f(k) = \int_0^n f(x)\,dx + \tfrac{1}{2}f(0) + \tfrac{1}{2}f(n) + \int_0^n P(x)f'(x)\,dx,$$

in which $P(x) = x - [x] - \frac{1}{2}$.

Proof. First write

$$\int_0^n P(x)f'(x)\,dx = \sum_{k=1}^{n} \int_{k-1}^{k} P(x)f'(x)\,dx.$$

Now

$$\int_{k-1}^{k} P(x)f'(x)\,dx = \int_{k-1}^{k} (x - k + \tfrac{1}{2})f'(x)\,dx,$$

and we integrate by parts to obtain

$$\int_{k-1}^{k} P(x)f'(x)\,dx = \Big[(x - k + \tfrac{1}{2})f(x)\Big]_{k-1}^{k} - \int_{k-1}^{k} f(x)\,dx$$

$$= \tfrac{1}{2}f(k) + \tfrac{1}{2}f(k - 1) - \int_{k-1}^{k} f(x)\,dx.$$

We may therefore write

$$\begin{aligned}\int_0^n P(x)f'(x)\,dx &= \tfrac{1}{2}\sum_{k=1}^{n} f(k) + \tfrac{1}{2}\sum_{k=1}^{n} f(k-1) - \sum_{k=1}^{n}\int_{k-1}^{k} f(x)\,dx \\ &= \tfrac{1}{2}\sum_{k=1}^{n} f(k) + \tfrac{1}{2}\sum_{k=0}^{n-1} f(k) - \int_0^n f(x)\,dx \\ &= \sum_{k=0}^{n} f(k) - \tfrac{1}{2}f(0) - \tfrac{1}{2}f(n) - \int_0^n f(x)\,dx,\end{aligned}$$

which is a simple rearrangement of the result in Theorem **11.**

Lemma 9. For $|\arg z| \leqq \pi - \delta$, *where* $\delta > 0$,

$$\begin{aligned}\sum_{k=0}^{n} \operatorname{Log}(z+k) &= (z + n + \tfrac{1}{2}) \operatorname{Log}(z+n) \\ &\quad - n - (z - \tfrac{1}{2}) \operatorname{Log} z + \int_0^n \frac{P(x)\,dx}{z+x}.\end{aligned}$$

Proof: Lemma 9 follows at once by applying Theorem 11 to the function $f(x) = \operatorname{Log}(z + x)$.

Let us next turn to the result

$$\Gamma(z) = \lim_{n\to\infty} \frac{(n-1)!\, n^z}{(z)_n}, \tag{1}$$

established on page 23. In (1) put $z = \frac{1}{2}$ and shift from n to $(n + 1)$ to obtain

$$\lim_{n\to\infty} \frac{n!\,(n+1)^{\frac{1}{2}}}{(\frac{1}{2})_{n+1}} = \Gamma(\tfrac{1}{2}),$$

which may be put in the form

$$\lim_{n\to\infty} \frac{n!\,(n+1)^{\frac{1}{2}} n!\, 2^{2n}}{(\frac{1}{2})(\frac{3}{2})_n(1)_n 2^{2n}} = \sqrt{\pi}.$$

Now Lemma 6, page 22, yields

$$2^{2n}\left(\frac{3}{2}\right)_n (1)_n = (2)_{2n} = (2n+1)!.$$

Therefore we have

$$\lim_{n\to\infty} \frac{2^{2n+1}[(n+1)!]^2(n+1)^{-\frac{3}{2}}}{(2n+1)!} = \sqrt{\pi}. \tag{2}$$

It is legitimate to take logarithms of each member of (2) and thus write

$$(3)\quad \lim_{n\to\infty}\left[(2n+1)\operatorname{Log}2-\tfrac{3}{2}\operatorname{Log}(1+n)+2\sum_{k=0}^{n}\operatorname{Log}(1+k)-\sum_{k=0}^{2n}\operatorname{Log}(1+k)\right]=\tfrac{1}{2}\operatorname{Log}\pi.$$

We shall apply the formula of Lemma 9 to the sums involved on the left in equation (3). The choice $z = 1$ in the result in Lemma 9 yields

$$(4)\quad \sum_{k=0}^{n}\operatorname{Log}(1+k)=\left(\frac{3}{2}+n\right)\operatorname{Log}(1+n)-n+\int_0^n\frac{P(x)\,dx}{1+x}.$$

By Lemma 9, with $z = 1$ and n replaced by $2n$, we get

$$\sum_{k=0}^{2n}\operatorname{Log}(1+k)=\left(\frac{3}{2}+2n\right)\operatorname{Log}(1+2n)-2n+\int_0^{2n}\frac{P(x)\,dx}{1+x}.$$

Equation (3) can now be put in the form

$$\lim_{n\to\infty}\left[(2n+1)\operatorname{Log}2+\left(2n+\frac{3}{2}\right)\operatorname{Log}\frac{1+n}{1+2n}+2\int_0^n\frac{P(x)\,dx}{1+x}-\int_0^{2n}\frac{P(x)\,dx}{1+x}\right]=\frac{1}{2}\operatorname{Log}\pi,$$

which leads* to

$$\lim_{n\to\infty}\left[-\frac{1}{2}\operatorname{Log}2+\left(2n+\frac{3}{2}\right)\operatorname{Log}\frac{2+2n}{1+2n}\right]+\int_0^\infty\frac{P(x)\,dx}{1+x}=\frac{1}{2}\operatorname{Log}\pi.$$

But

$$\lim_{n\to\infty}\left[\left(2n+\frac{3}{2}\right)\operatorname{Log}\frac{2+2n}{1+2n}\right]=1.$$

Therefore we arrive at the evaluation

$$(5)\qquad \int_0^\infty\frac{P(x)\,dx}{1+x}=-1+\frac{1}{2}\operatorname{Log}(2\pi).$$

22. The behavior of log $\Gamma(z)$ for large $|z|$. From formula (3), page 11, it follows that

$$\Gamma(z)=\lim_{n\to\infty}\frac{(n+1)!\,(n+1)^{z-1}}{(z)_{n+1}},$$

*For proof of convergence of $\int_0^\infty \frac{P(x)\,dx}{1+x}$, see the exercises at the end of this chapter.

and so also that

$$(1)\quad \log \Gamma(z) = \lim_{n\to\infty}\left[\sum_{k=0}^{n} \operatorname{Log}(1+k) + (z-1)\operatorname{Log}(1+n) - \sum_{k=0}^{n}\operatorname{Log}(z+k)\right].$$

Using equation (4) and Lemma 9 of the preceding section, we may now conclude that, if $|\arg(z)| \leqq \pi - \delta$, $\delta > 0$,

$$(2)\quad \log \Gamma(z) = \lim_{n\to\infty}\left[\left(z+n+\frac{1}{2}\right)\{\operatorname{Log}(1+n) - \operatorname{Log}(z+n)\} + \left(z-\frac{1}{2}\right)\operatorname{Log} z + \int_0^n \frac{P(x)\,dx}{1+x} - \int_0^n \frac{P(x)\,dx}{z+x}\right].$$

The elementary limit

$$\lim_{n\to\infty}\left[\left(z+n+\frac{1}{2}\right)\{\operatorname{Log}(1+n) - \operatorname{Log}(z+n)\}\right] = 1 - z,$$

together with equation (5) of Section 21, permits us to put (2) in the form

$$\log \Gamma(z) = 1 - z + \left(z - \frac{1}{2}\right)\operatorname{Log} z - 1 + \frac{1}{2}\operatorname{Log}(2\pi) - \int_0^\infty \frac{P(x)\,dx}{z+x}.$$

THEOREM 12. *If* $|\arg(z)| \leqq \pi - \delta$, *where* $\delta > 0$,

$$(3)\quad \log \Gamma(z) = \left(z - \frac{1}{2}\right)\operatorname{Log} z - z + \frac{1}{2}\operatorname{Log}(2\pi) - \int_0^\infty \frac{P(x)\,dx}{z+x},$$

in which $P(x) = x - [x] - \frac{1}{2}$, *as in Section 21.*

Let us next consider the integral on the right in (3). Since

$$\int P(x)\,dx = \tfrac{1}{2}P^2(x) + c,$$

we may use $c = -\frac{1}{24}$ and integrate by parts to find that

$$\int_0^\infty \frac{P(x)\,dx}{z+x} = \frac{1}{2}\left[\frac{P^2(x) - \frac{1}{12}}{z+x}\right]_0^\infty + \frac{1}{2}\int_0^\infty \frac{[P^2(x) - \frac{1}{12}]\,dx}{(z+x)^2}$$

$$= -\frac{1}{12z} + \frac{1}{2}\int_0^\infty \frac{[P^2(x) - \frac{1}{12}]\,dx}{(z+x)^2}.$$

Now the maximum value of $[P^2(x) - \frac{1}{12}]$ is $\frac{1}{6}$ and, in the region $|\arg z| \leqq \pi - \delta$, $\delta > 0$,

$$|z+x|^2 \geqq x^2 + |z|^2, \qquad \text{for } \operatorname{Re}(z) \geqq 0,$$

$$|z + x|^2 \geqq [x + \text{Re}(z)]^2 + |z|^2 \sin^2 \delta, \qquad \text{for } \text{Re}(z) < 0.$$

It follows that

$$\int_0^\infty \frac{[P^2(x) - \frac{1}{12}]\, dx}{(z + x)^2} = \text{O}\left(\frac{1}{|z|}\right),$$

as $|z| \to \infty$ in $|\arg z| \leqq \pi - \delta$, $\delta > 0$.

We have shown that as $|z| \to \infty$ in $|\arg z| \leqq \pi - \delta$, $\delta > 0$,

$$\text{(4)} \qquad \log \Gamma(z) = (z - \tfrac{1}{2}) \,\text{Log}\, z - z + \tfrac{1}{2} \,\text{Log}(2\pi) + \text{o}(1).$$

Indeed we showed a little more than that, but (4) is itself more precise than is needed later in this book.

From (4) we obtain at once the actual result to be employed in Chapter 5.

THEOREM 13. *As* $|z| \to \infty$ *in the region where* $|\arg z| \leqq \pi - \delta$ *and* $|\arg(z + a)| \leqq \pi - \delta$, $\delta > 0$,

$$\text{(5)} \qquad \log \Gamma(z + a) = (z + a - \tfrac{1}{2}) \,\text{Log}\, z - z + \text{O}(1).$$

EXERCISES

1. Start with $\dfrac{\Gamma'(z)}{\Gamma(z)} = -\gamma - \dfrac{1}{z} - \sum_{n=1}^{\infty}\left(\dfrac{1}{z+n} - \dfrac{1}{n}\right)$,

prove that

$$\frac{2\Gamma'(2z)}{\Gamma(2z)} - \frac{\Gamma'(z)}{\Gamma(z)} - \frac{\Gamma'(z + \frac{1}{2})}{\Gamma(z + \frac{1}{2})} = 2 \,\text{Log}\, 2,$$

and thus derive Legendre's duplication formula, page 24.

2. Show that $\Gamma'(\frac{1}{2}) = -(\gamma + 2 \,\text{Log}\, 2)\sqrt{\pi}$.

3. Use Euler's integral form $\Gamma(z) = \displaystyle\int_0^\infty e^{-t} t^{z-1}\, dt$ to show that $\Gamma(z + 1) = z\Gamma(z)$.

4. Show that $\Gamma(z) = \lim\limits_{n\to\infty} n^z B(z, n)$.

5. Derive the following properties of the Beta function:
   (a) $pB(p, q + 1) = qB(p + 1, q)$;
   (b) $B(p, q) = B(p + 1, q) + B(p, q + 1)$;
   (c) $(p + q)B(p, q + 1) = qB(p, q)$;
   (d) $B(p, q)B(p + q, r) = B(q, r)B(q + r, p)$.

6. Show that for positive integral n, $B(p, n + 1) = n!/(p)_{n+1}$.

7. Evaluate $\displaystyle\int_{-1}^{1} (1 + x)^{p-1}(1 - x)^{q-1}\, dx$.

Ans. $2^{p+q-1}B(p, q)$.

8. Show that for $0 \leqq k \leqq n$

$$(\alpha)_{n-k} = \frac{(-1)^k(\alpha)_n}{(1-\alpha-n)_k}.$$

Note particularly the special case $\alpha = 1$.

9. Show that if α is not an integer,

$$\frac{\Gamma(1-\alpha-n)}{\Gamma(1-\alpha)} = \frac{(-1)^n}{(\alpha)_n}.$$

In Exs. 10–14, the function $P(x)$ is that of Section 21.

10. Evaluate $\int_0^x P(y)\,dy$. *Ans.* $\frac{1}{2}P^2(x) - \frac{1}{8}$.

11. Use integration by parts and the result of Ex. 10 to show that

$$\left|\int_n^\infty \frac{P(x)\,dx}{1+x}\right| \leqq \frac{1}{8(1+n)}.$$

12. With the aid of Ex. 11 prove the convergence of $\int_0^\infty \frac{P(x)\,dx}{1+x}$.

13. Show that

$$\int_0^\infty \frac{P(x)\,dx}{1+x} = \sum_{n=0}^\infty \int_n^{n+1} \frac{P(x)\,dx}{1+x} = \sum_{n=0}^\infty \int_0^1 \frac{(y-\frac{1}{2})\,dy}{1+n+y}.$$

Then prove that

$$\lim_{n\to\infty} n^2 \int_0^1 \frac{(y-\frac{1}{2})\,dy}{1+n+y} = \frac{1}{12}$$

and thus conclude that $\int_0^\infty \frac{P(x)\,dx}{1+x}$ is convergent.

14. Apply Theorem 11, page 27, to the function $f(x) = (1+x)^{-1}$; let $n \to \infty$ and thus conclude that

$$\gamma = \tfrac{1}{2} - \int_1^\infty y^{-2}P(y)\,dy.$$

15. Use the relation $\Gamma(z)\Gamma(1-z) = \pi/\sin \pi z$ and the elementary result

$$\sin x \sin y = \tfrac{1}{2}\,[\cos(x-y) - \cos(x+y)]$$

to prove that

$$1 - \frac{\Gamma(c)\Gamma(1-c)\Gamma(c-a-b)\Gamma(a+b+1-c)}{\Gamma(c-a)\Gamma(a+1-c)\Gamma(c-b)\Gamma(b+1-c)}$$

$$= \frac{\Gamma(2-c)\Gamma(c-1)\Gamma(c-a-b)\Gamma(a+b+1-c)}{\Gamma(a)\Gamma(1-a)\Gamma(b)\Gamma(1-b)}.$$

CHAPTER 3

Asymptotic Series

23. Definition of an asymptotic expansion. Let us first recall the sense in which a convergent power series expansion represents the function being expanded. When a function $F(z)$, analytic at $z = 0$, is expanded in a power series about $z = 0$, we write

$$F(z) = \sum_{n=0}^{\infty} c_n z^n, \qquad |z| < r. \tag{1}$$

Define a partial sum of the series by

$$S_n(z) = \sum_{k=0}^{n} c_k z^k.$$

Then the series on the right in (1) represents $F(z)$ in the sense that

$$\lim_{n \to \infty} [F(z) - S_n(z)] = 0 \tag{2}$$

for each z in the region $|z| < r$. That is, for each fixed z the series in (1) can be made to approximate $F(z)$ as closely as desired by taking a sufficiently large number of terms of the series.

We now define an *asymptotic power series representation* of a function $f(z)$ as $z \to 0$ in some region R. We write

$$f(z) \sim \sum_{n=0}^{\infty} a_n z^n, \qquad z \to 0 \text{ in } R, \tag{3}$$

if and only if

$$\lim_{z\to 0 \text{ in } R} \frac{|f(z) - s_n(z)|}{|z|^n} = 0, \tag{4}$$

for each fixed n, with

$$s_n(z) = \sum_{k=0}^{n} a_k z^k. \tag{5}$$

By employing the order symbol defined in Section 13, we may write the condition (4) in the form

$$f(z) - s_n(z) = o(z^n), \qquad \text{as } z \to 0 \text{ in } R. \tag{6}$$

Here we see that the series in (3) represents the function $f(z)$ in the sense that for each fixed n, the sum of the terms out to the term $a_n z^n$ can be made to approximate $f(z)$ more closely than $|z|^n$ approximates zero, in the sense of (4), by choosing z sufficiently close to zero in the region R.

It is particularly noteworthy that in the definition of an asymptotic expansion, there is no requirement that the series converge. Indeed some authors include the additional restriction that the series in (3) diverge. Most asymptotic expansions do diverge, but it seems artificial to insist upon that behavior.

Asymptotic series are of great value in many computations. They play an important role in the solution of linear differential equations about irregular singular points. Such series were used by astronomers more than a century ago, long before the pertinent mathematical theory was developed.

Example: Show that

$$\int_0^\infty \frac{e^{-t}\,dt}{1 - xt} \sim \sum_{n=0}^{\infty} n!x^n, \qquad x \to 0 \text{ in } \operatorname{Re}(x) \leqq 0. \tag{7}$$

Let us put

$$s_n(x) = \sum_{k=0}^{n} k!x^k.$$

In the region $\operatorname{Re}(x) \leqq 0$, the integral on the left in (7) is absolutely and uniformly convergent. To see this, note that $t \geqq 0$ so that $\operatorname{Re}(1 - xt) \geqq 1$. Hence $|1 - xt| \geqq 1$, and we have

$$\left| \int_0^\infty \frac{e^{-t}\,dt}{1 - xt} \right| \leqq \int_0^\infty e^{-t}\,dt = 1.$$

For k a non-negative integer,

$$\int_0^\infty e^{-t} t^k \, dt = \Gamma(k+1) = k!. \tag{8}$$

Hence

$$\int_0^\infty \frac{e^{-t}\,dt}{1-xt} - s_n(x) = \int_0^\infty \frac{e^{-t}\,dt}{1-xt} - \sum_{k=0}^{n} \int_0^\infty e^{-t} t^k x^k \, dt$$

$$= \int_0^\infty e^{-t} \left[\frac{1}{1-xt} - \sum_{k=0}^{n} (xt)^k \right] dt.$$

From elementary algebra we have

$$\sum_{k=0}^{n} r^k = \frac{1-r^{n+1}}{1-r}, \qquad r \neq 1.$$

Therefore

$$\int_0^\infty \frac{e^{-t}\,dt}{1-xt} - s_n(x) = \int_0^\infty \frac{e^{-t}(xt)^{n+1}\,dt}{1-xt},$$

from which, since $|1 - xt| \geqq 1$, we obtain

$$\left| \int_0^\infty \frac{e^{-t}\,dt}{1-xt} - s_n(x) \right| \leqq |x|^{n+1} \int_0^\infty e^{-t} t^{n+1} \, dt, \qquad \text{in } \operatorname{Re}(x) \leqq 0.$$

We may conclude that

$$\left| \int_0^\infty \frac{e^{-t}\,dt}{1-xt} - s_n(x) \right| \leqq (n+1)! \, |x|^{n+1}, \qquad \text{in } \operatorname{Re}(x) \leqq 0. \tag{9}$$

From (9) it follows at once that the condition (4), page 34, is satisfied, which concludes the proof. Actually (9) gives more information than that. Let $E_n(x)$ be the error made in computing the sum function by discarding all terms after the term $n!x^n$. Then $|E_n(x)|$ is the left member of (9), and the inequality (9) shows that $|E_n(x)|$ is smaller than the magnitude of the first term omitted. This property, although not possessed by all asymptotic series, is one of frequent occurrence.

The preceding example gives little indication of methods for obtaining asymptotic expansions. Later we shall exhibit two common methods, successive integration by parts and term-by-term integration of power series.

Extension of the concept of an asymptotic expansion to one in which the variable approaches any specific point in the finite plane is direct. For finite z_0 we say that

$$f(z) \sim \sum_{n=0}^{\infty} a_n (z - z_0)^n, \qquad \text{as } z \to z_0 \text{ in } R,$$

if and only if, for each fixed n,

$$f(z) - s_n(z) = o([z - z_0]^n), \qquad \text{as } z \to z_0 \text{ in } R,$$

in which

$$s_n(z) = \sum_{k=0}^{n} a_k(z - z_0)^k.$$

24. Asymptotic expansions about infinity. Asymptotic series are often used for large $|z|$. We say that

$$(1) \qquad f(z) \sim \sum_{n=0}^{\infty} a_n z^{-n}, \text{ as } z \to \infty \text{ in } R,$$

if and only if, for each fixed n,

$$(2) \qquad f(z) - s_n(z) = o(z^{-n}), \qquad \text{as } z \to \infty \text{ in } R,$$

in which

$$(3) \qquad s_n(z) = \sum_{k=0}^{n} a_k z^{-k}.$$

At times, as in the subsequent example, we wish to work only along the axis of reals. We then use (1), (2), and (3) for a real variable x, with the region R replaced by a direction along the real axis.

One last extension of the term asymptotic expansion follows. It may be that $f(z)$ itself has no asymptotic expansion in the sense of the foregoing definitions. We do, however, write

$$(4) \qquad f(z) \sim h(z) + g(z) \sum_{n=0}^{\infty} a_n z^{-n}, \qquad \text{as } z \to \infty \text{ in } R,$$

if and only if

$$(5) \qquad \frac{f(z) - h(z)}{g(z)} \sim \sum_{n=0}^{\infty} a_n z^{-n}, \qquad \text{as } z \to \infty \text{ in } R,$$

and similarly for asymptotic expansions about a point in the finite plane.

EXAMPLE: Obtain, for real x, as $x \to \infty$, an asymptotic expansion of the error function

$$(6) \qquad \operatorname{erf}(x) = \frac{2}{\sqrt{\pi}} \int_0^x \exp(-t^2)\, dt.$$

From the fact that $\Gamma(\frac{1}{2}) = \sqrt{\pi}$, it follows at once that

$$\lim_{x\to\infty} \operatorname{erf}(x) = 1.$$

Let us write

$$\operatorname{erf}(x) = \frac{2}{\sqrt{\pi}}\int_0^\infty \exp(-t^2)\,dt - \frac{2}{\sqrt{\pi}}\int_x^\infty \exp(-t^2)\,dt$$

$$= 1 - \frac{2}{\sqrt{\pi}}\int_x^\infty \exp(-t^2)\,dt.$$

Now consider the function

$$B(x) = \int_x^\infty \exp(-t^2)\,dt$$

and integrate by parts to get

$$B(x) = -\tfrac{1}{2}\Big[t^{-1}\exp(-t^2)\Big]_x^\infty - \tfrac{1}{2}\int_x^\infty t^{-2}\exp(-t^2)\,dt$$

$$= \tfrac{1}{2}x^{-1}\exp(-x^2) - \tfrac{1}{2}\int_x^\infty t^{-2}\exp(-t^2)\,dt.$$

Iteration of the integration by parts soon yields

$B(x) =$

$$\exp(-x^2)\left[\frac{1}{2x} - \frac{1}{2^2x^3} + \frac{1\cdot3}{2^3x^5} - \frac{1\cdot3\cdot5}{2^4x^7} + \cdots + \frac{(-1)^n 1\cdot3\cdot5\cdots(2n-1)}{2^{n+1}x^{2n+1}}\right]$$

$$+\frac{(-1)^{n+1}1\cdot3\cdot5\cdots(2n+1)}{2^{n+1}}\int_x^\infty t^{-2n-2}\exp(-t^2)\,dt,$$

or

$$B(x) = \frac{1}{2}\exp(-x^2)\sum_{k=0}^{n}\frac{(-1)^k(\frac{1}{2})_k}{x^{2k+1}} + (-1)^{n+1}\left(\frac{1}{2}\right)_{n+1}\int_x^\infty t^{-2n-2}\exp(-t^2)\,dt. \tag{7}$$

Let

$$s_n(x) = \frac{1}{2}\sum_{k=0}^{n}\frac{(-1)^k(\frac{1}{2})_k}{x^{2k+1}}.$$

Then, from (7),

$$\exp(x^2)B(x) - s_n(x) = (-1)^{n+1}(\tfrac{1}{2})_{n+1}\exp(x^2)\int_x^\infty t^{-2n-2}\exp(-t^2)\,dt.$$

The variable of integration is never less than x. We replace the factor t^{-2n-2} in the integrand by tx^{-2n-3} and thus obtain

$$\left| \exp(x^2)B(x) - s_n(x) \right| < \frac{(\frac{1}{2})_{n+1}\exp(x^2)}{x^{2n+3}} \int_x^\infty t\exp(-t^2)\,dt,$$

from which it follows that

$$\left| \exp(x^2)B(x) - s_n(x) \right| < \frac{(\frac{1}{2})_{n+1}}{2x^{2n+3}}. \tag{8}$$

Hence

$$\exp(x^2)B(x) - s_n(x) = o(x^{-2n-2}), \qquad \text{as } x \to \infty,$$

which permits us to write the asymptotic expansion

$$\exp(x^2)B(x) \sim \sum_{n=0}^{\infty} \frac{(-1)^n(\frac{1}{2})_n}{2x^{2n+1}}, \qquad x \to \infty.$$

But $\operatorname{erf}(x) = 1 - \dfrac{2}{\sqrt{\pi}} B(x)$. Hence

$$\operatorname{erf}(x) \sim 1 - \frac{1}{\sqrt{\pi}} \exp(-x^2) \sum_{n=0}^{\infty} \frac{(-1)^n(\frac{1}{2})_n}{x^{2n+1}}, \qquad x \to \infty. \tag{9}$$

Note also the useful bound in (8).

25. Algebraic properties. Asymptotic expansions behave like convergent power series in many ways. We shall treat only expansions as $z \to \infty$ in some region R. The reader can easily extend the results to theorems in which $z \to z_0$ in the finite plane.

THEOREM 14. *If, as* $z \to \infty$ *in* R,

$$f(z) \sim \sum_{n=0}^{\infty} a_n z^{-n} \tag{1}$$

and

$$g(z) \sim \sum_{n=0}^{\infty} b_n z^{-n}, \tag{2}$$

then

$$f(z) + g(z) \sim \sum_{n=0}^{\infty} (a_n + b_n) z^{-n} \tag{3}$$

and

$$f(z)g(z) \sim \sum_{n=0}^{\infty} \sum_{k=0}^{n} a_k b_{n-k} z^{-n}. \tag{4}$$

Proof: Let

$$S_n(z) = \sum_{k=0}^{n} a_k z^{-k}, \qquad T_n(z) = \sum_{k=0}^{n} b_k z^{-k}.$$

From (1) and (2) we know that

$$f(z) - S_n(z) = o(z^{-n}), \tag{5}$$

$$g(z) - T_n(z) = o(z^{-n}), \tag{6}$$

from which

$$f(z) + g(z) - [S_n(z) + T_n(z)] = o(z^{-n}),$$

yielding (3).

To prove the validity of (4), first put

$$Q_n(z) = \sum_{k=0}^{n} \sum_{i=0}^{k} a_i b_{k-i} z^{-k},$$

which is the "nth partial sum" of the series on the right in (4). By direct multiplication,

$$S_n(z)T_n(z) = Q_n(z) + o(z^{-n}),$$

and by (5) and (6),

$$f(z)g(z) = S_n(z)T_n(z) + o(z^{-n}).$$

Hence

$$f(z)g(z) = Q_n(z) + o(z^{-n}),$$

which shows the validity of (4).

The right member in (4) is the ordinary Cauchy product of the series (1) and (2).

26. Term-by-term integration. Suppose that for real x we have

$$f(x) \sim \sum_{n=0}^{\infty} a_n x^{-n}, \qquad x \to \infty. \tag{1}$$

Surely we are interested here in large x, so that an integral which it is natural to consider is $\int_y^\infty f(x)\,dx$. But $\int_y^\infty a_0\,dx$ and $\int_y^\infty a_1 x^{-1}\,dx$ do not exist. Therefore we restrict ourselves to the consideration of an expansion

$$g(x) \sim \sum_{n=2}^{\infty} a_n x^{-n}, \qquad x \to \infty, \tag{2}$$

and seek $\int_y^\infty g(x)\,dx$. Of course $g(x) = f(x) - a_0 - a_1x^{-1}$.

Let

$$s_n(x) = \sum_{k=2}^{n} a_k x^{-k}.$$

Then

$$g(x) - s_n(x) = o(x^{-n}), \qquad x \to \infty,$$

and

$$\left|\int_y^\infty g(x)\,dx - \int_y^\infty s_n(x)\,dx\right| \leqq \int_y^\infty |g(x) - s_n(x)|\,dx$$

$$< \int_y^\infty |o(x^{-n})|\,dx$$

$$= o(y^{-n+1}).$$

But

$$\int_y^\infty s_n(x)\,dx = \sum_{k=2}^{n} a_k \int_y^\infty x^{-k}\,dx = \sum_{k=2}^{n} \frac{a_k y^{-k+1}}{(k-1)}.$$

Hence

$$\int_y^\infty g(x)\,dx \sim \sum_{n=2}^{\infty} \frac{a_n y^{-n+1}}{n-1}, \qquad y \to \infty, \tag{3}$$

the desired result.

27. Uniqueness. Since $e^{-x} = o(x^k)$, as $x \to \infty$, for any real k, whole classes of functions have the same asymptotic expansion. Surely if

$$f(x) \sim \sum_{n=0}^{\infty} A_n x^{-n},$$

then also

$$f(x) + ce^{-x} \sim \sum_{n=0}^{\infty} A_n x^{-n},$$

and numerous similar examples are easily concocted.

On the other hand a given function cannot have more than one asymptotic expansion as $z \to z_0$, finite or infinite. Let us use $z \to \infty$ in a region R as a representative example.

THEOREM 15. *If*

$$f(z) \sim \sum_{n=0}^{\infty} A_n z^{-n}, \qquad z \to \infty \text{ in } R, \tag{1}$$

and

(2) $$f(z) \sim \sum_{n=0}^{\infty} B_n z^{-n}, \qquad z \to \infty \text{ in } R,$$

then $A_n = B_n$.

Proof: From (1) and (2) we have

$$f(z) - \sum_{k=0}^{n} A_k z^{-k} = o(z^{-n}),$$

$$f(z) - \sum_{k=0}^{n} B_k z^{-k} = o(z^{-n}),$$

from which it follows that

$$\sum_{k=0}^{n} (A_k - B_k) z^{-k} = o(z^{-n}),$$

or its equivalent

$$\sum_{k=0}^{n} (A_k - B_k) z^{n-k} = o(1), \qquad z \to \infty \text{ in } R,$$

for each n. Therefore $A_k = B_k$ for each k. The expansion (1) associated with $z \to \infty$ in a particular region R is unique. The function $f(z)$ may, of course, have a different asymptotic expansion as $z \to \infty$ in some region other than R.

28. Watson's lemma. The following useful result due to Watson [1;236] gives conditions under which the term-by-term Laplace transform of a series yields an asymptotic representation for the transform of the sum of the series. For details on Laplace transforms see Churchill [1].

Since relatively complicated exponents appear in the following few pages, we shall simplify the printing by the introduction of a notation similar to the common one, exp $u = e^u$. The symbol $\exp_x(m)$ is defined by

$$\exp_x(m) = x^m.$$

Watson's Lemma. Let $F(t)$ *satisfy the following conditions:*

(1) $F(t) = \sum_{n=1}^{\infty} a_n \exp_t\left(\frac{n}{r} - 1\right)$, *in* $|t| \leqq a + \delta$, *with* $a, \delta, r > 0$;

(2) *There exist positive constants* K *and* b *such that*

$$|F(t)| < Ke^{bt}, \qquad \text{for } t \geqq a.$$

Then

$$(3)\qquad f(s) = \int_0^\infty e^{-st}F(t)\,dt \sim \sum_{n=1}^{\infty} \frac{a_n\Gamma(n/r)}{s^{n/r}},$$

as $|s| \to \infty$ *in the region* $|\arg s| \leqq \frac{1}{2}\pi - \Delta$, *for arbitrarily small positive* Δ.

Note that (1) implies that $F(t)$ is either analytic at $t = 0$ or has at most a certain type of branch point there.

Proof: It is not difficult to show (Exs. 1 and 2 at the end of this chapter) that under the conditions of Watson's lemma, there exist positive constants c and β such that for all $t \geqq 0$, whether $t \leqq a$ or $t > a$,

$$(4)\qquad \left|F(t) - \sum_{k=1}^{n} a_k \exp_t\left(\frac{k}{r} - 1\right)\right| < c\exp_t\left(\frac{n+1}{r} - 1\right)e^{\beta t}.$$

We know also the Laplace transform of a power of t,

$$(5)\qquad \int_0^\infty e^{-st}t^m\,dt = \Gamma(m+1)s^{-m-1}, \qquad m > -1,\ \mathrm{Re}(s) > 0.$$

In order to derive (3), we need to show that for each fixed n

$$\left|f(s) - \sum_{k=1}^{n} a_k\Gamma\left(\frac{k}{r}\right)s^{-k/r}\right| \cdot \left|s\right|^{n/r} = o(1),$$

as $|s| \to \infty$ in $|\arg s| \leqq \frac{1}{2}\pi - \Delta$, $\Delta > 0$.

Now

$$f(s) - \sum_{k=1}^{n} a_k\Gamma\left(\frac{k}{r}\right)s^{-k/r} = \int_0^\infty e^{-st}\left[F(t) - \sum_{k=1}^{n} a_k \exp_t\left(\frac{k}{r} - 1\right)\right]dt.$$

Hence, with the aid of (4),

$$|s|^{n/r}\left|f(s) - \sum_{k=1}^{n} a_k\Gamma\left(\frac{k}{r}\right)s^{-k/r}\right|$$

$$< |s|^{n/r}c\int_0^\infty |e^{-st}|\exp_t\left(\frac{n+1}{r} - 1\right)e^{\beta t}\,dt$$

$$< c|s|^{n/r}\int_0^\infty e^{-\mathrm{Re}(s)t}\exp_t\left(\frac{n+1}{r} - 1\right)e^{\beta t}\,dt$$

$$< c|s|^{n/r}\Gamma\left(\frac{n+1}{r}\right)[\mathrm{Re}(s) - \beta]^{-\frac{(n+1)}{r}},$$

if $\mathrm{Re}(s) > \beta$. In the region $|\arg s| \leqq \frac{1}{2}\pi - \Delta$, $\Delta > 0$, $\mathrm{Re}(s) > \beta$

as soon as we choose $|s| > \beta(\sin \Delta)^{-1}$. Therefore, as $|s| \to \infty$ in the region $|\arg s| \leqq \frac{1}{2}\pi - \Delta$,

$$|s|^{n/r}\left|f(s) - \sum_{k=1}^{n} a_k\Gamma\left(\frac{k}{r}\right)s^{-k/r}\right| = o(1),$$

as desired.

Example: Obtain an asymptotic expansion of

$$f(x) = \int_0^\infty \frac{e^{-xt}\,dt}{1+t^2}, \qquad |x| \to \infty \text{ in } |\arg x| \leqq \tfrac{1}{2}\pi - \Delta,\ \Delta > 0.$$

Note that the result will be valid in particular for real $x \to \infty$.

We shall apply Watson's lemma with $F(t) = 1/(1 + t^2)$. Then

$$F(t) = \sum_{n=0}^{\infty} (-1)^n t^{2n} = \sum_{n=1}^{\infty} (-1)^{n+1} t^{2n-2}, \qquad |t| < 1,$$

so that we may write

$$F(t) = \sum_{n=1}^{\infty} a_n t^{n-1}, \qquad \text{in } |t| \leqq \frac{5}{6},$$

in which $a_{2n} = 0$, $a_{2n-1} = (-1)^{n+1}$, and we have chosen $r = 1$, $a = \frac{1}{2}$, $\delta = \frac{1}{3}$ in the notation of Watson's lemma.

For $t \geqq \frac{1}{2}$, $e^t > 1$ and $1/(1 + t^2) < 1$, from which

$$F(t) = \frac{1}{1+t^2} < e^t.$$

We may therefore conclude from Watson's lemma that

$$\int_0^\infty \frac{e^{-xt}\,dt}{1+t^2} \sim \sum_{n=1}^{\infty} a_n\Gamma(n)x^{-n},$$

or

$$\int_0^\infty \frac{e^{-xt}\,dt}{1+t^2} \sim \sum_{n=1}^{\infty} \frac{(-1)^{n+1}\Gamma(2n-1)}{x^{2n-1}},$$

and finally that

$$\int_0^\infty \frac{e^{-xt}\,dt}{1+t^2} \sim \sum_{n=0}^{\infty} \frac{(-1)^n(2n)!}{x^{2n+1}}, \tag{6}$$

as $|x| \to \infty$ in $|\arg x| \leqq \frac{1}{2}\pi - \Delta$, $\Delta > 0$.

EXERCISES

1. With the assumptions of Watson's lemma, page 41, show, with the aid of the convergence of the series in (1), that for $0 \leqq t \leqq a$, there exists a positive constant c_1 such that

$$\left| F(t) - \sum_{k=1}^{n} a_k \exp_t\left(\frac{k}{r} - 1\right) \right| < c_1 \exp_t\left(\frac{n+1}{r} - 1\right).$$

2. With the assumptions of Watson's lemma, page 41, show that for $t > a$, there exist positive constants c_2 and β such that

$$\left| F(t) - \sum_{k=1}^{n} a_k \exp_t\left(\frac{k}{r} - 1\right) \right| < c_2 \exp_t\left(\frac{n+1}{r} - 1\right) e^{\beta t}.$$

3. Derive the asymptotic expansion (6) immediately preceding these exercises by applying Watson's lemma to the function

$$f'(x) = -\int_0^\infty \frac{te^{-xt}\,dt}{1+t^2}$$

and then integrating the resultant expansion term by term.

4. Establish (6), page 43, directly, first showing that

$$f(x) - \sum_{k=0}^{n} (-1)^k (2k)!\,x^{-2k-1} = (-1)^{n+1} \int_0^\infty \frac{e^{-xt}t^{2n+2}\,dt}{1+t^2},$$

and thus obtain not only (6) but also a bound on the error made in computing with the series involved.

5. Use integration by parts to establish that for real $x \to \infty$,

$$\int_x^\infty e^{-t}t^{-1}\,dt \sim e^{-x} \sum_{n=0}^{\infty} (-1)^n n!\,x^{-n-1}.$$

6. Let the Hermite polynomials $H_n(x)$ be defined by

$$\exp(2xt - t^2) = \sum_{n=0}^{\infty} \frac{H_n(x)t^n}{n!}$$

for all x and t, as in Chapter 11. Also let the complementary error function erfc x be defined by

$$\operatorname{erfc} x = 1 - \operatorname{erf} x = \frac{2}{\sqrt{\pi}} \int_x^\infty \exp(-\beta^2)\,d\beta.$$

Apply Watson's lemma to the function $F(t) = \exp(2xt - t^2)$; obtain

$$\exp(x - \tfrac{1}{2}s)^2 \int_{\frac{1}{2}s - x}^\infty \exp(-\beta^2)\,d\beta \sim \sum_{n=0}^{\infty} H_n(x)s^{-n-1}, \qquad s \to \infty,$$

and thus arrive at the result

$$\tfrac{1}{2}t^{-1}\sqrt{\pi}\exp[(\tfrac{1}{2}t^{-1} - x)^2]\operatorname{erfc}(\tfrac{1}{2}t^{-1} - x) \sim \sum_{n=0}^{\infty} H_n(x)t^n, \qquad t \to 0^+.$$

7. Use integration by parts to show that if $\operatorname{Re}(\alpha) > 0$, and if x is real,

$$\int_x^\infty e^{-t}t^{-\alpha}\,dt \sim x^{1-\alpha}e^{-x} \sum_{n=0}^{\infty} \frac{(-1)^n(\alpha)_n}{x^{n+1}}, \qquad x \to \infty,$$

of which Ex. 5 is the special case $\alpha = 1$.

CHAPTER 4

The Hypergeometric Function

29. The function $F(a, b; c; z)$. In the study of second-order linear differential equations with three regular singular points, there arises the function

$$F(a, b; c; z) = 1 + \sum_{n=1}^{\infty} \frac{(a)_n (b)_n z^n}{(c)_n n!}, \tag{1}$$

for c neither zero nor a negative integer. In (1) the notation

$$\begin{aligned} (\alpha)_n &= \alpha(\alpha+1)(\alpha+2)\cdots(\alpha+n-1), \qquad n \geqq 1, \\ (\alpha)_0 &= 1, \qquad \alpha \neq 0, \end{aligned} \tag{2}$$

of Section 18 is used. We are here concerned with various properties of the special functions under consideration; that (1) satisfies a certain differential equation is, for us, only one among many facts of interest.

Since

$$\begin{aligned} \operatorname*{Lim}_{n\to\infty} &\left| \frac{(a)_{n+1}(b)_{n+1}z^{n+1}}{(c)_{n+1}(n+1)!} \cdot \frac{(c)_n n!}{(a)_n (b)_n z^n} \right| \\ &= \operatorname*{Lim}_{n\to\infty} \left| \frac{(a+n)(b+n)z}{(c+n)(n+1)} \right| = |z|, \end{aligned}$$

so long as none of a, b, c is zero or a negative integer, the series in (1) has the circle $|z| < 1$ as its circle of convergence. If either or both of a and b is zero or a negative integer, the series terminates, and convergence does not enter the discussion.

On the boundary $|z| = 1$ of the region of convergence, a sufficient condition for absolute convergence of the series is $\text{Re}(c-a-b) > 0$. To prove this, let

$$\delta = \tfrac{1}{2}\,\text{Re}(c - a - b) > 0,$$

and compare terms of the series

$$1 + \sum_{n=1}^{\infty} \left| \frac{(a)_n(b)_n z^n}{(c)_n n!} \right| \tag{3}$$

with corresponding terms of the series

$$\sum_{n=1}^{\infty} \frac{1}{n^{1+\delta}}, \tag{4}$$

known to be convergent. Since $|z| = 1$ and

$$\begin{aligned}
\lim_{n\to\infty} \left| \frac{n^{1+\delta}(a)_n(b)_n}{(c)_n n!} \right|
&= \lim_{n\to\infty} \left| \frac{(a)_n}{(n-1)!n^a} \cdot \frac{(b)_n}{(n-1)!n^b} \cdot \frac{(n-1)!n^c}{(c)_n} \cdot \frac{(n-1)!n^{1+\delta}}{n!n^{c-a-b}} \right| \\
&= \left| \frac{1}{\Gamma(a)} \cdot \frac{1}{\Gamma(b)} \cdot \frac{\Gamma(c)}{1} \right| \lim_{n\to\infty} \left| \frac{1}{n^{c-a-b-\delta}} \right| = 0,
\end{aligned}$$

because $\text{Re}(c - a - b - \delta) = 2\delta - \delta > 0$, the series in (1) is absolutely convergent on $|z| = 1$ when $\text{Re}(c - a - b) > 0$.

A mild variation of the notation $F(a, b; c; z)$ is often used; it is

$$F\left[\begin{matrix} a, b; \\ \\ c; \end{matrix} \quad z \right], \tag{5}$$

which is sometimes more convenient for printing and which has the advantage of exhibiting the numerator parameters a and b above the denominator parameter c, thus making it easy to remember the respective roles of a, b, and c. When we come to the generalized hypergeometric functions, we shall frequently use a notation like that in (5).

The series on the right in (1) or in

$$F\left[\begin{matrix} a, b; \\ \\ c; \end{matrix} \quad z \right] = \sum_{n=0}^{\infty} \frac{(a)_n(b)_n z^n}{(c)_n n!} \tag{6}$$

is called the *hypergeometric series*. The special case $a = c$, $b = 1$ yields the elementary geometric series $\sum_{n=0}^{\infty} z^n$; hence the term *hypergeometric*. The function in (6) or in (1) is correspondingly called the *hypergeometric function*. Although Euler obtained many properties of the function $F(a, b; c; z)$, we owe much of our knowledge of the subject to the more systematic and detailed study made by Gauss.

30. A simple integral form. If n is a non-negative integer,

$$\frac{(b)_n}{(c)_n} = \frac{\Gamma(b+n)\,\Gamma(c)}{\Gamma(c+n)\,\Gamma(b)} = \frac{\Gamma(c)}{\Gamma(b)\,\Gamma(c-b)} \cdot \frac{\Gamma(b+n)\,\Gamma(c-b)}{\Gamma(c+n)}.$$

If $\text{Re}(c) > \text{Re}(b) > 0$, we know from Theorem 7, page 19, and the integral definition of the Beta function, that

$$\frac{\Gamma(b+n)\,\Gamma(c-b)}{\Gamma(c+n)} = \int_0^1 t^{b+n-1}(1-t)^{c-b-1}\,dt.$$

Therefore, for $|z| < 1$,

$$F(a, b; c; z) = \frac{\Gamma(c)}{\Gamma(b)\,\Gamma(c-b)} \sum_{n=0}^{\infty} \frac{(a)_n}{n!} \int_0^1 t^{b+n-1}(1-t)^{c-b-1} z^n\,dt$$

$$= \frac{\Gamma(c)}{\Gamma(b)\,\Gamma(c-b)} \int_0^1 t^{b-1}(1-t)^{c-b-1} \sum_{n=0}^{\infty} \frac{(a)_n (zt)^n\,dt}{n!}.$$

The binomial theorem states that

$$(1-y)^{-a} = \sum_{n=0}^{\infty} \frac{(-a)(-a-1)(-a-2)\cdots(-a-n+1)(-1)^n y^n}{n!},$$

which may be written

$$(1-y)^{-a} = \sum_{n=0}^{\infty} \frac{a(a+1)(a+2)\cdots(a+n-1)y^n}{n!}.$$

Therefore, in factorial function notation,

$$(1-y)^{-a} = \sum_{n=0}^{\infty} \frac{(a)_n y^n}{n!},$$

which we use with $y = zt$ to obtain the following result.

THEOREM 16. *If* $|z| < 1$ *and if* $\text{Re}(c) > \text{Re}(b) > 0$,

$$F(a, b; c; z) = \frac{\Gamma(c)}{\Gamma(b)\,\Gamma(c-b)} \int_0^1 t^{b-1}(1-t)^{c-b-1}(1-tz)^{-a}\,dt.$$

31. $F(a, b; c; 1)$ as a function of the parameters. We know already that if c is neither zero nor a negative integer and if $\text{Re}(c - a - b) > 0$, the series

$$(1) \qquad F(a, b; c; 1) = 1 + \sum_{n=1}^{\infty} \frac{(a)_n(b)_n}{(c)_n n!}$$

is absolutely convergent.

Let δ be any positive number. We shall show that in the region $\text{Re}(c - a - b) \geqq 2\delta > 0$, the series (1) for $F(a, b; c; 1)$ is uniformly convergent. To fix the ideas, it may be desirable to think of $\text{Re}(c - a - b) \geqq 2\delta > 0$ as a region in the c-plane, with a and b chosen first. It is not necessary to look on the region in that way.

The series of positive constants

$$(2) \qquad \sum_{n=1}^{\infty} \frac{1}{n^{1+\delta}}$$

is convergent because $\delta > 0$. We show that for n sufficiently large and for all a, b, c in the region $\text{Re}(c - a - b) \geqq 2\delta > 0$, with c neither zero nor a negative integer,

$$(3) \qquad \left| \frac{(a)_n(b)_n}{(c)_n n!} \right| < \frac{1}{n^{1+\delta}}.$$

Now (see page 46)

$$\lim_{n\to\infty} \left| \frac{(a)_n(b)_n n^{1+\delta}}{(c)_n n!} \right| = \left| \frac{\Gamma(c)}{\Gamma(a)\Gamma(b)} \right| \lim_{n\to\infty} \left| \frac{1}{n^{c-a-b-\delta}} \right| = 0,$$

since $\text{Re}(c - a - b - \delta) \geqq 2\delta - \delta = \delta > 0$. Hence (3) is true for n sufficiently large, and the Weierstrass M-test can be applied to the series in equation (1).

THEOREM 17. *If c is neither zero nor a negative integer and $\text{Re}(c - a - b) > 0$, $F(a, b; c; 1)$ is an analytic function of a, b, c.*

32. Evaluation of $F(a, b; c; 1)$. If $\text{Re}(c - a - b) > 0$, Theorem 17 permits us to extend the integral form for $F(a, b; c; z)$, page 47, to the point $z = 1$ in the following manner. Since $\text{Re}(c - a - b) > 0$, we may write

$$F(a, b; c; 1) = \sum_{n=0}^{\infty} \frac{(a)_n(b)_n}{(c)_n n!}.$$

If we also stipulate that $\text{Re}(c) > \text{Re}(b) > 0$, it follows by the technique of Section 30 that

$$F(a, b; c; 1) = \frac{\Gamma(c)}{\Gamma(b)\Gamma(c-b)} \sum_{n=0}^{\infty} \frac{(a)_n}{n!} \int_0^1 t^{b+n-1}(1-t)^{c-b-1}\, dt$$

$$= \frac{\Gamma(c)}{\Gamma(b)\Gamma(c-b)} \int_0^1 t^{b-1}(1-t)^{c-b-1}(1-t)^{-a}\, dt.$$

Therefore, if $\text{Re}(c - a - b) > 0$, if $\text{Re}(c) > \text{Re}(b) > 0$, and since c is neither zero nor a negative integer,

$$F(a, b; c; 1) = \frac{\Gamma(c)}{\Gamma(b)\Gamma(c-b)} \int_0^1 t^{b-1}(1-t)^{c-a-b-1}\, dt$$

$$= \frac{\Gamma(c)}{\Gamma(b)\Gamma(c-b)} \cdot \frac{\Gamma(b)\Gamma(c-a-b)}{\Gamma(c-a)}$$

$$= \frac{\Gamma(c)\Gamma(c-a-b)}{\Gamma(c-a)\Gamma(c-b)}.$$

We now resort to Theorem 17 and analytic continuation to conclude that the foregoing evaluation of $F(a, b; c; 1)$ is valid without the condition $\text{Re}(c) > \text{Re}(b) > 0$.

THEOREM 18. *If* $\text{Re}(c - a - b) > 0$ *and if c is neither zero nor a negative integer,*

$$F(a, b; c; 1) = \frac{\Gamma(c)\Gamma(c-a-b)}{\Gamma(c-a)\Gamma(c-b)}.$$

The value of $F(a, b; c; 1)$ will play a vital role in many of the results to be obtained in this and later chapters. Theorem 18 can be proved without the aid of the integral in Theorem 16. For such a proof see Whittaker and Watson [1;281–282].

EXAMPLE: Show that if $\text{Re}(b) > 0$ and if n is a non-negative integer,

$$F\left[\begin{matrix} -\tfrac{1}{2}n, \ -\tfrac{1}{2}n + \tfrac{1}{2}; \\ b + \tfrac{1}{2}; \end{matrix} \quad 1\right] = \frac{2^n (b)_n}{(2b)_n}.$$

By Theorem 18 we get

$$F\left[\begin{matrix} -\tfrac{1}{2}n, \ -\tfrac{1}{2}n + \tfrac{1}{2}; \\ b + \tfrac{1}{2}; \end{matrix} \quad 1\right] = \frac{\Gamma(b + \tfrac{1}{2})\Gamma(b + n)}{\Gamma(b + \tfrac{1}{2}n)\Gamma(b + \tfrac{1}{2}n + \tfrac{1}{2})}$$

$$= \frac{(b)_n \Gamma(b)\Gamma(b + \tfrac{1}{2})}{\Gamma(b + \tfrac{1}{2}n)\Gamma(b + \tfrac{1}{2}n + \tfrac{1}{2})}.$$

Legendre's duplication formula for the Gamma function, page 24, yields

$$\Gamma(b)\Gamma(b+\tfrac{1}{2}) = 2^{1-2b}\sqrt{\pi}\,\Gamma(2b),$$

$$\Gamma(b+\tfrac{1}{2}n)\Gamma(b+\tfrac{1}{2}n+\tfrac{1}{2}) = 2^{1-2b-n}\sqrt{\pi}\,\Gamma(2b+n).$$

Therefore

$$F\left[\begin{matrix} -\tfrac{1}{2}n,\ -\tfrac{1}{2}n+\tfrac{1}{2}; & \\ & 1 \\ b+\tfrac{1}{2}; & \end{matrix}\right] = \frac{(b)_n 2^n \Gamma(2b)}{\Gamma(2b+n)} = \frac{2^n (b)_n}{(2b)_n},$$

as desired.

33. The contiguous function relations. Gauss defined as *contiguous* to $F(a, b; c; z)$ each of the six functions obtained by increasing or decreasing one of the parameters by unity. For simplicity in printing, we use the notations

$$F = F(a, b; c; z), \tag{1}$$

$$F(a+) = F(a+1, b; c; z), \tag{2}$$

$$F(a-) = F(a-1, b; c; z), \tag{3}$$

together with similar notations $F(b+)$, $F(b-)$, $F(c+)$, $F(c-)$ for the other four of the six functions contiguous to F.

Gauss proved, and we shall follow his technique, that between F and any two of its contiguous functions, there exists a linear relation with coefficients at most linear in z. The proof is one of remarkable directness; we prove that the relations exist by obtaining them. There are, of course, fifteen (six things taken two at a time) such relations.

Put

$$\delta_n = \frac{(a)_n (b)_n z^n}{(c)_n n!}$$

so that

$$F = \sum_{n=0}^{\infty} \delta_n \tag{4}$$

and

$$F(a+) = \sum_{n=0}^{\infty} \frac{(a+1)_n\, \delta_n}{(a)_n}.$$

Since $a(a+1)_n = (a+n)\ (a)_n$, we may write the six functions contiguous to F in the form

(5) $$F(a+) = \sum_{n=0}^{\infty} \frac{a+n}{a}\,\delta_n, \qquad F(a-) = \sum_{n=0}^{\infty} \frac{a-1}{a-1+n}\,\delta_n,$$

$$F(b+) = \sum_{n=0}^{\infty} \frac{b+n}{b}\,\delta_n, \qquad F(b-) = \sum_{n=0}^{\infty} \frac{b-1}{b-1+n}\,\delta_n,$$

$$F(c+) = \sum_{n=0}^{\infty} \frac{c}{c+n}\,\delta_n, \qquad F(c-) = \sum_{n=0}^{\infty} \frac{c-1+n}{c-1}\,\delta_n.$$

We also employ the differential operator $\theta = z\left(\frac{d}{dz}\right)$. This operator has the particularly pleasant property that $\theta z^n = nz^n$, thus making it handy to use on power series.

Since

(6) $$(\theta + a)F = \sum_{n=0}^{\infty} (a+n)\,\delta_n,$$

it can be seen with the aid of (5) that

(7) $$(\theta + a)F = aF(a+),$$

(8) $$(\theta + b)F = bF(b+),$$

(9) $$(\theta + c - 1)F = (c-1)F(c-).$$

From (7), (8), and (9), it follows at once that

(10) $$(a - b)F = aF(a+) - bF(b+),$$

and

(11) $$(a - c + 1)F = aF(a+) - (c-1)F(c-),$$

two of the simplest of the contiguous function relations.

Next consider

$$\theta F = \sum_{n=1}^{\infty} \frac{n(a)_n(b)_n z^n}{(c)_n n!} = z \sum_{n=0}^{\infty} \frac{(a)_{n+1}(b)_{n+1} z^n}{(c)_{n+1} n!},$$

from which

(12) $$\theta F = z \sum_{n=0}^{\infty} \frac{(a+n)(b+n)\,\delta_n}{c+n}.$$

Now

$$\frac{(a+n)(b+n)}{c+n} = n + (a + b - c) + \frac{(c-a)(c-b)}{c+n}.$$

Hence equation (12) yields

$$\theta F = z\sum_{n=0}^{\infty} n\,\delta_n + (a+b-c)z\sum_{n=0}^{\infty}\delta_n + \frac{(c-a)(c-b)z}{c}\sum_{n=0}^{\infty}\frac{c}{c+n}\,\delta_n,$$

or

$$(13)\quad (1-z)\theta F = (a+b-c)zF + c^{-1}(c-a)(c-b)zF(c+).$$

From (7) we obtain

$$(1-z)\theta F = -a(1-z)F + a(1-z)F(a+),$$

which combines with (13) to yield another of the contiguous function relations,

$$(14)\qquad [(1-z)a+(a+b-c)z]F = a(1-z)F(a+) - c^{-1}(c-a)(c-b)zF(c+).$$

The coefficient of F on the left in (14) is in a form desirable for certain later developments. Equation (14) may be replaced by

$$(15)\quad [a+(b-c)z]F = a(1-z)F(a+) - c^{-1}(c-a)(c-b)zF(c+).$$

Next let us operate with θ on the series defining $F(a-)$. We thus obtain

$$\theta F(a-) = \sum_{n=1}^{\infty}\frac{(a-1)_n(b)_n z^n}{(c)_n(n-1)!} = \sum_{n=0}^{\infty}\frac{(a-1)_{n+1}(b)_{n+1}z^{n+1}}{(c)_{n+1}n!},$$

or

$$(16)\qquad \theta F(a-) = (a-1)z\sum_{n=0}^{\infty}\frac{b+n}{c+n}\,\delta_n.$$

But

$$\frac{b+n}{c+n} = 1 - \frac{c-b}{c+n},$$

so that (16) becomes

$$\theta F(a-) = (a-1)z\sum_{n=0}^{\infty}\delta_n - \frac{(a-1)(c-b)z}{c}\sum_{n=0}^{\infty}\frac{c}{c+n}\,\delta_n,$$

which, in view of (5), yields

$$(17)\qquad \theta F(a-) = (a-1)zF - c^{-1}(a-1)(c-b)zF(c+).$$

We return to (7) and replace a by $(a-1)$ in it to get

$$(18)\qquad \theta F(a-) = (a-1)F - (a-1)F(a-).$$

From (17) and (18) it follows that

$$(19)\qquad (1-z)F = F(a-) - c^{-1}(c-b)zF(c+).$$

Similarly, since a and b may be interchanged without affecting the hypergeometric series, we may write

$$(20) \qquad (1 - z)F = F(b-) - c^{-1}(c - a)zF(c+).$$

We now have five contiguous function relations, (19) and (20) together with

$$(10) \qquad (a - b)F = aF(a+) - bF(b+),$$

$$(11) \qquad (a - c + 1)F = aF(a+) - (c - 1)F(c-),$$

and

$$(15) \qquad [a+(b-c)z]F = a(1-z)F(a+) - c^{-1}(c-a)(c-b)zF(c+).$$

From these five relations the remaining ten may be obtained by performing suitable eliminations. See Ex. 21 at the end of this chapter.

34. The hypergeometric differential equation. The operator $\theta = z\left(\dfrac{d}{dz}\right)$, already used in the derivation of the contiguous function relations, is helpful in deriving a differential equation satisfied by

$$(1) \qquad w = F(a, b; c; z) = \sum_{n=0}^{\infty} \frac{(a)_n(b)_n z^n}{(c)_n n!}.$$

From (1) we obtain

$$\theta(\theta + c - 1)w = \sum_{n=0}^{\infty} \frac{n(n + c - 1)(a)_n(b)_n z^n}{(c)_n n!}$$

$$= \sum_{n=1}^{\infty} \frac{(a)_n(b)_n z^n}{(c)_{n-1}(n - 1)!}.$$

A shift of index yields

$$\theta(\theta + c - 1)w = \sum_{n=0}^{\infty} \frac{(a)_{n+1}(b)_{n+1} z^{n+1}}{(c)_n n!}$$

$$= z \sum_{n=0}^{\infty} \frac{(a + n)(b + n)(a)_n(b)_n z^n}{(c)_n n!}$$

$$= z(\theta + a)(\theta + b)w.$$

We have shown that $w = F(a, b; c; z)$ is a solution of the differential equation

$$(2) \qquad [\theta(\theta + c - 1) - z(\theta + a)(\theta + b)]w = 0. \qquad \theta = z\frac{d}{dz}.$$

Equation (2) is easily put in the form

$$(3) \qquad z(1 - z)w'' + [c - (a + b + 1)z]w' - abw = 0$$

by employing the relations $\theta w = zw'$ and $\theta(\theta - 1)w = z^2w''$.

The second-order linear differential equation (3) is treated in many texts* and therefore we omit details here. In order to avoid tedious repetition, we shall, in this chapter only, refer to the text mentioned in the footnote as IDE.

In IDE, pages 144–148, it is shown that if c is nonintegral, two linearly independent solutions of (3) in $|z| < 1$ are

$$(4) \qquad w_1 = F(a, b; c; z)$$

and

$$(5) \qquad w_2 = z^{1-c}F(a + 1 - c, b + 1 - c; 2 - c; z).$$

We shall also make free use of Kummer's 24 solutions of equation (3) as listed in IDE, pages 157–158. In any specific instance, however, previous knowledge of Kummer's 24 solutions is not necessary; the desired solution can be obtained directly with the aid of simple changes of variable performed on the differential equation (3). See Ex. 12 at the end of this chapter.

35. Logarithmic solutions of the hypergeometric equation. If c is not an integer, the hypergeometric equation

$$(1) \qquad z(1 - z)w'' + [c - (a + b + 1)z]w' - abw = 0$$

always has in $|z| < 1$ the two power series solutions (4) and (5) of the preceding section. If c is an integer, one solution may or may not, depending on the values of a and b, become logarithmic. In this book we are primarily interested in the functions rather than in the differential equations. We shall, whenever it is feasible, avoid discussion of logarithmic solutions.

If c is a positive integer and neither a nor b is an integer, two linearly independent solutions of (1) are as listed below. These solutions may be obtained by standard elementary techniques (see Rainville [1], Sections 92 and 94), and the details are therefore omitted here.

If $c = 1$ and neither a nor b is zero or a negative integer, two linearly independent solutions of (1) valid in $0 < |z| < 1$ are

*See, for example, Chapter 6 of Rainville [2].

(2) $$w_1 = F(a, b; 1; z),$$

(3) $$w_3 = F(a, b; 1; z)\log z + \sum_{n=1}^{\infty} \frac{(a)_n(b)_n z^n}{(n!)^2}\{H(a, n) + H(b, n) - 2H(1, n)\},$$

in which

(4) $$H(a, n) = \sum_{k=1}^{n} \frac{1}{a + k - 1},$$

including the ordinary harmonic sum $H(1, n) = H_n$.

If c is an integer >1 and neither a nor b is an integer, two linearly independent solutions of (1) valid in $0 < |z| < 1$ are

(5) $$w_1 = F(a, b; c; z)$$

and

(6) $$w_4 = F(a, b; c; z)\log z + \sum_{n=1}^{\infty} \frac{(a)_n(b)_n z^n}{(c)_n n!}\{H(a, n) + H(b, n) - H(c, n) - H(1, n)\} - \sum_{n=0}^{c-2} \frac{n!(1 - c)_{n+1}}{(1 - a)_{n+1}(1 - b)_{n+1} z^{n+1}}.$$

If c is an integer, $c \leqq 0$, equation (1) may be transformed by using $w = z^{1-c}y$ into a hypergeometric equation for y with new parameters $a' = a + 1 - c$, $b' = b + 1 - c$, and $c' = 2 - c$. If neither a' nor b' is an integer, the y-equation can be solved by using (5) and (6).

36. $F(a, b; c; z)$ as a function of its parameters. We have already noted that the series in

(1) $$F(a, b; c; z) = \sum_{n=0}^{\infty} \frac{(a)_n(b)_n z^n}{(c)_n n!}$$

is absolutely convergent (ratio test) for $|z| < 1$, independent of the choice of a, b, c as long as c is neither zero nor a negative integer. Recall that $(c)_n = \Gamma(c + n)/\Gamma(c)$. Consider the function

(2) $$\frac{F(a, b; c; z)}{\Gamma(c)} = \sum_{n=0}^{\infty} \frac{(a)_n(b)_n z^n}{\Gamma(c + n)n!},$$

in which the possibility of zero denominators on the right has been removed. In any closed region in the finite a, b, and c planes,

$$\lim_{n\to\infty}\left|\frac{(a)_n(b)_n z^{\frac{1}{2}n}}{\Gamma(c+n)n!}\right|$$

$$= \lim_{n\to\infty}\left|\frac{(a)_n}{(n-1)!n^a}\frac{(b)_n}{(n-1)!n^b}\frac{(n-1)!n^c}{\Gamma(c+n)}\frac{z^{\frac{1}{2}n}}{n^{1+c-a-b}}\right|$$

$$= \frac{1}{\Gamma(a)\Gamma(b)}\lim_{n\to\infty}\left|\frac{z^{\frac{1}{2}n}}{n^{1+c-a-b}}\right| = 0, \qquad \text{for } |z| < 1.$$

Therefore, for any fixed z in $|z| < 1$, there exists a constant K independent of a, b, c and such that

$$\left|\frac{(a)_n(b)_n z^n}{\Gamma(c+n)n!}\right| < K|z|^{\frac{1}{2}n}.$$

Since $\sum_{n=0}^{\infty} K|z|^{\frac{1}{2}n}$ is convergent, the series on the right in (2) is absolutely and uniformly convergent for all finite a, b, c as long as $|z| < 1$.

We know the location and character of the singular points of $\Gamma(c)$ and are now able to stipulate the behavior, with regard to analyticity, of the hypergeometric function with z fixed, $|z| < 1$, and a, b, c as variables.

THEOREM 19. *For $|z| < 1$ the function $F(a, b; c; z)$ is analytic in a, b, and c for all finite a, b, and c except for simple poles at c = zero and c = each negative integer.*

Reference to Theorem 19 will enable us to simplify many proofs in later work.

37. Elementary series manipulations. A common tool to be used in much of our later work is the rearrangement of terms in iterated series. Here we prove two basic lemmas of the kind needed. When convergent power series are involved, the infinite rearrangements can be justified in the elementary sense. In our study of generating functions of sets of polynomials, we sometimes deal with divergent power series. For such series the identities of this section may be considered as purely formal, but we shall find that the manipulative techniques are fully as useful as when applied to convergent series.

Lemma 10.

$$(1) \qquad \sum_{n=0}^{\infty}\sum_{k=0}^{\infty} A(k, n) = \sum_{n=0}^{\infty}\sum_{k=0}^{n} A(k, n-k),$$

and

$$\sum_{n=0}^{\infty}\sum_{k=0}^{n} B(k, n) = \sum_{n=0}^{\infty}\sum_{k=0}^{\infty} B(k, n + k). \tag{2}$$

Proof: Consider the series

$$\sum_{n=0}^{\infty}\sum_{k=0}^{\infty} A(k, n)t^{n+k} \tag{3}$$

in which t^{n+k} has been inserted for convenience and will be removed later by placing $t = 1$. Let us collect the powers of t in (3). We introduce new indices of summation j and m by

$$k = j, \qquad n = m - j, \tag{4}$$

so that the exponent $(n + k)$ in (3) becomes m. The old indices n and k in (3) are restricted, as indicated in the summation symbol, by the inequalities

$$n \geqq 0, \qquad k \geqq 0. \tag{5}$$

Because of (4) the inequalities (5) become

$$m - j \geqq 0, \qquad j \geqq 0,$$

or $0 \leqq j \leqq m$ with m, the exponent on t, restricted only in that it must be a non-negative integer. Thus we arrive at

$$\sum_{n=0}^{\infty}\sum_{k=0}^{\infty} A(k, n)t^{n+k} = \sum_{m=0}^{\infty}\sum_{j=0}^{m} A(j, m - j)t^{m}, \tag{6}$$

and the identity (1) of Lemma 10 follows by putting $t = 1$ and replacing the dummy indices j and m on the right by dummy indices k and n.

There is no need to use k and n for indices in both members of (1), but neither is there harm in it once a small degree of sophistication is acquired. We frequently employ many parameters and prefer to keep to a minimum the number of different symbols used.

In Lemma 10, equation (2) is merely (1) written in reverse; it needs no separate derivation.

Lemma 11.

$$\sum_{n=0}^{\infty}\sum_{k=0}^{\infty} A(k, n) = \sum_{n=0}^{\infty}\sum_{k=0}^{[n/2]} A(k, n - 2k), \tag{7}$$

and

$$\sum_{n=0}^{\infty}\sum_{k=0}^{[n/2]} B(k, n) = \sum_{n=0}^{\infty}\sum_{k=0}^{\infty} B(k, n + 2k). \tag{8}$$

Proof: Consider the series

$$\sum_{n=0}^{\infty}\sum_{k=0}^{\infty} A(k, n)t^{n+2k} \tag{9}$$

and in it collect powers of t, introducing new indices by

$$k = j, \qquad n = m - 2j, \tag{10}$$

so that $n + 2k = m$. Since

$$n \geqq 0, \qquad k \geqq 0, \tag{11}$$

we conclude that

$$m - 2j \geqq 0, \qquad j \geqq 0,$$

from which $0 \leqq 2j \leqq m$ and $m \geqq 0$. Since $0 \leqq j \leqq \frac{1}{2}m$ and j is integral, the index j runs from 0 to the greatest integer in $\frac{1}{2}m$. Thus we obtain

$$\sum_{n=0}^{\infty}\sum_{k=0}^{\infty} A(k, n)t^{n+2k} = \sum_{m=0}^{\infty}\sum_{j=0}^{[\frac{1}{2}m]} A(j, m - 2j)t^{m}, \tag{12}$$

from which (7) follows by placing $t = 1$ and making the proper change of letters for the dummy indices on the right in (12). Equation (8) is (7) written in reverse order.

There is no bound to the number of such identities. The reader should now be able to obtain whatever lemmas he needs along these lines.

Note also that a combination of Lemmas 10 and 11 yields

$$\sum_{n=0}^{\infty}\sum_{k=0}^{n} C(k, n) = \sum_{n=0}^{\infty}\sum_{k=0}^{[n/2]} C(k, n - k). \tag{13}$$

38. Simple transformations. It is convenient for us to write the ordinary binomial expansion with the factorial function notation,

$$(1 - z)^{-a} = \sum_{n=0}^{\infty} \frac{(a)_n z^n}{n!} \tag{1}$$

and to recall the result of Ex. 8, page 32,

$$(\alpha)_{n-k} = \frac{(-1)^k(\alpha)_n}{(1 - \alpha - n)_k}, \qquad 0 \leqq k \leqq n. \tag{2}$$

In particular $\alpha = 1$ in (2) yields

$$(n - k)! = \frac{(-1)^k n!}{(-n)_k}, \qquad 0 \leqq k \leqq n. \tag{3}$$

Consider now the product

$$(1-z)^{-a}F\left[\begin{matrix}a, c-b; & \\ & \dfrac{-z}{1-z} \\ c; & \end{matrix}\right] = \sum_{k=0}^{\infty} \frac{(a)_k(c-b)_k(-1)^k z^k}{(c)_k k!(1-z)^{k+a}}.$$

With the aid of (1) we may write

$$(1-z)^{-a}F\left[\begin{matrix}a, c-b; & \\ & \dfrac{-z}{1-z} \\ c; & \end{matrix}\right] = \sum_{k=0}^{\infty}\sum_{n=0}^{\infty} \frac{(a)_k(c-b)_k(a+k)_n(-1)^k z^{n+k}}{(c)_k k! n!}.$$

Now $(a)_k(a+k)_n = (a)_{n+k}$, so that

$$(1-z)^{-a}F\left[\begin{matrix}a, c-b; & \\ & \dfrac{-z}{1-z} \\ c; & \end{matrix}\right] = \sum_{n=0}^{\infty}\sum_{k=0}^{\infty} \frac{(c-b)_k(a)_{n+k}(-1)^k z^{n+k}}{(c)_k k! n!}$$

and we collect powers of z to obtain

$$(1-z)^{-a}F\left[\begin{matrix}a, c-b; & \\ & \dfrac{-z}{1-z} \\ c; & \end{matrix}\right] = \sum_{n=0}^{\infty}\sum_{k=0}^{n} \frac{(c-b)_k(a)_n(-1)^k z^n}{(c)_k k!(n-k)!}$$

$$= \sum_{n=0}^{\infty}\sum_{k=0}^{n} \frac{(-n)_k(c-b)_k}{(c)_k k!} \cdot \frac{(a)_n z^n}{n!},$$

by (3). The inner sum on the right is a terminating hypergeometric series. Hence

$$(1-z)^{-a}F\left[\begin{matrix}a, c-b; & \\ & \dfrac{-z}{1-z} \\ c; & \end{matrix}\right] = \sum_{n=0}^{\infty} F\left[\begin{matrix}-n, c-b; & \\ & 1 \\ c; & \end{matrix}\right]\frac{(a)_n z^n}{n!}.$$

Since $F(-n, c-b; c; 1)$ terminates, we may write

$$(1-z)^{-a}F\left[\begin{matrix}a, c-b; & \\ & \dfrac{-z}{1-z} \\ c; & \end{matrix}\right] = \sum_{n=0}^{\infty} \frac{\Gamma(c)\Gamma(b+n)(a)_n z^n}{\Gamma(c+n)\Gamma(b) n!}$$

$$= \sum_{n=0}^{\infty} \frac{(b)_n(a)_n z^n}{(c)_n n!}$$

$$= F(a, b; c; z),$$

a result valid where both $|z| < 1$ and $|z/(1-z)| < 1$ (for which see Figure 3, page 60).

THEOREM 20. *If* $|z| < 1$ *and* $|z/(1-z)| < 1$,

$$(4)\quad F\left[\begin{matrix} a, b; & \\ & z \\ c; & \end{matrix}\right] = (1-z)^{-a} F\left[\begin{matrix} a, c-b; & \\ & \dfrac{-z}{1-z} \\ c; & \end{matrix}\right].$$

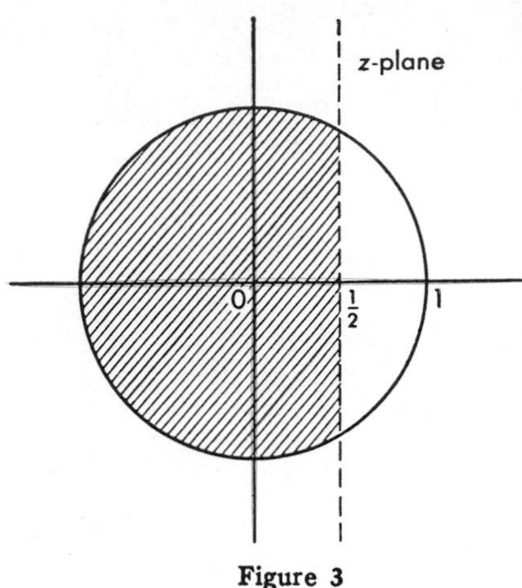

Figure 3

The roles of a and b may be interchanged in (4).

The type of series manipulations involved above in arriving at the identity (4) will be used frequently throughout the remainder of this book, and such steps will be taken hereafter without detailed explanation.

Let us use Theorem 20 on the hypergeometric function on the right in (4). Put

$$y = \frac{-z}{1-z}.$$

Then

$$F\left[\begin{matrix} a, c-b; & \\ & y \\ c; & \end{matrix}\right] = (1-y)^{-c+b} F\left[\begin{matrix} c-a, c-b; & \\ & \dfrac{-y}{1-y} \\ c; & \end{matrix}\right].$$

But $1 - y = (1-z)^{-1}$ and $-y/(1-y) = z$. Hence

$$F\left[\begin{matrix} a, c-b; & \\ & \dfrac{-z}{1-z} \\ c; & \end{matrix}\right] = (1-z)^{c-b} F\left[\begin{matrix} c-a, c-b; & \\ & z \\ c; & \end{matrix}\right],$$

which combines with Theorem 20 to yield the following result due to Euler.

THEOREM 21. *If* $|z| < 1$,

$$(5)\qquad F(a, b; c; z) = (1-z)^{c-a-b} F(c-a, c-b; c; z).$$

The identities in Theorems 20 and 21 are statements of equality among certain of the 24 Kummer solutions of the hypergeometric differential equation. In the terminology of IDE, pages 157–158, we have shown that IIIa = Va = IIIb. Alternate proofs of Theorems 20 and 21, making use of the differential equation, are left as exercises.

39. Relation between functions of z and $1 - z$. The hypergeometric differential equation

$$(1) \qquad z(1-z)w'' + [c - (a+b+1)z]w' - abw = 0$$

has, in $|1 - z| < 1$, the solution

$$(2) \qquad w = F(a, b; a+b+1-c; 1-z),$$

as indicated in IDE, page 157, formula IVa. The solution (2) can be obtained independently by placing $z = 1 - y$ in the differential equation (1) and observing that the transformed equation is also hypergeometric with parameters $a' = a$, $b' = b$, $c' = a + b + 1 - c$, and argument $y = 1 - z$.

We already know that in $|z| < 1$, the equation (1) has the linearly independent solutions

$$(3) \qquad w_1 = F(a, b; c; z),$$

$$(4) \qquad w_2 = z^{1-c}F(a+1-c, b+1-c; 2-c; z).$$

Then there must exist constants A and B such that

$$(5) \qquad F(a, b; a+b+1-c; 1-z) = AF(a, b; c; z) + Bz^{1-c}F(a+1-c, b+1-c; 2-c; z)$$

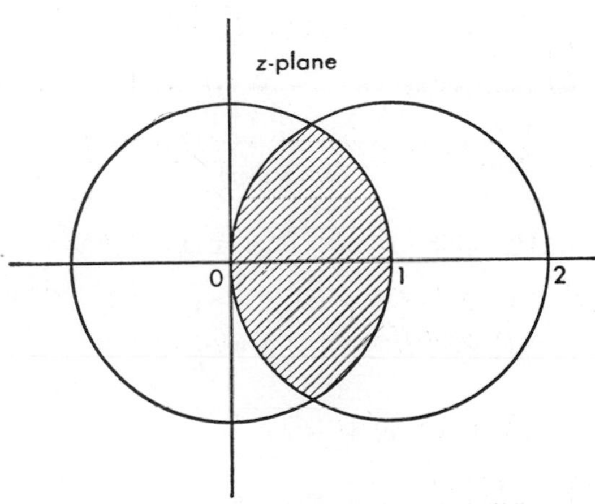

Figure 4

is an identity in the region (Figure 4) where both $|z| < 1$ and $|1 - z| < 1$. If we insist that $\text{Re}(1 - c) > 0$ and let $z \to 0$ from within the pertinent region, (5) yields

$$F(a, b; a+b+1-c; 1) = A \cdot 1 + B \cdot 0,$$

from which, by Theorem 18, page 49,

$$(6) \qquad A = \frac{\Gamma(a+b+1-c)\Gamma(1-c)}{\Gamma(a+1-c)\Gamma(b+1-c)}.$$

Again from (5) if we let $z \to 1$ from inside the region and insist that $\text{Re}(c-a-b) > 0$, we obtain

$$1 = AF(a, b; c; 1) + BF(a+1-c, b+1-c; 2-c; 1),$$

or, by Theorem 18,

$$(7) \qquad \frac{\Gamma(c)\Gamma(c-a-b)A}{\Gamma(c-a)\Gamma(c-b)} + \frac{\Gamma(2-c)\Gamma(c-a-b)B}{\Gamma(1-a)\Gamma(1-b)} = 1.$$

Employing (6) in (7) leads to

$$(8) \qquad \frac{\Gamma(2-c)\Gamma(c-a-b)B}{\Gamma(1-a)\Gamma(1-b)}$$

$$= 1 - \frac{\Gamma(a+b+1-c)\Gamma(1-c)\Gamma(c)\Gamma(c-a-b)}{\Gamma(a+1-c)\Gamma(b+1-c)\Gamma(c-a)\Gamma(c-b)}.$$

In Ex. 15, page 32, we showed that the right member of (8) is equal to

$$(9) \qquad \frac{\Gamma(2-c)\Gamma(c-1)\Gamma(c-a-b)\Gamma(a+b+1-c)}{\Gamma(a)\Gamma(1-a)\Gamma(b)\Gamma(1-b)}.$$

From (8) and (9) we obtain

$$(10) \qquad B = \frac{\Gamma(a+b+1-c)\Gamma(c-1)}{\Gamma(a)\Gamma(b)},$$

which completes the proof of the following statement.

Theorem 22. *If* $|z| < 1$ *and* $|1-z| < 1$, *if* $\text{Re}(c) < 1$ *and* $\text{Re}\,(c-a-b) > 0$, *and if none of* $a, b, c, c-a, c-b, c-a-b$ *is an integer,*

$$(11) \quad F(a, b; a+b+1-c; 1-z)$$

$$= \frac{\Gamma(a+b+1-c)\Gamma(1-c)}{\Gamma(a+1-c)\Gamma(b+1-c)} \cdot F(a, b; c; z)$$

$$+ \frac{\Gamma(a+b+1-c)\Gamma(c-1)}{\Gamma(a)\Gamma(b)} \cdot z^{1-c}F(a+1-c, b+1-c; 2-c; z).$$

The restrictions on a, b, c can be relaxed somewhat, if desired. The expression of $F(a, b; c; z)$ as a linear combination of functions of $(1 - z)$ is left as an exercise; see Ex. 15 at the end of this chapter.

Theorems 20, 21, and 22 exhibit only three of the numerous relations among the 24 Kummer solutions. For other such relations see volume one of the Bateman Manuscript Project work, Erdélyi [1; 106–109].

40. A quadratic transformation. In any detailed study of the hypergeometric differential equation

$$z(1 - z)w'' + [c - (a + b + 1)z]w' - abw = 0, \tag{1}$$

the derivation of the 24 Kummer solutions is a natural result of the study of transformations of equation (1) into itself under linear fractional transformations on the independent variable. It is reasonable to attempt to use a quadratic transformation on the independent variable for the same purpose. Such a study shows that the parameters a, b, c need to be related for the new equation to be of hypergeometric character. Since differential equations are not our primary interest, we bypass the fairly simple determination of all such quadratic transformations and corresponding relations among a, b, and c. Here we move directly to the particular change of variables which leads to the relation we need for our later work.

In equation (1) put $c = 2b$ to get

$$z(1 - z)w'' + [2b - (a + b + 1)z]w' - abw = 0, \tag{2}$$

of which one solution is $w = F(a, b; 2b; z)$. Next let

$$z = \frac{4x}{(1 + x)^2}, \tag{3}$$

and obtain, after the usual labor involved in changing independent variables, the equation

$$x(1 - x)(1 + x)^2\frac{d^2w}{dx^2} + 2(1 + x)(b - 2ax + bx^2 - x^2)\frac{dw}{dx} - 4ab(1 - x)w = 0, \tag{4}$$

of which one solution is therefore

$$w = F\left[\begin{matrix} a, b; \\ 2b; \end{matrix} \quad \frac{4x}{(1 + x)^2}\right]. \tag{5}$$

In (4) put $w = (1 + x)^{2a}y$ to obtain the equation

$$x(1-x^2)y''+2[b-(2a-b+1)x^2]y'-2ax(1+2a-2b)y=0, \tag{6}$$

of which one solution is

$$(7) \qquad y = (1+x)^{-2a} F\left[\begin{matrix} a, b; & \\ & \dfrac{4x}{(1+x)^2} \\ 2b; & \end{matrix}\right].$$

The differential equation (6) is invariant under a change from x to $(-x)$. Hence we introduce a new independent variable $v = x^2$. The equation in y and v is found to be

$$(8) \quad v(1-v)\frac{d^2y}{dv^2} + \left[b + \frac{1}{2} - \left(2a - b + \frac{3}{2}\right)v\right]\frac{dy}{dv} - a\left(a - b + \frac{1}{2}\right)y = 0,$$

which has, in $|v| < 1$, the general solution

$$(9) \quad y = AF\left[\begin{matrix} a, a-b+\frac{1}{2}; & \\ & v \\ b+\frac{1}{2}; & \end{matrix}\right] + Bv^{\frac{1}{2}-b}F\left[\begin{matrix} a-b+\frac{1}{2}, a+1-2b; & \\ & v \\ \frac{3}{2}-b; & \end{matrix}\right].$$

We now have the following situation. The differential equation (6) has a solution (7) valid in

$$\left|\frac{4x}{(1+x)^2}\right| < 1$$

as long as $2b$ is neither zero nor a negative integer. At the same time, equation (6) has the general solution* (9) with $v = x^2$, this solution valid in $|x| < 1$.

Therefore, if both $|x| < 1$ and $\left|\dfrac{4x}{(1+x)^2}\right| < 1$ and if $2b$ is neither zero nor a negative integer, there exist constants A and B such that

$$(10) \quad (1+x)^{-2a}F\left[\begin{matrix} a, b; & \\ & \dfrac{4x}{(1+x)^2} \\ 2b; & \end{matrix}\right] = AF\left[\begin{matrix} a, a-b+\frac{1}{2}; & \\ & x^2 \\ b+\frac{1}{2}; & \end{matrix}\right]$$

$$+ Bx^{1-2b}F\left[\begin{matrix} a-b+\frac{1}{2}, a+1-2b; & \\ & x^2 \\ \frac{3}{2}-b; & \end{matrix}\right].$$

In (10) the left member and the first term on the right are analytic at $x = 0$, but the last term is not analytic at $x = 0$ because of the

*If $2b$ is a positive integer, the second term on the right in (9) may or may not need to be replaced by a logarithmic solution. If such a logarithmic solution is involved in (9), reasoning parallel to that following equation (10) shows again that $B = 0$.

factor x^{1-2b}. Hence $B = 0$ and A is easily determined by using $x = 0$ in the resultant identity

$$(11)\quad (1+x)^{-2a}F\left[\begin{matrix} a, b; & \\ & \dfrac{4x}{(1+x)^2} \\ 2b; & \end{matrix}\right] = AF\left[\begin{matrix} a, a-b+\frac{1}{2}; & \\ & x^2 \\ b+\frac{1}{2}; & \end{matrix}\right].$$

Thus $A = 1$, and we are led to the following result due to Gauss.

THEOREM 23. *If $2b$ is neither zero nor a negative integer, and if both $|x| < 1$ and $|4x(1+x)^{-2}| < 1$,*

$$(12)\quad (1+x)^{-2a}F\left[\begin{matrix} a, b; & \\ & \dfrac{4x}{(1+x)^2} \\ 2b; & \end{matrix}\right] = F\left[\begin{matrix} a, a-b+\frac{1}{2}; & \\ & x^2 \\ b+\frac{1}{2}; & \end{matrix}\right].$$

41. Other quadratic transformations. For variety of technique we shall now prove the following theorem without recourse to the differential equation.

THEOREM 24. *If $2b$ is neither zero nor a negative integer and if $|y| < \frac{1}{2}$ and $|y/(1-y)| < 1$,*

$$(1)\quad (1-y)^{-a}F\left[\begin{matrix} \frac{1}{2}a, \frac{1}{2}a+\frac{1}{2}; & \\ & \dfrac{y^2}{(1-y)^2} \\ b+\frac{1}{2}; & \end{matrix}\right] = F\left[\begin{matrix} a, b; & \\ & 2y \\ 2b; & \end{matrix}\right].$$

Proof: Let ψ denote the left member of (1). Then

$$\psi = \sum_{k=0}^{\infty} \frac{(\frac{1}{2}a)_k(\frac{1}{2}a+\frac{1}{2})_k y^{2k}}{(b+\frac{1}{2})_k k!(1-y)^{a+2k}} = \sum_{k=0}^{\infty} \frac{(a)_{2k} y^{2k}}{2^{2k}(b+\frac{1}{2})_k k!(1-y)^{a+2k}}$$

with the aid of Lemma 5, page 22. Also

$$(1-y)^{-a-2k} = \sum_{n=0}^{\infty} \frac{(a+2k)_n y^n}{n!}$$

and $(a)_{2k}\,(a+2k)_n = (a)_{n+2k}$. Hence

$$\psi = \sum_{n,k=0}^{\infty} \frac{(a)_{n+2k} y^{n+2k}}{2^{2k}(b+\frac{1}{2})_k k!n!}.$$

Using Lemma 11, page 57, we may collect powers of y and obtain

$$\psi = \sum_{n=0}^{\infty} \sum_{k=0}^{[n/2]} \frac{(a)_n y^n}{2^{2k}(b+\frac{1}{2})_k k!(n-2k)!}.$$

We know that $(n-2k)! = n!/(-n)_{2k}$ and that

$$(-n)_{2k} = 2^{2k}(-\tfrac{1}{2}n)_k(-\tfrac{1}{2}n+\tfrac{1}{2})_k.$$

Therefore we have

$$\psi = \sum_{n=0}^{\infty}\sum_{k=0}^{[n/2]}\frac{(-\tfrac{1}{2}n)_k(-\tfrac{1}{2}n+\tfrac{1}{2})_k}{(b+\tfrac{1}{2})_k k!}\cdot\frac{(a)_n y^n}{n!}$$

$$= \sum_{n=0}^{\infty} F\left[\begin{matrix}-\tfrac{1}{2}n, -\tfrac{1}{2}n+\tfrac{1}{2}; \\ b+\tfrac{1}{2};\end{matrix}\; 1\right]\frac{(a)_n y^n}{n!}.$$

In the example on page 49 we found that the terminating hypergeometric function above has the value $2^n(b)_n/(2b)_n$. Hence

$$\psi = \sum_{n=0}^{\infty}\frac{2^n(b)_n(a)_n y^n}{(2b)_n n!} = F\left[\begin{matrix}a, b; \\ 2b;\end{matrix}\; 2y\right],$$

which completes the proof of Theorem 24.

In Theorem 24 put $y = 2x/(1+x)^2$. Then

$$1 - y = \frac{1+x^2}{(1+x)^2}, \qquad \frac{y}{1-y} = \frac{2x}{1+x^2}$$

and we may write

$$(1+x^2)^{-a}(1+x)^{2a}F\left[\begin{matrix}\tfrac{1}{2}a, \tfrac{1}{2}a+\tfrac{1}{2}; \\ b+\tfrac{1}{2};\end{matrix}\; \frac{4x^2}{(1+x^2)^2}\right] = F\left[\begin{matrix}a, b; \\ 2b;\end{matrix}\; \frac{4x}{(1+x)^2}\right].$$

In view of Theorem 23 we may now conclude that

$$(1+x^2)^{-a}F\left[\begin{matrix}\tfrac{1}{2}a, \tfrac{1}{2}a+\tfrac{1}{2}; \\ b+\tfrac{1}{2};\end{matrix}\; \frac{4x^2}{(1+x^2)^2}\right] = F\left[\begin{matrix}a, a-b+\tfrac{1}{2}; \\ b+\tfrac{1}{2};\end{matrix}\; x^2\right].$$

Now put $x^2 = z$ and replace b by $(\tfrac{1}{2}+a-b)$ to obtain

$$(2)\quad (1+z)^{-a}F\left[\begin{matrix}\tfrac{1}{2}a, \tfrac{1}{2}a+\tfrac{1}{2}; \\ 1+a-b;\end{matrix}\; \frac{4z}{(1+z)^2}\right] = F\left[\begin{matrix}a, b; \\ 1+a-b;\end{matrix}\; z\right].$$

By Theorem 20, page 60, with appropriate substitutions for the a, b, c and z of the theorem,

$$F\left[\begin{matrix}\tfrac{1}{2}a, \tfrac{1}{2}a+\tfrac{1}{2}; \\ 1+a-b;\end{matrix}\; \frac{4z}{(1+z)^2}\right] = \left(\frac{1-z}{1+z}\right)^{-a}F\left[\begin{matrix}\tfrac{1}{2}a, \tfrac{1}{2}+\tfrac{1}{2}a-b; \\ 1+a-b;\end{matrix}\; \frac{-4z}{(1-z)^2}\right].$$

We may therefore rewrite (2) in the form

$$(3)\quad (1-z)^{-a}F\left[\begin{matrix}\tfrac{1}{2}a, \tfrac{1}{2}+\tfrac{1}{2}a-b; & \\ & \dfrac{-4z}{(1-z)^2}\\ 1+a-b; & \end{matrix}\right] = F\left[\begin{matrix}a, b; & \\ & z\\ 1+a-b; & \end{matrix}\right],$$

which will be useful in Section 42.

Let us next return to the differential equation to establish one more relation involving a quadratic transformation.

THEOREM 25. *If $a + b + \frac{1}{2}$ is neither zero nor a negative integer, and if $|x| < 1$ and $|4x(1-x)| < 1$,*

$$(4)\quad F\left[\begin{matrix}a, b; & \\ & 4x(1-x)\\ a+b+\tfrac{1}{2}; & \end{matrix}\right] = F\left[\begin{matrix}2a, 2b; & \\ & x\\ a+b+\tfrac{1}{2}; & \end{matrix}\right].$$

The function

$$(5)\quad y = F\left[\begin{matrix}a, b; & \\ & z\\ a+b+\tfrac{1}{2}; & \end{matrix}\right]$$

is a solution of the differential equation

$$(6)\quad z(1-z)\frac{d^2y}{dz^2} + [a+b+\tfrac{1}{2} - (a+b+1)z]\frac{dy}{dz} - aby = 0.$$

In (6) put $z = 4x(1-x)$, and with some labor thus obtain

$$(7)\quad x(1-x)y'' + [a+b+\tfrac{1}{2} - (2a+2b+1)x]y' - 4aby = 0.$$

Equation (7) is hypergeometric in character and has the general solution

$$y = AF\left[\begin{matrix}2a, 2b; & \\ & x\\ a+b+\tfrac{1}{2}; & \end{matrix}\right] + Bx^{\frac{1}{2}-a-b}F\left[\begin{matrix}\tfrac{1}{2}+a-b, \tfrac{1}{2}+b-a; & \\ & x\\ \tfrac{3}{2}-a-b; & \end{matrix}\right],$$

as well as the solution

$$y = F\left[\begin{matrix}a, b; & \\ & 4x(1-x)\\ a+b+\tfrac{1}{2}; & \end{matrix}\right]$$

from (5) above. By the usual argument it is easy to conclude the validity of (4).

42. A theorem due to Kummer. Let us return to equation (3) of the preceding section and let $z \to -1$. The result is

$$2^{-a}F\left[\begin{matrix}\tfrac{1}{2}a, \tfrac{1}{2} + \tfrac{1}{2}a - b; \\ 1 + a - b;\end{matrix} \quad 1\right] = F\left[\begin{matrix}a, b; \\ 1 + a - b;\end{matrix} \quad -1\right].$$

We can sum the series on the left and thus obtain

$$(1) \qquad F\left[\begin{matrix}a, b; \\ 1 + a - b;\end{matrix} \quad -1\right] = \frac{\Gamma(1 + a - b)\Gamma(\tfrac{1}{2})}{2^{a}\Gamma(1 + \tfrac{1}{2}a - b)\Gamma(\tfrac{1}{2} + \tfrac{1}{2}a)}.$$

Legendre's duplication formula, page 24, yields

$$\Gamma(\tfrac{1}{2})\Gamma(1 + a) = 2^{a}\Gamma(\tfrac{1}{2} + \tfrac{1}{2}a)\Gamma(1 + \tfrac{1}{2}a),$$

which may be used on the right in (1).

THEOREM 26. *If* $(1 + a - b)$ *is neither zero nor a negative integer, and* $\text{Re}(b) < 1$ *for convergence,*

$$(2) \qquad F\left[\begin{matrix}a, b; \\ 1 + a - b;\end{matrix} \quad -1\right] = \frac{\Gamma(1 + a - b)\Gamma(1 + \tfrac{1}{2}a)}{\Gamma(1 + \tfrac{1}{2}a - b)\Gamma(1 + a)}.$$

43. Additional properties. Further results applying to special hypergeometric functions appear in later chapters, where we shall find that the polynomials of Legendre, Jacobi, Gegenbauer, and others are terminating hypergeometric series.

We now obtain one more identity as an example of those resulting from combinations of the theorems proved earlier in this chapter. In the identity of Theorem 25, page 67, replace a by $(\tfrac{1}{2}c - \tfrac{1}{2}a)$ and b by $(\tfrac{1}{2}c + \tfrac{1}{2}a - \tfrac{1}{2})$ to get

$$F\left[\begin{matrix}\tfrac{1}{2}c - \tfrac{1}{2}a, \tfrac{1}{2}c + \tfrac{1}{2}a - \tfrac{1}{2}; \\ c;\end{matrix} \quad 4x(1 - x)\right] = F\left[\begin{matrix}c - a, c + a - 1; \\ c;\end{matrix} \quad x\right].$$

Theorem 21, page 60, yields

$$F\left[\begin{matrix}c - a, c + a - 1; \\ c;\end{matrix} \quad x\right] = (1 - x)^{1-c}F\left[\begin{matrix}a, 1 - a; \\ c;\end{matrix} \quad x\right],$$

which leads to the desired result.

THEOREM 27. *If c is neither zero nor a negative integer, and if both* $|x| < 1$ *and* $|4x(1 - x)| < 1$,

$$F\left[\begin{matrix} a, 1-a; \\ c; \end{matrix} \; x\right] = (1-x)^{c-1} F\left[\begin{matrix} \tfrac{1}{2}c - \tfrac{1}{2}a, \tfrac{1}{2}c + \tfrac{1}{2}a - \tfrac{1}{2}; \\ c; \end{matrix} \; 4x(1-x)\right].$$

EXERCISES

1. Show that

$$\frac{d}{dx} F\left[\begin{matrix} a, b; \\ c; \end{matrix} \; x\right] = \frac{ab}{c} F\left[\begin{matrix} a+1, b+1; \\ c+1; \end{matrix} \; x\right].$$

2. Show that

$$F\left[\begin{matrix} 2a, 2b; \\ a+b+\tfrac{1}{2}; \end{matrix} \; \tfrac{1}{2}\right] = \frac{\Gamma(a+b+\tfrac{1}{2})\Gamma(\tfrac{1}{2})}{\Gamma(a+\tfrac{1}{2})\Gamma(b+\tfrac{1}{2})}.$$

3. Show that

$$F\left[\begin{matrix} a, 1-a; \\ c; \end{matrix} \; \tfrac{1}{2}\right] = \frac{2^{1-c}\Gamma(c)\Gamma(\tfrac{1}{2})}{\Gamma(\tfrac{1}{2}c + \tfrac{1}{2}a)\Gamma(\tfrac{1}{2}c - \tfrac{1}{2}a + \tfrac{1}{2})}.$$

4. Obtain the result

$$F\left[\begin{matrix} -n, b; \\ c; \end{matrix} \; 1\right] = \frac{(c-b)_n}{(c)_n}.$$

5. Obtain the result

$$F\left[\begin{matrix} -n, a+n; \\ c; \end{matrix} \; 1\right] = \frac{(-1)^n(1+a-c)_n}{(c)_n}.$$

6. Show that

$$F\left[\begin{matrix} -n, 1-b-n; \\ a; \end{matrix} \; 1\right] = \frac{(a+b-1)_{2n}}{(a)_n(a+b-1)_n}.$$

7. Prove that if $g_n = F(-n, \alpha; 1 + \alpha - n; 1)$ and α is not an integer, then $g_n = 0$ for $n \geqq 1$, $g_0 = 1$.

8. Show that

$$\frac{d^n}{dx^n}[x^{a-1+n}F(a, b; c; x)] = (a)_n x^{a-1}F(a+n, b; c; x).$$

9. Use equation (2), page 66, with $z = -x$, $b = -n$, in which n is a non-negative integer, to conclude that

$$F\left[\begin{matrix} -n, a; \\ 1+a+n; \end{matrix} -x\right] = (1-x)^{-a} F\left[\begin{matrix} \frac{1}{2}a, \frac{1}{2}a+\frac{1}{2}; \\ 1+a+n; \end{matrix} \frac{-4x}{(1-x)^2}\right].$$

10. In Theorem 23, page 65, put $b = \gamma$, $a = \gamma + \frac{1}{2}$, $4x(1+x)^{-2} = z$ and thus prove that

$$F\left[\begin{matrix} \gamma, \gamma+\frac{1}{2}; \\ 2\gamma; \end{matrix} z\right] = (1-z)^{-\frac{1}{2}}\left[\frac{2}{1+\sqrt{1-z}}\right]^{2\gamma-1}$$

and further that

$$F\left[\begin{matrix} \gamma, \gamma-\frac{1}{2}; \\ 2\gamma; \end{matrix} z\right] = \left(\frac{2}{1+\sqrt{1-z}}\right)^{2\gamma-1}.$$

11. Use Theorem 27, page 69, to show that

$$(1-x)^{1-c} F\left[\begin{matrix} a, 1-a; \\ c; \end{matrix} x\right] = (1-2x)^{a-c} F\left[\begin{matrix} \frac{1}{2}c-\frac{1}{2}a, \frac{1}{2}c-\frac{1}{2}a+\frac{1}{2}; \\ c; \end{matrix} \frac{4x(x-1)}{(1-2x)^2}\right].$$

12. In the differential equation (3), page 54, for

$$w = F(a, b; c; z)$$

introduce a new dependent variable u by $w = (1-z)^{-a}u$, thus obtaining

$$z(1-z)^2u'' + (1-z)[c + (a-b-1)z]u' + a(c-b)u = 0.$$

Next change the independent variable to x by putting $x = -z/(1-z)$. Show that the equation for u in terms of x is

$$x(1-x)\frac{d^2u}{dx^2} + [c - (a+c-b+1)x]\frac{du}{dx} - a(c-b)u = 0,$$

and thus derive the solution

$$w = (1-z)^{-a} F\left[\begin{matrix} a, c-b; \\ c; \end{matrix} \frac{-z}{1-z}\right].$$

13. Use the result of Ex. 12 and the method of Section 40 to prove Theorem 20, page 60.

14. Prove Theorem 21, page 60, by the method suggested by Exs. 12 and 13.

15. Use the method of Section 39 to prove that if both $|z| < 1$ and $|1-z| < 1$, and if a, b, c are suitably restricted,

$$F\left[\begin{matrix} a, b; \\ c; \end{matrix} z\right] = \frac{\Gamma(c)\Gamma(c-a-b)}{\Gamma(c-a)\Gamma(c-b)} F\left[\begin{matrix} a, b; \\ a+b+1-c; \end{matrix} 1-z\right]$$

$$+ \frac{\Gamma(c)\Gamma(a+b-c)(1-z)^{c-a-b}}{\Gamma(a)\Gamma(b)} F\left[\begin{matrix} c-a, c-b; \\ c-a-b+1; \end{matrix} 1-z\right].$$

16. In a common notation for the Laplace transform

$$L\{F(t)\} = \int_0^\infty e^{-st}F(t)\,dt = f(s); \qquad L^{-1}\{f(s)\} = F(t).$$

Show that

$$L^{-1}\left\{\frac{1}{s}F\left[\begin{matrix} a, b; \\ \\ s+1; \end{matrix} \quad z\right]\right\} = F\left[\begin{matrix} a, b; \\ \\ 1; \end{matrix} \quad z(1-e^{-t})\right].$$

17. With the notation of Ex. 16 show that

$$L\{t^n \sin at\} = \frac{a\Gamma(n+2)}{s^{n+2}}F\left[\begin{matrix} 1+\frac{1}{2}n, \frac{3}{2}+\frac{1}{2}n; \\ \\ \frac{3}{2}; \end{matrix} \quad -\frac{a^2}{s^2}\right].$$

18. Obtain the results

$$\begin{aligned} \text{Log}\,(1+x) &= xF(1, 1; 2; -x), \\ \text{Arcsin}\, x &= xF(\tfrac{1}{2}, \tfrac{1}{2}; \tfrac{3}{2}; x^2), \\ \text{Arctan}\, x &= xF(\tfrac{1}{2}, 1; \tfrac{3}{2}; -x^2). \end{aligned}$$

19. The complete elliptic integral of the first kind is

$$K = \int_0^{\frac{1}{2}\pi} \frac{d\varphi}{\sqrt{1-k^2\sin^2\varphi}}.$$

Show that $K = \frac{1}{2}\pi F(\frac{1}{2}, \frac{1}{2}; 1; k^2)$.

20. The complete elliptic integral of the second kind is

$$E = \int_0^{\frac{1}{2}\pi} \sqrt{1-k^2\sin^2\theta}\,d\theta.$$

Show that $E = \frac{1}{2}\pi F(\frac{1}{2}, -\frac{1}{2}; 1; k^2)$.

21. From the contiguous function relations

(1) $(a-b)F = aF(a+) - bF(b+)$,
(2) $(a-c+1)F = aF(a+) - (c-1)F(c-)$,
(3) $[a+(b-c)z]F = a(1-z)F(a+) - c^{-1}(c-a)(c-b)zF(c+)$,
(4) $(1-z)F = F(a-) - c^{-1}(c-b)zF(c+)$,
(5) $(1-z)F = F(b-) - c^{-1}(c-a)zF(c+)$, derived in Section 33,

obtain the remaining ten such relations:

(6) $[2a-c+(b-a)z]F = a(1-z)F(a+) - (c-a)F(a-)$,
(7) $(a+b-c)F = a(1-z)F(a+) - (c-b)F(b-)$,
(8) $(c-a-b)F = (c-a)F(a-) - b(1-z)F(b+)$,
(9) $(b-a)(1-z)F = (c-a)F(a-) - (c-b)F(b-)$,
(10) $[1-a+(c-b-1)z]F = (c-a)F(a-) - (c-1)(1-z)F(c-)$,
(11) $[2b-c+(a-b)z]F = b(1-z)F(b+) - (c-b)F(b-)$,
(12) $[b+(a-c)z]F = b(1-z)F(b+) - c^{-1}(c-a)(c-b)zF(c+)$,
(13) $(b-c+1)F = bF(b+) - (c-1)F(c-)$,
(14) $[1-b+(c-a-1)z]F = (c-b)F(b-) - (c-1)(1-z)F(c-)$,
(15) $[c-1+(a+b+1-2c)z]F = (c-1)(1-z)F(c-) - c^{-1}(c-a)(c-b)zF(c+)$.

22. The notation used in Ex. 21 and in Section 33 is often extended as in the examples

$$F(a-, b+) = F(a - 1, b + 1; c; z),$$
$$F(b+, c+) = F(a, b + 1; c + 1; z).$$

Use the relations (4) and (5) of Ex. 21 to obtain

$$F(a-) - F(b-) + c^{-1}(b - a)zF(c+) = 0$$

and from it, by changing b to $(b + 1)$ arrive at

$$F = F(a-, b+) + c^{-1}(b + 1 - a)zF(b+, c+),$$

a relation we wish to use in Chapter 16.

23. In equation (9) of Ex. 21 shift b to $(b + 1)$ to obtain the relation

$$(c - 1 - b)F = (c - a)F(a-, b+) + (a - 1 - b)(1 - z)F(b+),$$

or

$$(c - 1 - b)F(a, b; c; z)$$
$$= (c - a)F(a - 1, b + 1; c; z) + (a - 1 - b)(1 - z)F(a, b + 1; c; z),$$

another relation we wish to use in Chapter 16.

CHAPTER 5

Generalized Hypergeometric Functions

44. The function $_pF_q$. The hypergeometric function

$$(1)\qquad F\left[\begin{matrix} a, b; \\ \\ c; \end{matrix}\ z\right] = \sum_{n=0}^{\infty} \frac{(a)_n (b)_n z^n}{(c)_n n!}$$

studied in Chapter 4 has two numerator parameters, a and b, and one denominator parameter, c. It is a natural generalization to move from the definition (1) to a similar function with any number of numerator and denominator parameters.

We define a generalized hypergeometric function by

$$(2)\qquad {}_pF_q\left[\begin{matrix} \alpha_1, \alpha_2, \cdots, \alpha_p; \\ \\ \beta_1, \beta_2, \cdots, \beta_q; \end{matrix}\ z\right] = 1 + \sum_{n=1}^{\infty} \frac{\prod_{i=1}^{p} (\alpha_i)_n}{\prod_{j=1}^{q} (\beta_j)_n} \cdot \frac{z^n}{n!},$$

in which no denominator parameter β_j is allowed to be zero or a negative integer. If any numerator parameter α_i in (2) is zero or a negative integer, the series terminates.

An application of the elementary ratio test to the power series on the right in (2) shows at once that:

(a) If $p \leqq q$, the series converges for all finite z;

(b) If $p = q + 1$, the series converges for $|z| < 1$ and diverges for $|z| > 1$;

(c) If $p > q + 1$, the series diverges for $z \neq 0$. If the series terminates, there is no question of convergence, and the conclusions (b) and (c) do not apply.

If $p = q + 1$, the series in (2) is absolutely convergent on the circle $|z| = 1$ if

$$\text{Re}\left(\sum_{j=1}^{q} \beta_j - \sum_{i=1}^{p} \alpha_i\right) > 0.$$

The proof can be made to parallel that used in Section 29 for the corresponding result on the ordinary hypergeometric series. The proof is left as an exercise.

When we wish to indicate the number of numerator parameters and of denominator parameters but not to specify them, we use the notation ${}_pF_q$. For instance, the ordinary hypergeometric function of Chapter 4 is a ${}_2F_1$.

We permit p or q, or both, to be zero. The absence of parameters is emphasized by a dash. The most general ${}_0F_1$, for example, is

$$(3) \qquad {}_0F_1(-;b;z) = \sum_{n=0}^{\infty} \frac{z^n}{(b)_n n!}.$$

We shall see later that the function in (3) is essentially a Bessel function.

For more results on the ${}_pF_q$ see Erdélyi [1; 182–247] and Bailey [1]. Additional references may also be found on pages 103–108 of Bailey [1] and on pages 199–201 and 246–247 of Erdélyi [1].

45. The exponential and binomial functions. Two elementary instances of the ${}_pF_q$ follow. If no numerator or denominator parameters are present, the result is

$$(1) \qquad {}_0F_0(-;-;z) = \sum_{n=0}^{\infty} \frac{z^n}{n!} = \exp(z).$$

If we use one numerator parameter and no denominator parameter, we obtain

$$(2) \qquad {}_1F_0(a;-;z) = \sum_{n=0}^{\infty} \frac{(a)_n z^n}{n!}.$$

Hence, by the argument on page 47,

$$(3) \qquad {}_1F_0(a;-;z) = (1-z)^{-a}.$$

46. A differential equation. Recall that the ordinary hypergeometric function $F(a, b; c; z)$ satisfies the differential equation

$$z(1-z)\frac{d^2w}{dz^2} + [c - (a+b+1)z]\frac{dw}{dz} - abw = 0, \tag{1}$$

or, in terms of the differential operator $\theta \equiv z\dfrac{d}{dz}$, the differential equation

$$[\theta(\theta + c - 1) - z(\theta + a)(\theta + b)]w = 0. \tag{2}$$

With the suggestion of equation (2) before us, we can proceed as follows. Let

$$w = {}_pF_q = \sum_{k=0}^{\infty} \frac{(a_1)_k(a_2)_k\cdots(a_p)_k}{(b_1)_k(b_2)_k\cdots(b_q)_k}\frac{z^k}{k!}.$$

Since $\theta z^k = kz^k$, it follows that

$$\theta \prod_{j=1}^{q}(\theta + b_j - 1)w = \sum_{k=1}^{\infty} \frac{k\prod_{j=1}^{q}(k + b_j - 1)\prod_{i=1}^{p}(a_i)_k}{\prod_{j=1}^{q}(b_j)_k}\frac{z^k}{k!}$$

$$= \sum_{k=1}^{\infty}\frac{\prod_{i=1}^{p}(a_i)_k}{\prod_{j=1}^{q}(b_j)_{k-1}}\frac{z^k}{(k-1)!}.$$

Now we replace k by $(k+1)$ and have

$$\theta \prod_{j=1}^{q}(\theta + b_j - 1)w = \sum_{k=0}^{\infty}\frac{\prod_{i=1}^{p}(a_i)_{k+1}}{\prod_{j=1}^{q}(b_j)_k}\frac{z^{k+1}}{k!}$$

$$= \sum_{k=0}^{\infty}\frac{\prod_{i=1}^{p}(a_i + k)\prod_{i=1}^{p}(a_i)_k}{\prod_{j=1}^{q}(b_j)_k}\frac{z^{k+1}}{k!}$$

$$= z\prod_{i=1}^{p}(\theta + a_i)w.$$

Thus we have shown that $w = {}_pF_q$ is a solution of the differential equation

$$\left[\theta\prod_{j=1}^{q}(\theta + b_j - 1) - z\prod_{i=1}^{p}(\theta + a_i)\right]w = 0, \tag{3}$$

when no b_j is a nonpositive integer. The solution is valid for all finite z when $p \leqq q$. If $p = q + 1$, the solution is valid in $|z| < 1$. We are not concerning ourselves with the case $p > q + 1$, when the series for ${}_pF_q$ has a zero radius of convergence unless the series terminates.

47. Other solutions of the differential equation. If $p \leqq q + 1$, the equation (3) of the preceding section is a linear differential equation of order $(q + 1)$. We have, in the neighborhood of the origin, one solution,

$$w = w_0 = F(a_1, a_2, \cdots, a_p; b_1, b_2, \cdots, b_q; z).$$

Naturally we wish to obtain q other solutions near $z = 0$.

Let us turn for the moment to the one case, the ${}_2F_1$, about which we already have quite a bit of information. The differential equation for $p = 2$, $q = 1$, is

$$[\theta(\theta + b_1 - 1) - z(\theta + a_1)(\theta + a_2)]w = 0. \tag{1}$$

We know that if b_1 is not integral, the general solution of (1) in the neighborhood of $z = 0$ is

$$w = Aw_0 + Bw_1,$$

where A and B are arbitrary constants and w_0 and w_1 may be taken to be

$$\begin{aligned} w_0 &= F(a_1, a_2; b_1; z), \\ w_1 &= z^{1-b_1}F(a_1 - b_1 + 1, a_2 - b_1 + 1; 2 - b_1; z). \end{aligned}$$

When we have determined just what it is that makes the w_1 satisfy equation (1), we shall be in a better position to move on to the generalized case.

Now

$$w_1 = \sum_{k=0}^{\infty} \frac{(a_1 - b_1 + 1)_k(a_2 - b_1 + 1)_k}{(2 - b_1)_k} \frac{z^{k+1-b_1}}{k!}$$

so that

$$\begin{aligned} &\theta(\theta + b_1 - 1)w_1 \\ &\quad = \sum_{k=0}^{\infty} \frac{(k+1-b_1)(k+1-b_1+b_1-1)(a_1-b_1+1)_k(a_2-b_1+1)_k}{(2 - b_1)_k} \frac{z^{k+1-b_1}}{k!} \\ &\quad = \sum_{k=1}^{\infty} \frac{k(k + 1 - b_1)(a_1 - b_1 + 1)_k(a_2 - b_1 + 1)_k}{(2 - b_1)_k} \frac{z^{k+1-b_1}}{k!}. \end{aligned}$$

But the last factor in $(2 - b_1)_k$ is $2 - b_1 + k - 1 = k + 1 - b_1$, so that

$$\theta(\theta + b_1 - 1)w_1 = \sum_{k=1}^{\infty} \frac{(a_1 - b_1 + 1)_k(a_2 - b_1 + 1)_k}{(2 - b_1)_{k-1}} \frac{z^{k+1-b_1}}{(k-1)!}$$

$$= \sum_{k=0}^{\infty} \frac{(a_1 - b_1 + 1)_{k+1}(a_2 - b_1 + 1)_{k+1}}{(2 - b_1)_k} \frac{z^{k+2-b_1}}{k!}.$$

The last factor in $(a_1 - b_1 + 1)_{k+1}$ is $a_1 - b_1 + 1 + k + 1 - 1 = k + 1 - b_1 + a_1$. Hence

$$\theta(\theta + b_1 - 1)w_1$$

$$= \sum_{k=0}^{\infty} \frac{(k+1-b_1+a_1)(k+1-b_1+a_2)(a_1-b_1+1)_k(a_2-b_1+1)_k}{(2 - b_1)_k} \frac{z^{k+2-b_1}}{k!}$$

$$= z(\theta + a_1)(\theta + a_2) \sum_{k=0}^{\infty} \frac{(a_1 - b_1 + 1)_k(a_2 - b_1 + 1)_k}{(2 - b_1)_k} \frac{z^{k+1-b_1}}{k!}$$

$$= z(\theta + a_1)(\theta + a_2)w_1.$$

Therefore w_1 is a solution of equation (1).

Notice the way in which the operator θ introduced a factor $(k + 1 - b_1)$ which canceled the last factor in $(2 - b_1)_k$ and the way in which the operator $(\theta + b_1 - 1)$ introduced a factor k which canceled the factor k in $k!$. If we were dealing with the generalized case, other operators in the form $(\theta + b_j - 1)$ would occur. Such an operator would introduce a factor

$$k + 1 - b_1 + b_j - 1 = b_j - b_1 + 1 + k - 1,$$

which is the last factor in the product $(b_j - b_1 + 1)_k$. Hence, in the solutions we seek, a denominator parameter b_j in the ${}_pF_q$ should be replaced by the parameter $(b_j - b_1 + 1)$ for the solution corresponding to the parameter b_1 in the original ${}_pF_q$. Of course each of the b_j should play a role like that of b_1, each yielding a different solution of the differential equation.

Thus we are led to the desired generalization. That is, the generalized hypergeometric equation,

$$(2) \qquad \left[\theta \prod_{j=1}^{q} (\theta + b_j - 1) - z \prod_{i=1}^{p} (\theta + a_i)\right] w = 0; p \leqq q + 1,$$

has, when no b_j is a nonpositive integer and no two b_j's differ by an integer, the solution

(3) $$w = \sum_{m=0}^{q} c_m w_m,$$

where the c_m are arbitrary constants, where

$$w_0 = F(a_1, a_2, \cdots, a_p; b_1\, b_2, \cdots, b_q; z),$$

and where, for $m = 1, 2, \cdots, q$,

(4) $$w_m = z^{1-b_m}$$

$$\times F\left[\begin{matrix} a_1 - b_m + 1, a_2 - b_m + 1, \cdots, a_p - b_m + 1; \\ b_1 - b_m + 1, \cdots, b_{m-1} - b_m + 1, 2 - b_m, b_{m+1} - b_m + 1, \cdots, b_q - b_m + 1; \end{matrix} z\right]$$

$$= \sum_{k=0}^{\infty} \frac{\prod_{i=1}^{p} (a_i - b_m + 1)_k}{\prod_{j=1}^{q} (b_j - b_m + 1)_k} \frac{z^{k+1-b_m}}{(2 - b_m)_k}.$$

Whenever, in addition to the above restrictions, no b_j is a positive integer, then the linear combination (3) is the general solution of equation (2) around $z = 0$. Note also that if $p \leqq q$, then the series for w_m; $m = 0, 1, 2, \cdots, q$, converges for all finite z and that for $p = q + 1$, the series for w_m converges for $|z| < 1$.

It is important to realize that the procedure we are now using is not necessary. All $(q + 1)$ solutions of the differential equation (2) can be obtained by the standard series methods, the method of Frobenius, keeping in mind that, for $p \leqq q + 1$, the equation (2) has at worst a regular singular point at the origin. The whole point to the method of the present section is an attempt to gain some insight into the way in which the parameters in ${}_pF_q$ must be altered to preserve its property of being a solution of equation (2) and, perhaps most of all, the method is an attempt to obtain some small practice in the technique of manipulating generalized hypergeometric series.

We shall now verify that w_m satisfies equation (2). Let, as before, for m any one of the integers $1, 2, 3, \cdots, q$,

(5) $$w_m = \sum_{k=0}^{\infty} \frac{\prod_{i=1}^{p} (a_i - b_m + 1)_k}{\prod_{j=1}^{q} (b_j - b_m + 1)_k} \frac{z^{k+1-b_m}}{(2 - b_m)_k}.$$

Note that the mth factor in the product

$$\prod_{j=1}^{q} (b_j - b_m + 1)_k$$

is $(1)_k$, which furnishes the factor $k!$ needed to make the right side of (5) a hypergeometric series.

From (5) we get at once

$$\theta \prod_{j=1}^{q} (\theta + b_j - 1) w_m$$

$$= \sum_{k=0}^{\infty} \frac{(k+1-b_m) \prod_{j=1}^{q} (k+1-b_m+b_j-1) \prod_{i=1}^{p} (a_i - b_m + 1)_k}{\prod_{j=1}^{q} (b_j - b_m + 1)_k} \frac{z^{k+1-b_m}}{(2-b_m)_k}.$$

Now $(b_j - b_m + k)$ is the last factor in the product $(b_j - b_m + 1)_k$ and $(k + 1 - b_m)$ is the last factor in the product $(2 - b_m)_k$. Hence

$$\theta \prod_{j=1}^{q} (\theta + b_j - 1) w_m = \sum_{k=1}^{\infty} \frac{\prod_{i=1}^{p} (a_i - b_m + 1)_k}{\prod_{j=1}^{q} (b_j - b_m + 1)_{k-1}} \frac{z^{k+1-b_m}}{(2 - b_m)_{k-1}},$$

where the $k = 0$ term dropped out because of the numerator factor $k + 1 - b_m + b_m - 1 = k$, which was later canceled.

Next, with a shift of index from k to $(k + 1)$, we get

$$\theta \prod_{j=1}^{q} (\theta+b_j-1) w_m = \sum_{k=0}^{\infty} \frac{\prod_{i=1}^{p} (a_i - b_m + 1)_{k+1}}{\prod_{j=1}^{q} (b_j - b_m + 1)_k} \frac{z^{k+2-b_m}}{(2-b_m)_k}$$

$$= z \sum_{k=0}^{\infty} \frac{\prod_{i=1}^{p} (a_i - b_m + 1 + k) \prod_{i=1}^{p} (a_i - b_m + 1)_k}{\prod_{j=1}^{q} (b_j - b_m + 1)_k} \frac{z^{k+1-b_m}}{(2-b_m)_k}.$$

Therefore

$$\theta \prod_{j=1}^{q} (\theta + b_j - 1) w_m = z \prod_{i=1}^{p} (\theta + a_i) w_m,$$

so that w_m; $m = 1, 2, \cdots, q$, satisfies the generalized hypergeometric equation.

If the exponents of the differential equation at $z = 0$ are such that one or more of their differences are integral, then the presence of logarithmic terms is to be expected in the general solution. For $p = q + 1$ the logarithmic solutions are obtained in F. C. Smith [1].

48. The contiguous function relations.* In this section the parameters are fixed, and the work is concerned only with the function ${}_pF_q$ and its contiguous functions. Hence, we are able to use an abbreviated notation illustrated by the following:

$$\begin{aligned} F &= {}_pF_q(\alpha_1, \alpha_2, \cdots, \alpha_p; \beta_1, \beta_2, \cdots, \beta_q; x), \\ F(\alpha_1 +) &= {}_pF_q(\alpha_1 + 1, \alpha_2, \cdots, \alpha_p; \beta_1, \beta_2, \cdots, \beta_q; x), \\ F(\beta_1 -) &= {}_pF_q(\alpha_1, \alpha_2, \cdots, \alpha_p; \beta_1 - 1, \beta_2, \cdots, \beta_q; x). \end{aligned} \tag{1}$$

There are, of course, $(2p+2q)$ functions contiguous to F. Corresponding to Gauss' five independent relations in the case of ${}_2F_1$, there is for F a set of $(2p+q)$ linearly independent relations, which we shall obtain. The canonical form into which we put this basic set may be described as follows:

First, there are $(p+q-1)$ relations, each containing F and two of its contiguous functions. These will be called the *simple* relations. Each simple relation connects F, $F(\alpha_1+)$, and $F(\alpha_k+)$ for $k = 2, 3, \cdots, p$, or it connects F, $F(\alpha_1+)$, and $F(\beta_j-)$ for $j = 1, 2, \cdots, q$. The simple relations are immediate extensions of two of Gauss' five relations and are not novel in any way.

Second, there are $(p+1)$ less simple relations, each containing F and $(q+1)$ of its contiguous functions. In our canonical form we shall select these so that one of them connects F, $F(\alpha_1+)$ and all the functions $F(\beta_j+)$ for $j = 1, 2, \cdots, q$. Each of the other p relations will contain F, all the functions $F(\beta_j+)$; $j = 1, 2, \cdots, q$, and one of the functions $F(\alpha_k-)$ for $k = 1, 2, \cdots, p$. The less simple relations are generalizations of three of Gauss' five relations but differ from them in one essential aspect in that each relation contains F and $(q + 1)$ of its contiguous functions. For Gauss' case $q + 1 = 2$, and the less simple relations contain the same number of contiguous functions as do the simple ones.

Since we shall actually exhibit the $(2p+q)$ relations, it will be evident upon looking at them that, just as in the case of the ordinary hypergeometric function, the coefficients are polynomials at most linear in x.

*This section is taken from an article (Rainville [3]) published in the *Bulletin* of the American Mathematical Society. Reproduced here by permission.

It is convenient to use the following notations:

(2) $$\prod_{s=1,(k)}^{m} A_s = \prod_{s=1}^{k-1} A_s \cdot \prod_{s=k+1}^{m} A_s,$$

a symbol denoting a product with a particular factor deleted,

(3) $$U_j = \frac{\prod_{s=1}^{p} (\alpha_s - \beta_j)}{\beta_j \prod_{s=1,(j)}^{q} (\beta_s - \beta_j)},$$

(4) $$W_{j,k} = \frac{\prod_{s=1,(k)}^{p} (\alpha_s - \beta_j)}{\beta_j \prod_{s=1,(j)}^{q} (\beta_s - \beta_j)},$$

(5) $$c_n = \frac{(\alpha_1)_n(\alpha_2)_n \cdots (\alpha_p)_n}{(\beta_1)_n(\beta_2)_n \cdots (\beta_q)_n},$$

(6) $$S_n = \frac{(\alpha_1 + n)(\alpha_2 + n) \cdots (\alpha_p + n)}{(\beta_1 + n)(\beta_2 + n) \cdots (\beta_q + n)},$$

(7) $$\tau_{n,k} = \frac{S_n}{\alpha_k + n},$$

(8) $$A = \sum_{s=1}^{p} \alpha_s,$$

(9) $$B = \sum_{s=1}^{q} \beta_s.$$

An examination of (5) and (6) shows that

(10) $$c_{n+1} = S_n c_n.$$

The relation $\alpha(\alpha+1)_n = (\alpha+n)(\alpha)_n$, together with the definitions of the contiguous functions, yields the formulas:

$$F = \sum_{n=0}^{\infty} \frac{c_n x^n}{n!},$$

(11) $$F(\alpha_k +) = \sum_{n=0}^{\infty} \frac{\alpha_k + n}{\alpha_k} \frac{c_n x^n}{n!}, \quad F(\alpha_k -) = \sum_{n=0}^{\infty} \frac{\alpha_k - 1}{\alpha_k + n - 1} \frac{c_n x^n}{n!},$$

$$F(\beta_k +) = \sum_{n=0}^{\infty} \frac{\beta_k}{\beta_k + n} \frac{c_n x^n}{n!}, \quad F(\beta_k -) = \sum_{n=0}^{\infty} \frac{\beta_k + n - 1}{\beta_k - 1} \frac{c_n x^n}{n!}.$$

Using the operator $\theta = x(d/dx)$, we see that

$$(\theta + \alpha_k)F = \sum_{n=0}^{\infty} (\alpha_k + n) \frac{c_n x^n}{n!}.$$

Hence, with the aid of (11),

$$(12) \qquad (\theta + \alpha_k)F = \alpha_k F(\alpha_k +); \qquad k = 1, 2, \cdots, p.$$

Similarly, it follows that

$$(13) \qquad (\theta + \beta_k - 1)F = (\beta_k - 1)F(\beta_k -); \qquad k = 1, 2, \cdots, q.$$

The $(p+q)$ equations (12) and (13) lead at once by elimination of θF to $(p+q-1)$ linear algebraic relations between F and pairs of its contiguous functions. Let us use $F(\alpha_1+)$ as an element in each equation. The result is the set of simple relations,

$$(14) \qquad (\alpha_1 - \alpha_k)F = \alpha_1 F(\alpha_1 +) - \alpha_k F(\alpha_k +); \qquad k = 2, 3, \cdots, p,$$

and

$$(15) \qquad (\alpha_1 - \beta_k + 1)F = \alpha_1 F(\alpha_1 +) - (\beta_k - 1)F(\beta_k -); \qquad k = 1, 2, \cdots, q.$$

From

$$F = \sum_{n=0}^{\infty} \frac{c_n x^n}{n!},$$

it follows that

$$\theta F = \sum_{n=1}^{\infty} \frac{n c_n x^n}{n!} = x \sum_{n=0}^{\infty} \frac{c_{n+1} x^n}{n!}.$$

Thus, because of (10) and (6),

$$\theta F = x \sum_{n=0}^{\infty} S_n \frac{c_n x^n}{n!},$$

where

$$S_n = \frac{(\alpha_1 + n)(\alpha_2 + n) \cdots (\alpha_p + n)}{(\beta_1 + n)(\beta_2 + n) \cdots (\beta_q + n)}.$$

Now, if $p<q$, then the degree of the numerator of S_n is lower than the degree of the denominator, and the elementary theory of rational fraction expansions yields, for no two β's equal,

$$S_n = \sum_{j=1}^{q} \frac{\beta_j U_j}{\beta_j + n}, \qquad p < q,$$

in which the U_j is as defined in (3).

Therefore,

$$\theta F = x \sum_{n=0}^{\infty} \sum_{j=1}^{q} \frac{\beta_j U_j}{\beta_j + n} \frac{c_n x^n}{n!},$$

which in view of (11) becomes

$$\theta F = x \sum_{j=1}^{q} U_j F(\beta_j +). \tag{16}$$

The elimination of θF, using (16) and the case $k = 1$ of (12), leads to

$$\alpha_1 F = \alpha_1 F(\alpha_1 +) - x \sum_{j=1}^{q} U_j F(\beta_j +), \qquad p < q. \tag{17}$$

If $p = q$, the degree of the numerator of S_n equals that of the denominator. However, when $p = q$,

$$S_n = 1 + \frac{\prod_{s=1}^{p} (\alpha_s + n) - \prod_{s=1}^{q} (\beta_s + n)}{\prod_{s=1}^{q} (\beta_s + n)},$$

in which the fraction on the right has the desired property that its numerator is of lower degree than its denominator. Thus

$$S_n = 1 + \sum_{j=1}^{q} \frac{\beta_j U_j}{\beta_j + n}, \qquad p = q,$$

and it is easy to see that in this case

$$\theta F = xF + x \sum_{j=1}^{q} U_j F(\beta_j +),$$

so that (17) is replaced by the relation

$$(\alpha_1 + x)F = \alpha_1 F(\alpha_1 +) - x \sum_{j=1}^{q} U_j F(\beta_j +), \qquad p = q. \tag{18}$$

If $p = q+1$, then with the notation of (8) and (9) we may write

$$S_n = n + A - B + \frac{\prod_{s=1}^{p} (\alpha_s + n) - (n + A - B) \prod_{s=1}^{q} (\beta_s + n)}{\prod_{s=1}^{q} (\beta_s + n)},$$

in which the fraction on the right has the desired rational fraction expansion, so that

$$S_n = n + A - B + \sum_{j=1}^{q} \frac{\beta_j U_j}{\beta_j + n}.$$

Thus we conclude that, when $p=q+1$,

$$\theta F = x\theta F + (A - B)xF + x\sum_{j=1}^{q} U_j F(\beta_j +),$$

so that (17) is this time to be replaced by the relation

$$[(1 - x)\alpha_1 + (A - B)x]F = (1 - x)\alpha_1 F(\alpha_1 +)$$

$$(19) \qquad - x\sum_{j=1}^{q} U_j F(\beta_j +); \qquad p = q + 1.$$

From equation (11) we obtain

$$\theta F(\alpha_k -) = \sum_{n=1}^{\infty} \frac{\alpha_k - 1}{\alpha_k + n - 1}\frac{nc_n x^n}{n!},$$

or

$$\theta F(\alpha_k -) = x\sum_{n=0}^{\infty} \frac{\alpha_k - 1}{\alpha_k + n}\frac{c_{n+1} x^n}{n!}.$$

Now, with the notation of (7),

$$\frac{c_{n+1}}{\alpha_k + n} = c_n \tau_{n,k},$$

where $\tau_{n,k}$ has, for $p \leqq q$, its numerator of lower degree than its denominator. Thus

$$\theta F(\alpha_k -) = (\alpha_k - 1)x\sum_{n=0}^{\infty} \tau_{n,k}\frac{c_n x^n}{n!},$$

where

$$\tau_{n,k} = \sum_{j=1}^{q} \frac{\beta_j W_{j,k}}{\beta_j + n}$$

in which the $W_{j,k}$ is as defined in (4).

We now have

$$\theta F(\alpha_k -) = (\alpha_k - 1)x\sum_{j=1}^{q} W_{j,k} F(\beta_j +); \qquad p \leqq q;\ k = 1, 2, \cdots, p.$$

But, from (12),

$$\theta F(\alpha_k -) = (\alpha_k - 1)[F - F(\alpha_k -)].$$

The elimination of $\theta F(\alpha_k -)$ from the preceding two formulas yields the p relations:

$$(20) \quad F = F(\alpha_k -) + x\sum_{j=1}^{q} W_{j,k} F(\beta_j +); \qquad p \leqq q;\ k = 1, 2, \cdots, p.$$

When $p=q+1$, the fraction $\tau_{n,k}$ has its numerator and denominator of equal degree. Then we write

$$\tau_{n,k} = 1 + \frac{\prod_{s=1,(k)}^{p} (\alpha_s + n) - \prod_{s=1}^{q} (\beta_s + n)}{\prod_{s=1}^{q} (\beta_s + n)}$$

and conclude in the same manner as before that

$$\theta F(\alpha_k -) = (\alpha_k - 1)xF + (\alpha_k - 1)x \sum_{j=1}^{q} W_{j,k}F(\beta_j +).$$

Therefore, for $p=q+1$, (20) is to be replaced by

$$(21) \quad (1 - x)F = F(\alpha_k -) + x \sum_{j=1}^{q} W_{j,k}F(\beta_j +); \qquad k = 1, 2, \cdots, p.$$

We have shown that for

$$F(\alpha_1, \alpha_2, \cdots, \alpha_p; \beta_1, \beta_2, \cdots, \beta_q; x)$$

in which no two β's are equal and no β is a nonpositive integer, a canonical set of $(2p+q)$ contiguous function relations is as described below.

If $p<q$, (14), (15), (17), and (20) hold.

If $p=q$, the relations are (14), (15), and (20) together with (18) to replace (17).

If $p=q+1$, the relations are (14), (15), (19), and (21).

49. A simple integral. It is an easy matter to extend the work of Section 30 to obtain the following result.

THEOREM 28. *If* $p \leqq q + 1$, *if* $\text{Re}(b_1) > \text{Re}(a_1) > 0$, *if no one of* $b_1, b_2 \cdots, b_q$ *is zero or a negative integer, and if* $|z| < 1$,

$${}_pF_q\left[\begin{matrix} a_1, a_2, \cdots, a_p; \\ \\ b_1, b_2, \cdots, b_q; \end{matrix} \; z\right]$$

$$= \frac{\Gamma(b_1)}{\Gamma(a_1)\Gamma(b_1 - a_1)} \int_0^1 t^{a_1-1}(1-t)^{b_1-a_1-1} {}_{p-1}F_{q-1}\left[\begin{matrix} a_2, \cdots, a_p; \\ \\ b_2, \cdots, b_q; \end{matrix} \; zt\right] dt.$$

If $p \leqq q$, *the condition* $|z| < 1$ *may be omitted.*

50. The ${}_pF_q$ with unit argument. We have already found much use for the evaluation of ${}_2F_1(a, b; c; 1)$. We know that if no denominator parameter is zero or a negative integer, the series

$$(1) \qquad {}_pF_q\left[\begin{matrix} a_1, a_2, \cdots, a_p; \\ b_1, b_2, \cdots, b_q; \end{matrix} \quad 1\right]$$

is absolutely convergent whenever $p \leqq q$ and for $p = q + 1$ is absolutely convergent if

$$\text{Re}\left(\sum_{j=1}^{q} b_j - \sum_{i=1}^{p} a_i\right) > 0.$$

It is natural to seek the value of (1), at least for $p = q + 1$. For $p \leqq q$, the point $z = 1$ is not a singular point of the differential equation for ${}_pF_q$, and there is less reason to hope that (1) will have a relatively simple value. Compare Theorem 28 with Theorem 16, page 47, to see why the method of Section 32 does not carry over to the more general case. In several special instances the desired evaluation has been accomplished. This book contains a few of the more widely used theorems along these lines.

51. Saalschütz' theorem. From Theorem 21, page 60, we see that

$$(1) \qquad F(c - a, c - b; c; z) = (1 - z)^{a+b-c}F(a, b; c; z).$$

Equation (1) may be interpreted as an identity involving three power series:

$$\sum_{n=0}^{\infty} \frac{(c-a)_n(c-b)_n z^n}{(c)_n n!} = \left[\sum_{n=0}^{\infty} \frac{(c-a-b)_n z^n}{n!}\right]\left[\sum_{n=0}^{\infty} \frac{(a)_n(b)_n z^n}{(c)_n n!}\right]$$

$$= \sum_{n=0}^{\infty}\sum_{k=0}^{n} \frac{(a)_k(b)_k(c-a-b)_{n-k}z^n}{(c)_k k!(n-k)!}.$$

We know that

$$(\alpha)_{n-k} = \frac{(-1)^k(\alpha)_n}{(1-\alpha-n)_k},$$

which we now use both for $\alpha = c - a - b$ and $\alpha = 1$. Thus we find that

$$\sum_{n=0}^{\infty} \frac{(c-a)_n(c-b)_n z^n}{(c)_n n!} = \sum_{n=0}^{\infty}\sum_{k=0}^{n} \frac{(a)_k(b)_k(-n)_k}{(c)_k(1-c+a+b-n)_k k!} \frac{(c-a-b)_n z^n}{n!}$$

$$= \sum_{n=0}^{\infty} {}_3F_2\left[\begin{matrix} -n, a, b; \\ c, 1-c+a+b-n; \end{matrix} \quad 1\right] \frac{(c-a-b)_n z^n}{n!}.$$

THEOREM 29. *If n is a non-negative integer and if a, b, c are independent of n,*

$$ {}_3F_2\left[\begin{matrix} -n, a, b; \\ c, 1-c+a+b-n; \end{matrix} \quad 1\right] = \frac{(c-a)_n(c-b)_n}{(c)_n(c-a-b)_n}. $$

The ${}_3F_2$ of Theorem 29 has the property that the sum of its denominator parameters exceeds the sum of its numerator parameters by unity. Any ${}_{q+1}F_q$ with that property is called *Saalschützian.*

The proof given above for Theorem 29 requires that a, b, c be independent of n. Bailey [1; 21] gives a general form for the Saalschütz theorem. Here we need only the simple special cases exhibited in the preceding Theorem 29 and the subsequent Theorem 30. See, however, the discussion at the end of this section.

THEOREM 30. *If n is a non-negative integer and if a and b are independent of n,*

$$ {}_3F_2\left[\begin{matrix} -n, a+n, \tfrac{1}{2}+\tfrac{1}{2}a-b; \\ 1+a-b, \tfrac{1}{2}a+\tfrac{1}{2}; \end{matrix} \quad 1\right] = \frac{(b)_n}{(1+a-b)_n}. $$

Proof: On page 67 we showed that

$$ {}_2F_1\left[\begin{matrix} a, b; \\ 1+a-b; \end{matrix} \quad z\right] = (1-z)^{-a}\,{}_2F_1\left[\begin{matrix} \tfrac{1}{2}a, \tfrac{1}{2}+\tfrac{1}{2}a-b; \\ 1+a-b; \end{matrix} \quad \frac{-4z}{(1-z)^2}\right]. $$

Then

$$ \sum_{n=0}^{\infty} \frac{(a)_n(b)_n z^n}{n!(1+a-b)_n} = \sum_{k=0}^{\infty} \frac{(\tfrac{1}{2}a)_k(\tfrac{1}{2}+\tfrac{1}{2}a-b)_k(-1)^k 2^{2k} z^k}{k!(1+a-b)_k(1-z)^{a+2k}} $$

$$ = \sum_{n,k=0}^{\infty} \frac{(-1)^k(\tfrac{1}{2}a)_k(\tfrac{1}{2}+\tfrac{1}{2}a-b)_k 2^{2k}(a+2k)_n z^{n+k}}{k!n!(1+a-b)_k} $$

$$ = \sum_{n,k=0}^{\infty} \frac{(\tfrac{1}{2}a)_k(\tfrac{1}{2}+\tfrac{1}{2}a-b)_k(a)_{n+2k}(-1)^k 2^{2k} z^{n+k}}{k!n!(1+a-b)_k(a)_{2k}} $$

$$ = \sum_{n,k=0}^{\infty} \frac{(\tfrac{1}{2}+\tfrac{1}{2}a-b)_k(a)_{n+2k}(-1)^k z^{n+k}}{k!n!(1+a-b)_k(\tfrac{1}{2}a+\tfrac{1}{2})_k} $$

$$ = \sum_{n=0}^{\infty}\sum_{k=0}^{n} \frac{(-1)^k(\tfrac{1}{2}+\tfrac{1}{2}a-b)_k(a)_{n+k} z^n}{k!(n-k)!(1+a-b)_k(\tfrac{1}{2}a+\tfrac{1}{2})_k}. $$

Now $(a)_{n+k} = (a)_n(a+n)_k$ and $(-1)^k/(n-k)! = (-n)_k/n!$. Therefore we may write

$$\sum_{n=0}^{\infty} \frac{(a)_n (b)_n z^n}{n!(1+a-b)_n} = \sum_{n=0}^{\infty} \sum_{k=0}^{n} \frac{(-n)_k (a+n)_k (\frac{1}{2}+\frac{1}{2}a-b)_k}{k!(1+a-b)_k (\frac{1}{2}a+\frac{1}{2})_k} \cdot \frac{(a)_n z^n}{n!}$$

$$= \sum_{n=0}^{\infty} {}_3F_2\left[\begin{matrix} -n, a+n, \frac{1}{2}+\frac{1}{2}a-b; \\ 1+a-b, \frac{1}{2}a+\frac{1}{2}; \end{matrix} \; 1\right] \frac{(a)_n z^n}{n!},$$

from which Theorem 30 follows.

Results such as those in Theorems 29 and 30 may be extended in the following way. Suppose that we have established the identity

$$f(a, n) = g(a, n) \tag{2}$$

for an arbitrary a which is independent of n and for every non-negative integral n.

Now consider an a which depends upon n, $a = h(n)$. Equation (2) yields

$$f(a, 0) = g(a, 0)$$

for arbitrary a. Therefore

$$f[h(0), 0] = g[h(0), 0].$$

Again from (2) we obtain

$$f(a, 1) = g(a, 1),$$

from which

$$f[h(1), 1] = g[h(1), 1],$$

etc. It follows readily that (2) leads to

$$f[h(n), n] = g[h(n), n], \tag{3}$$

so that (2) is also true for a dependent upon n.

52. Whipple's theorem. Later we shall need the following theorem due to Whipple[1].

THEOREM 31. *If n is a non-negative integer and if b and c are independent of n,*

$${}_3F_2\left[\begin{matrix} -n, b, c; \\ 1-b-n, 1-c-n; \end{matrix} \; x\right]$$

$$= (1-x)^n \, {}_3F_2\left[\begin{matrix} -\frac{1}{2}n, -\frac{1}{2}n+\frac{1}{2}, 1-b-c-n; \\ 1-b-n, 1-c-n; \end{matrix} \; \frac{-4x}{(1-x)^2}\right].$$

Proof: Consider the ordinary hypergeometric function

$$
{}_2F_1\left[\begin{matrix} b, c; \\ b+c; \end{matrix} \quad t(1-x+xt)\right]
$$

$$
= \sum_{n=0}^{\infty} \frac{(b)_n(c)_n[(1-x)+xt]^n t^n}{n!(b+c)_n}
$$

$$
= \sum_{n=0}^{\infty}\sum_{k=0}^{n} \frac{(b)_n(c)_n x^k(1-x)^{n-k}t^{n+k}}{k!(n-k)!(b+c)_n}
$$

$$
= \sum_{n=0}^{\infty}\sum_{k=0}^{[n/2]} \frac{(b)_{n-k}(c)_{n-k}x^k(1-x)^{n-2k}t^n}{k!(n-2k)!(b+c)_{n-k}}
$$

$$
= \sum_{n=0}^{\infty}\sum_{k=0}^{[n/2]} \frac{(-n)_{2k}(b)_n(c)_n(1-b-c-n)_k(-1)^k x^k(1-x)^{n-2k}t^n}{k!n!(1-b-n)_k(1-c-n)_k(b+c)_n}
$$

$$
= \sum_{n=0}^{\infty} (1-x)^n \, {}_3F_2\left[\begin{matrix} -\tfrac{1}{2}n, -\tfrac{1}{2}n+\tfrac{1}{2}, 1-b-c-n; \\ 1-b-n, 1-c-n; \end{matrix} \quad \frac{-4x}{(1-x)^2}\right] \frac{(b)_n(c)_n t^n}{n!(b+c)_n}.
$$

We next expand the same hypergeometric function in powers of t in another way:

$$
{}_2F_1\left[\begin{matrix} b, c; \\ b+c; \end{matrix} \quad t(1-x+xt)\right]
$$

$$
= \sum_{n=0}^{\infty} \frac{(b)_n(c)_n[1-x(1-t)]^n t^n}{n!(b+c)_n}
$$

$$
= \sum_{n=0}^{\infty}\sum_{k=0}^{n} \frac{(b)_n(c)_n(-1)^k x^k(1-t)^k t^n}{k!(n-k)!(b+c)_n}
$$

$$
= \sum_{n,k=0}^{\infty} \frac{(b)_{n+k}(c)_{n+k}(-1)^k x^k(1-t)^k t^{n+k}}{k!n!(b+c)_{n+k}}
$$

$$
= \sum_{k=0}^{\infty}\sum_{n=0}^{\infty} \frac{(b+k)_n(c+k)_n t^n}{n!(b+c+k)_n} \cdot \frac{(b)_k(c)_k(-1)^k x^k(1-t)^k t^k}{k!(b+c)_k}
$$

$$
= \sum_{k=0}^{\infty} {}_2F_1\left[\begin{matrix} b+k, c+k; \\ b+c+k; \end{matrix} \quad t\right] \frac{(b)_k(c)_k(-1)^k x^k(1-t)^k t^k}{k!(b+c)_k}
$$

$$= \sum_{k=0}^{\infty} (1-t)^{-k} {}_2F_1\left[\begin{matrix} c, b; \\ b+c+k; \end{matrix} \; t\right] \frac{(b)_k(c)_k(-1)^k x^k (1-t)^k t^k}{k!(b+c)_k}$$

$$= \sum_{n,k=0}^{\infty} \frac{(c)_n(b)_n(b)_k(c)_k(-1)^k x^k t^{k+n}}{k!n!(b+c)_k(b+c+k)_n}$$

$$= \sum_{n,k=0}^{\infty} \frac{(b)_k(c)_k(b)_n(c)_n(-1)^k x^k t^{n+k}}{k!n!(b+c)_{n+k}}$$

$$= \sum_{n=0}^{\infty} \sum_{k=0}^{n} \frac{(b)_k(c)_k(b)_{n-k}(c)_{n-k}(-1)^k x^k t^n}{k!(n-k)!(b+c)_n}$$

$$= \sum_{n=0}^{\infty} \sum_{k=0}^{n} \frac{(-n)_k(b)_k(c)_k(b)_n(c)_n x^k t^n}{k!n!(b+c)_n(1-b-n)_k(1-c-n)_k}$$

$$= \sum_{n=0}^{\infty} {}_3F_2\left[\begin{matrix} -n, b, c; \\ 1-b-n, 1-c-n; \end{matrix} \; x\right] \frac{(b)_n(c)_n t^n}{n!(b+c)_n}.$$

Theorem 31 now follows by equating coefficients of t^n in the preceding two expansions.

In Whipple's work [1] the ${}_3F_2$'s involved are not necessarily terminating. We now prove Whipple's theorem for nonterminating series.

THEOREM 32. *If neither* $(a-b)$ *nor* $(a-c)$ *nor* a *is a negative integer,*

$$(1) \qquad {}_3F_2\left[\begin{matrix} a, b, c; \\ 1+a-b, 1+a-c; \end{matrix} \; x\right]$$

$$= (1-x)^{-a} {}_3F_2\left[\begin{matrix} \tfrac{1}{2}a, \tfrac{1}{2}a+\tfrac{1}{2}, 1+a-b-c; \\ 1+a-b, 1+a-c; \end{matrix} \; \frac{-4x}{(1-x)^2}\right].$$

Proof:

$${}_3F_2\left[\begin{matrix} a, b, c; \\ 1+a-b, 1+a-c; \end{matrix} \; x\right]$$

$$= \sum_{k=0}^{\infty} \frac{(a)_k(b)_k(c)_k x^k}{k!(1+a-b)_k(1+a-c)_k}$$

$$= \frac{\Gamma(1+a-b)\Gamma(1+a-c)}{\Gamma(1+a)\Gamma(1+a-b-c)} \sum_{k=0}^{\infty} \frac{(a)_k(b)_k(c)_k x^k}{k!(1+a)_{2k}} \cdot \frac{\Gamma(1+a+2k)\Gamma(1+a-b-c)}{\Gamma(1+a-b+k)\Gamma(1+a-c+k)}$$

$$= \frac{\Gamma(1+a-b)\Gamma(1+a-c)}{\Gamma(1+a)\Gamma(1+a-b-c)} \sum_{k=0}^{\infty} {}_2F_1\left[\begin{matrix} b+k, c+k; \\ 1+a+2k; \end{matrix}\; 1\right] \cdot \frac{(a)_k(b)_k(c)_k x^k}{k!(1+a)_{2k}}$$

$$= \frac{\Gamma(1+a-b)\Gamma(1+a-c)}{\Gamma(1+a)\Gamma(1+a-b-c)} \sum_{n,k=0}^{\infty} \frac{(a)_k(b)_{n+k}(c)_{n+k}x^k}{k!n!(1+a)_{n+2k}}$$

$$= \frac{\Gamma(1+a-b)\Gamma(1+a-c)}{\Gamma(1+a)\Gamma(1+a-b-c)} \sum_{n=0}^{\infty} \sum_{k=0}^{n} \frac{(a)_k(b)_n(c)_n x^k}{k!(n-k)!(1+a)_{n+k}}$$

$$= \frac{\Gamma(1+a-b)\Gamma(1+a-c)}{\Gamma(1+a)\Gamma(1+a-b-c)} \cdot \sum_{n=0}^{\infty} {}_2F_1\left[\begin{matrix} -n, a; \\ 1+a+n; \end{matrix}\; -x\right] \frac{(b)_n(c)_n}{n!(1+a)_n}.$$

By Ex. 9, page 69, we know that

$${}_2F_1\left[\begin{matrix} -n, a; \\ 1+a+n; \end{matrix}\; -x\right] = (1-x)^{-a}\, {}_2F_1\left[\begin{matrix} \tfrac{1}{2}a, \tfrac{1}{2}a+\tfrac{1}{2}; \\ 1+a+n; \end{matrix}\; \frac{-4x}{(1-x)^2}\right].$$

Therefore we may write

$${}_3F_2\left[\begin{matrix} a, b, c; \\ 1+a-b, 1+a-c; \end{matrix}\; x\right]$$

$$= \frac{\Gamma(1+a-b)\Gamma(1+a-c)}{\Gamma(1+a)\Gamma(1+a-b-c)} \sum_{n=0}^{\infty} {}_2F_1\left[\begin{matrix} \tfrac{1}{2}a, \tfrac{1}{2}a+\tfrac{1}{2}; \\ 1+a+n; \end{matrix}\; \frac{-4x}{(1-x)^2}\right] \cdot \frac{(b)_n(c)_n}{n!(1+a)_n(1-x)^a}$$

$$= \frac{\Gamma(1+a-b)\Gamma(1+a-c)}{\Gamma(1+a)\Gamma(1+a-b-c)} \sum_{n,k=0}^{\infty} \frac{(-1)^k(a)_{2k}(b)_n(c)_n x^k}{k!n!(1+a)_{n+k}(1-x)^{a+2k}}$$

$$= \frac{\Gamma(1+a-b)\Gamma(1+a-c)}{\Gamma(1+a)\Gamma(1+a-b-c)} \sum_{k=0}^{\infty} {}_2F_1\left[\begin{matrix} b, c; \\ 1+a+k; \end{matrix} \; 1\right] \cdot \frac{(-1)^k (a)_{2k} x^k}{k!(1+a)_k (1-x)^{a+2k}}$$

$$= \frac{\Gamma(1+a-b)\Gamma(1+a-c)}{\Gamma(1+a)\Gamma(1+a-b-c)} \cdot \sum_{k=0}^{\infty} \frac{\Gamma(1+a+k)\Gamma(1+a-b-c+k)(-1)^k (a)_{2k} x^k}{\Gamma(1+a-b+k)\Gamma(1+a-c+k)k!(1+a)_k(1-x)^{a+2k}}$$

$$= (1-x)^{-a} \sum_{k=0}^{\infty} \frac{(-1)^k (a)_{2k} (1+a-b-c)_k x^k}{k!(1+a-b)_k (1+a-c)_k (1-x)^{2k}}$$

$$= (1-x)^{-a} \, {}_3F_2\left[\begin{matrix} \tfrac{1}{2}a, \tfrac{1}{2}a+\tfrac{1}{2}, 1+a-b-c; \\ 1+a-b, 1+a-c; \end{matrix} \; \frac{-4x}{(1-x)^2}\right],$$

which completes the proof of Whipple's theorem.

53. Dixon's theorem. We shall obtain one more theorem on a special ${}_3F_2$. The series

$${}_{q+1}F_q\left[\begin{matrix} a_0, a_1, a_2, \cdots, a_q; \\ b_1, b_2, \cdots, b_q; \end{matrix} \; x\right]$$

is said to be well poised if

$$1 + a_0 = a_1 + b_1 = a_2 + b_2 = \cdots = a_q + b_q.$$

Note that the left members of the relations in Theorems 31 and 32 are well poised ${}_3F_2$'s.

Dixon [1] summed the well poised ${}_3F_2$ with unit argument.

THEOREM 33. *The following is an identity if a, b, and c are so restricted that each of the functions involved exists:*

$${}_3F_2\left[\begin{matrix} a, b, c; \\ 1+a-b, 1+a-c; \end{matrix} \; 1\right]$$

$$= \frac{\Gamma(1+\tfrac{1}{2}a)\Gamma(1+a-b)\Gamma(1+a-c)\Gamma(1+\tfrac{1}{2}a-b-c)}{\Gamma(1+a)\Gamma(1+\tfrac{1}{2}a-b)\Gamma(1+\tfrac{1}{2}a-c)\Gamma(1+a-b-c)}.$$

Proof: Let

$$\psi = {}_3F_2\left[\begin{matrix} a, b, c; \\ 1+a-b, 1+a-c; \end{matrix} \quad 1\right].$$

Then

$$\psi = \sum_{k=0}^{\infty} \frac{(a)_k(b)_k(c)_k \Gamma(1+a-b)\Gamma(1+a-c)}{k!\Gamma(1+a-b+k)\Gamma(1+a-c+k)}$$

$$= \frac{\Gamma(1+a-b)\Gamma(1+a-c)}{\Gamma(1+a-b-c)\Gamma(1+a)} \sum_{k=0}^{\infty} \frac{(a)_k(b)_k(c)_k}{k!(1+a)_{2k}} \cdot \frac{\Gamma(1+a+2k)\Gamma(1+a-b-c)}{\Gamma(1+a-b+k)\Gamma(1+a-c+k)}.$$

But

$$\frac{\Gamma(1+a+2k)\Gamma(1+a-b-c)}{\Gamma(1+a-b+k)\Gamma(1+a-c+k)} = {}_2F_1\left[\begin{matrix} b+k, c+k; \\ 1+a+2k; \end{matrix} \quad 1\right]$$

$$= \sum_{n=0}^{\infty} \frac{(b+k)_n(c+k)_n}{n!(1+a+2k)_n}.$$

Hence

$$\psi = \frac{\Gamma(1+a-b)\Gamma(1+a-c)}{\Gamma(1+a-b-c)\Gamma(1+a)} \sum_{n,k=0}^{\infty} \frac{(a)_k(b)_{n+k}(c)_{n+k}}{k!n!(1+a)_{n+2k}}$$

$$= \frac{\Gamma(1+a-b)\Gamma(1+a-c)}{\Gamma(1+a-b-c)\Gamma(1+a)} \sum_{n=0}^{\infty}\sum_{k=0}^{n} \frac{(a)_k(b)_n(c)_n}{k!(n-k)!(1+a)_{n+k}}.$$

Therefore

$$\psi = \frac{\Gamma(1+a-b)\Gamma(1+a-c)}{\Gamma(1+a-b-c)\Gamma(1+a)} \sum_{n=0}^{\infty}\sum_{k=0}^{n} \frac{(-1)^k(-n)_k(a)_k(b)_n(c)_n}{k!n!(1+a+n)_k(1+a)_n}$$

$$= \frac{\Gamma(1+a-b)\Gamma(1+a-c)}{\Gamma(1+a-b-c)\Gamma(1+a)} \sum_{n=0}^{\infty} {}_2F_1\left[\begin{matrix} -n, a; \\ 1+a+n; \end{matrix} \quad -1\right] \frac{(b)_n(c)_n}{n!(1+a)_n}.$$

By Theorem 26, page 68, with b replaced by $(-n)$,

$${}_2F_1\left[\begin{matrix} -n, a; \\ 1+a+n: \end{matrix} \quad -1\right] = \frac{\Gamma(1+a+n)\Gamma(1+\tfrac{1}{2}a)}{\Gamma(1+\tfrac{1}{2}a+n)\Gamma(1+a)} = \frac{(1+a)_n}{(1+\tfrac{1}{2}a)_n}.$$

Therefore we have

$$\psi = \frac{\Gamma(1+a-b)\Gamma(1+a-c)}{\Gamma(1+a-b-c)\Gamma(1+a)} \sum_{n=0}^{\infty} \frac{(b)_n(c)_n}{n!(1+\frac{1}{2}a)_n}$$

$$= \frac{\Gamma(1+a-b)\Gamma(1+a-c)}{\Gamma(1+a-b-c)\Gamma(1+a)}\, {}_2F_1\left[\begin{matrix} b, c; \\ 1+\frac{1}{2}a; \end{matrix} \; 1\right]$$

$$= \frac{\Gamma(1+a-b)\Gamma(1+a-c)\Gamma(1+\frac{1}{2}a)\Gamma(1+\frac{1}{2}a-b-c)}{\Gamma(1+a-b-c)\Gamma(1+a)\Gamma(1+\frac{1}{2}a-b)\Gamma(1+\frac{1}{2}a-c)},$$

which completes the proof of Dixon's theorem.

For another form of Dixon's theorem see Ex. 3 at the end of this chapter.

54. Contour integrals of Barnes' type. Consider the integral

(1) $$\frac{1}{2\pi i}\int_B \frac{\left[\prod_{m=1}^{p} \Gamma(a_m+s)\right]\Gamma(-s)(-z)^s\, ds}{\prod_{j=1}^{q} \Gamma(b_j+s)},$$

where B is a Barnes path of integration: that is, B starts at $-i\infty$ and runs to $+i\infty$ in the s-plane, curving if necessary to put the poles of $\Gamma(a_m+s)$; $m = 1, 2, \cdots, p$, to the left of the path and to put the poles of $\Gamma(-s)$ to the right of the path. The assumption that there exists such a path precludes the possibility that any a_m is zero or a negative integer. A representative Barnes path with $m = 3$ is shown in Figure 5.

Let the integrand in (1) be $\psi(s)$,

(2) $$\psi(s) = \frac{\Gamma(-s)(-z)^s \prod_{m=1}^{p} \Gamma(a_m+s)}{\prod_{j=1}^{q} \Gamma(b_j+s)}.$$

We wish to determine the behavior of $\psi(s)$ as s recedes from the origin along the upper portion of the pure imaginary axis and along its lower portion. We know from Theorem 13, page 31, that

(3) $$\log\Gamma(\alpha+s) = (s+\alpha-\tfrac{1}{2})\,\mathrm{Log}\, s - s + O\,(1)$$

as $|s| \to \infty$ in a region in which we have both $|\arg s| \leqq \pi - \delta$ and $|\arg(s+\alpha)| \leqq \pi - \delta$, $\delta > 0$. Also, by (2) above,

(4) $\psi(s)$

$$= \exp\left[\sum_{m=1}^{p} \log \Gamma(a_m+s) - \sum_{j=1}^{q} \log \Gamma(b_j+s) + \log \Gamma(-s) + s \log(-z)\right].$$

For the moment let s be on the upper imaginary axis. Then $s = it$, $|s| = t$, $\arg s = \frac{1}{2}\pi$. Using (3) we have, as $t \to \infty$,

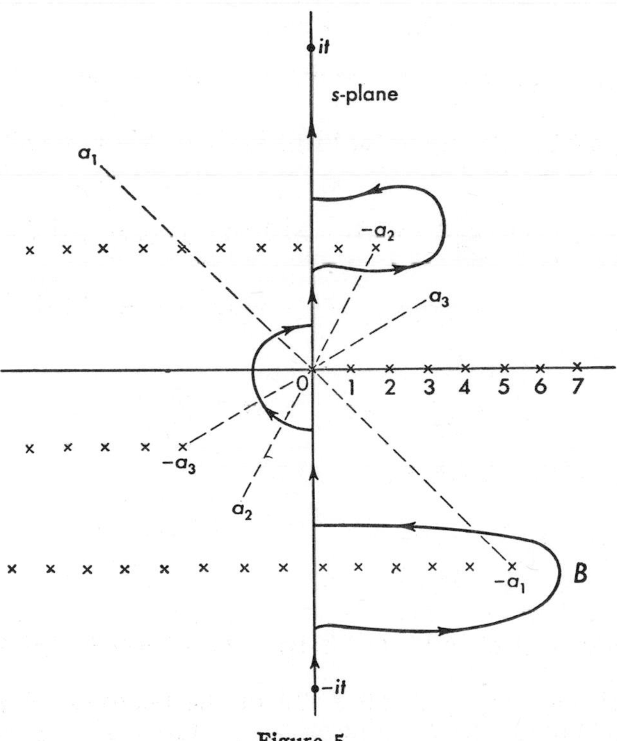

Figure 5

$$\log \Gamma(a_m + s) = (it + a_m - \tfrac{1}{2}) \operatorname{Log}(it) - it + O(1)$$
$$= (it + a_m - \tfrac{1}{2})(\operatorname{Log} t + \tfrac{1}{2}\pi i) - it + O(1).$$

Since $|\exp(iy)| = 1$ for real y, we are completely uninterested in any pure imaginary terms involved in the exponent in $\psi(s)$. Hence any such terms will hereafter be lumped together and designated as *PI* (pure imaginary).

We now have, as $t \to \infty$, $s = it$,

$$\log \Gamma(a_m + s) = [\operatorname{Re}(a_m) - \tfrac{1}{2}] \operatorname{Log} t - \tfrac{1}{2}\pi t + PI + O(1).$$

Then also

$$\log \Gamma\,(b_j + s) = [\operatorname{Re}(b_j) - \tfrac{1}{2}] \operatorname{Log} t - \tfrac{1}{2}\pi t + PI + O\,(1)$$

and

$$\begin{aligned}\log \Gamma\,(-s) &= (-it - \tfrac{1}{2})(\operatorname{Log} t - \tfrac{1}{2}\pi i) + it + O\,(1)\\ &= -\tfrac{1}{2} \operatorname{Log} t - \tfrac{1}{2}\pi t + PI + O\,(1).\end{aligned}$$

Finally,

$$\begin{aligned}s \log(-z) &= it\,[\operatorname{Log}|z| + i \arg\,(-z)]\\ &= -t \arg(-z) + \mathrm{PI}.\end{aligned}$$

We are now prepared to examine $\psi(s)$ as $s \to i\infty$. Indeed,

$$\begin{aligned}\log \psi(s) = \sum_{m=1}^{p} &[\operatorname{Re}(a_m) - \tfrac{1}{2}] \operatorname{Log} t - \tfrac{1}{2}p\pi t + PI + O(1)\\ &- \sum_{j=1}^{q} [\operatorname{Re}(b_j) - \tfrac{1}{2}] \operatorname{Log} t + \tfrac{1}{2}q\pi t + PI + O(1)\\ &- \tfrac{1}{2} \operatorname{Log} t - \tfrac{1}{2}\pi t - t \arg(-z) + PI + O(1).\end{aligned}$$

Let

$$A = \sum_{m=1}^{p} \operatorname{Re}(a_m), \qquad B = \sum_{j=1}^{q} \operatorname{Re}(b_j).$$

Then, for $s = it$, $t \to \infty$,

$$\begin{aligned}(5)\quad \log \psi(s) = [A-B+\tfrac{1}{2}(q-p-1)] \operatorname{Log} t + \tfrac{1}{2}(q-p-1)\pi t - t \arg(-z)\\ +\mathrm{PI}+O(1),\end{aligned}$$

or

$$(6)\qquad \psi(s) = O[t^{A-B+\frac{1}{2}(q-p-1)} \exp\{-\,t \arg(-z) + \tfrac{1}{2}\pi t(q-p-1)\}].$$

We turn next to a consideration of the behavior of $\psi(s)$ on the lower portion of the imaginary axis. Let $s = -it$, $|s| = t$, and $\arg s = -\tfrac{1}{2}\pi$. Then as $t \to \infty$, $s = -it$,

$$\log \Gamma(a_m + s) = (-it + a_m - \tfrac{1}{2})(\operatorname{Log} t - \tfrac{1}{2}\pi i) + it + O(1),$$

so that

$$\begin{aligned}\log \Gamma(a_m + s) &= [\operatorname{Re}(a_m) - \tfrac{1}{2}] \operatorname{Log} t - \tfrac{1}{2}\pi t + PI + O(1),\\ \log \Gamma(b_j + s) &= [\operatorname{Re}(b_j) - \tfrac{1}{2}] \operatorname{Log} t - \tfrac{1}{2}\pi t + PI + O(1),\\ \log \Gamma(-s) &= (it - \tfrac{1}{2})(\operatorname{Log} t + \tfrac{1}{2}\pi i) - it + O(1)\\ &= -\tfrac{1}{2} \operatorname{Log} t - \tfrac{1}{2}\pi t + PI + O(1),\\ s \log(-z) &= -it[\operatorname{Log}|z| + i \arg(-z)]\\ &= t \arg(-z) + PI.\end{aligned}$$

Therefore, as $t \to \infty$, $s = -it$, we have

$$\log \psi(s) = [A - B + \tfrac{1}{2}(q-p-1)] \operatorname{Log} t + \tfrac{1}{2}(q-p-1)\pi t + t \arg(-z) + PI + O(1), \tag{7}$$

or

$$\psi(s) = O[t^{A-B+\frac{1}{2}(q-p-1)} \exp\{t \arg(-z) + \tfrac{1}{2}\pi t(q-p-1\}]. \tag{8}$$

We wish to impose conditions which will insure that $\psi(s)$ be dominated by an exponential with negative exponent. Equations (6) and (8) show that we need both

$$\arg(-z) + \tfrac{1}{2}\pi(q-p-1) < 0 \tag{9}$$

and

$$-\arg(-z) + \tfrac{1}{2}\pi(q-p-1) < 0. \tag{10}$$

If (9) and (10) are satisfied,

$$\psi(s) = O(e^{-c|s|}), \qquad |s| \to \infty \text{ on } B,\ c > 0.$$

Then in any z-region in which (9) and (10) are satisfied

$$\frac{1}{2\pi i}\int_B \psi(s)\, ds$$

represents an analytic function of z.

If $p = q + 1$, $\frac{1}{2}\pi(q - p - 1) = -\pi$, and we require only that $|\arg(-z)| \leqq \pi - \delta$, $\delta > 0$ to obtain (9) and (10).

If $p = q$, $\frac{1}{2}\pi(q - p - 1) = -\frac{1}{2}\pi$, and we proceed to choose $|\arg(-z)| \leqq \frac{1}{2}\pi - \delta$, $\delta > 0$ to obtain (9) and (10).

If $p < q$, $\frac{1}{2}\pi(q - p - 1) \geqq 0$ (since p and q are integers), and there is no region in the z-plane for which both (9) and (10) are true.

THEOREM 34. *If $p = q + 1$ and no a_m is zero or a negative integer,*

$$\frac{1}{2\pi i}\int_B \frac{\Gamma(-s)(-z)^s \prod_{m=1}^{p} \Gamma(a_m + s)\, ds}{\prod_{j=1}^{q} \Gamma(b_j + s)} \tag{11}$$

is an analytic function of z in the cut plane $|\arg(-z)| < \pi$. If $p = q$ and no a_m is zero or a negative integer, (11) is an analytic function of z in the open half-plane $|\arg(-z)| < \frac{1}{2}\pi$ [i.e., in $\operatorname{Re}(z) < 0$]. The contour B is to be a Barnes contour as in Figure 5, page 95.

55. The Barnes integrals and the function ${}_pF_q$. We next relate the Barnes integrals of Theorem 34 with the function ${}_pF_q$ for $p = q+1$ and for $p = q$.

Once more let

$$\psi(s) = \frac{\Gamma(-s)(-z)^s \prod_{m=1}^{p} \Gamma(a_m + s)}{\prod_{j=1}^{q} \Gamma(b_j + s)}. \tag{1}$$

Let n be a non-negative integer, and consider the integral

$$\frac{1}{2\pi i} \int_{B_n + C_n} \psi(s)\, ds \tag{2}$$

over the closed path $(B_n + C_n)$ shown in Figure 6. The semicircle C_n is defined by $s = (n + \frac{1}{2})e^{i\theta}$, $-\frac{1}{2}\pi \leqq \theta \leqq \frac{1}{2}\pi$. The path B_n is

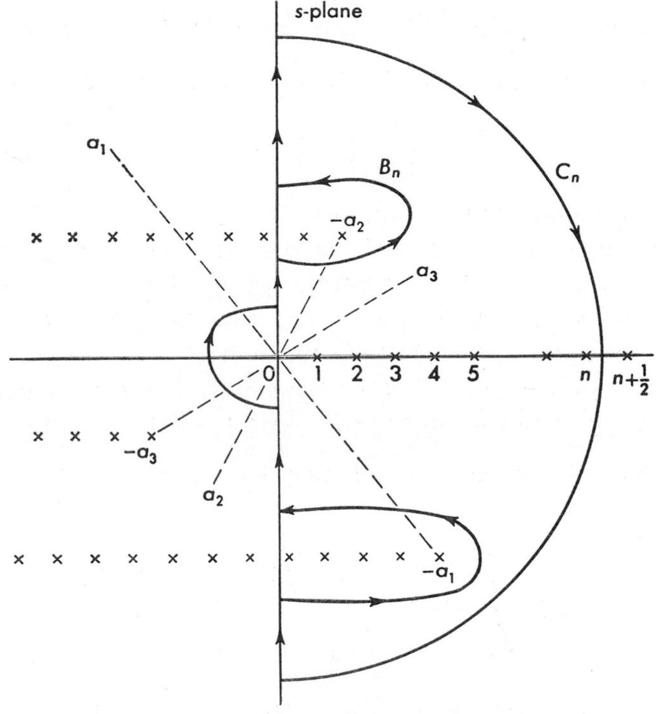

Figure 6

one of Barnes' type but is terminated at $s = (n + \frac{1}{2})i$ and $s = -(n + \frac{1}{2})i$. As $n \to \infty$, $B_n \to B$, the Barnes path of Figure 5.

The value of the integral (2) is the negative of the sum of the residues of $\psi(s)$ at the simple poles $s = k$, $k = 0, 1, 2, \cdots, n$ within the closed contour. Now

$$\Gamma(-s) = \frac{\Gamma(-s)\Gamma(1+s)}{\Gamma(1+s)} = \frac{1}{\Gamma(1+s)} \frac{-\pi}{\sin \pi s},$$

and it is convenient to write $\psi(s)$ in the form

$$\psi(s) = \frac{-\pi}{\sin \pi s} \cdot \frac{(-z)^s \prod_{m=1}^{p} \Gamma(a_m + s)}{\Gamma(1+s) \prod_{j=1}^{q} \Gamma(b_j + s)}. \tag{3}$$

Since $\lim_{s \to k} \pi(s-k)/\sin \pi s = (-1)^k$, the negative of the residue of $\psi(s)$ at $s = k$ is

$$\frac{(-1)^k(-z)^k \prod_{m=1}^{p} \Gamma(a_m + k)}{\Gamma(1+k) \prod_{j=1}^{q} \Gamma(b_j + k)} = \frac{\prod_{m=1}^{p} \Gamma(a_m)(a_m)_k}{\prod_{j=1}^{q} \Gamma(b_j)(b_j)_k} \cdot \frac{z^k}{k!}. \tag{4}$$

Therefore

$$\frac{1}{2\pi i}\int_{B_n} \psi(s)\,ds + \frac{1}{2\pi i}\int_{C_n} \psi(s)\,ds = \frac{\prod_{m=1}^{p} \Gamma(a_m)}{\prod_{j=1}^{q} \Gamma(b_j)} \sum_{k=0}^{n} \frac{\prod_{m=1}^{p} (a_m)_k}{\prod_{j=1}^{q} (b_j)_k} \cdot \frac{z^k}{k!}. \tag{5}$$

As $n \to \infty$, the first integral in (5) approaches the Barnes integral of the preceding section, and we now show that the second integral in (5) approaches zero.

Let $\rho = n + \frac{1}{2}$ so that the path C_n becomes $s = \rho e^{i\theta}$, $-\frac{1}{2}\pi \leqq \theta \leqq \frac{1}{2}\pi$. Let us determine the behavior of the $\psi(s)$ of (3) for s on C_n and n large. Since $s = \rho \cos \theta + i\rho \sin \theta$, we have, as $\rho \to \infty$,

$$\log \Gamma(a_m + s) = (\rho \cos \theta + i\rho \sin \theta + a_m - \tfrac{1}{2})(\operatorname{Log} \rho + i\theta) - \rho \cos \theta - i\rho \sin \theta + O(1),$$

or

$$\log \Gamma(a_m + s) = [\rho \cos \theta + \operatorname{Re}(a_m) - \tfrac{1}{2}] \operatorname{Log} \rho - \rho \cos \theta - \rho\theta \sin \theta + PI + O(1).$$

Again we use PI to denote all pure imaginary terms. The real term $-\theta\,\mathrm{Im}(a_m)$ is bounded by the constant $\frac{1}{2}\pi\,\mathrm{Im}(a_m)$ and is therefore included in $O(1)$. Of course

$$\log\Gamma(b_j+s)=[\rho\cos\theta+\mathrm{Re}(b_j)-\tfrac{1}{2}]\,\mathrm{Log}\,\rho-\rho\cos\theta-\rho\theta\sin\theta+PI+O(1),$$

$$\log\,\Gamma(1+s)=[\rho\cos\theta+1-\tfrac{1}{2}]\,\mathrm{Log}\rho-\rho\cos\theta-\rho\theta\sin\theta+PI+O(1),$$

and

$$\log(-z)^s = \rho\cos\theta\,\mathrm{Log}|z| - \rho\sin\theta\arg(-z) + PI.$$

It is an elementary matter (consider $|\sin\pi s|^{-1}$) to show also that

$$\log\frac{\pi}{\sin\pi s} = -\pi\rho\,|\sin\theta| + PI + O(1).$$

For the $\psi(s)$ of (3) it follows that

$$(6)\quad \log\psi(s)=[A-B-1+(\rho\cos\theta-\tfrac{1}{2})(p-q-1)]\,\mathrm{Log}\,\rho$$
$$-\rho(\cos\theta+\theta\sin\theta)(p-q-1)+\rho\cos\theta\,\mathrm{Log}|z|-\rho\sin\theta\arg(-z)$$
$$-\pi\rho|\sin\theta|+PI+O(1),$$

in which

$$A = \sum_{m=1}^{p}\mathrm{Re}(a_m),\qquad B = \sum_{j=1}^{q}\mathrm{Re}(b_j).$$

Because of Theorem 34, page 97, we are interested only in $p = q + 1$ and in $p = q$. First let $p = q + 1$. Then (6) yields

$$\log\psi(s) = (A - B - 1)\,\mathrm{Log}\,\rho + \rho\cos\theta\,\mathrm{Log}\,|z| - \rho\sin\theta\arg(-z)$$
$$-\pi\rho|\sin\theta| + PI + O(1),$$

from which

$$(7)\ \psi(s)=O[\rho^{A-B-1}\exp\{\rho\cos\theta\,\mathrm{Log}|z|-\rho\sin\theta\arg(-z)-\pi\rho|\sin\theta|\}].$$

Since $-\frac{1}{2}\pi \leqq \theta \leqq \frac{1}{2}\pi$, $\cos\theta \geqq 0$. If we choose $|z| < 1$, $\mathrm{Log}|z| < 0$. For $|\arg(-z)| \leqq \pi-\delta$, $\delta > 0$, the term $-\pi\rho|\sin\theta|$, never positive, dominates the term $-\rho\sin\theta\arg(-z)$. Hence for $p=q+1$, $|z|<1$, $|\arg(-z)| < \pi$, $\psi(s) = O[\exp(-c\rho)]$, $c > 0$, as $\rho\to\infty$, and therefore

$$(8)\qquad \frac{1}{2\pi i}\int_{C_n}\psi(s)\,ds \to 0 \text{ as } n\to\infty.$$

Then reference to equation (5) leads us to the following result.

THEOREM 35. *If* $|z| < 1$, *if* $|\arg(-z)| \leqq \pi - \delta$, $\delta > 0$, *and if no* a_m *or* b_j *is zero or a negative integer,*

$$(9)\qquad \frac{1}{2\pi i}\int_B \frac{(-z)^s\Gamma(-s)\prod_{m=1}^{q+1}\Gamma(a_m+s)\,ds}{\prod_{j=1}^{q}\Gamma(b_j+s)}$$

$$= \frac{\prod_{m=1}^{q+1}\Gamma(a_m)}{\prod_{j=1}^{q}\Gamma(b_j)}\cdot {}_{q+1}F_q\left[\begin{matrix} a_1, a_2, \cdots, a_{q+1}; \\ b_1, b_2, \cdots, b_q; \end{matrix}\; z\right],$$

in which B is the Barnes path of integration, Figure 5, page 95.

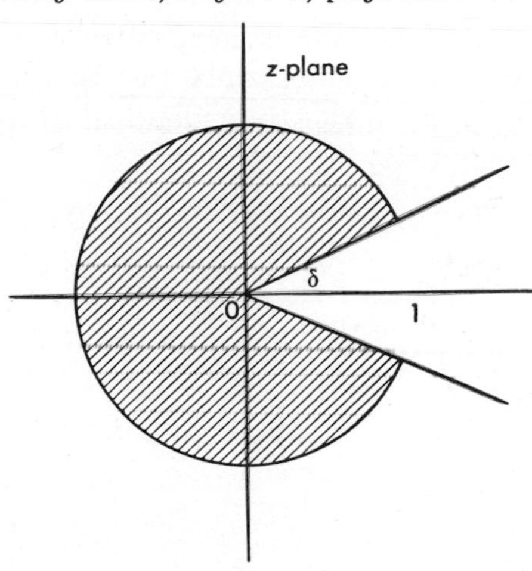

Figure 7

Theorem 35 states the equality of the two members of (9) in the region shown in Figure 7, a region in which both members are analytic. Since the left member of (9) is analytic in the larger region, the cut plane with the non-negative axis of reals deleted, that left member of (9) furnishes an analytic continuation of the right member of (9).

Let us now return to equation (6), page 100, and put $p = q$. The result is that as $\rho \to \infty$, s on C_n,

$$(10)\qquad \log \psi(s) = (A - B - \tfrac{1}{2} - \rho\cos\theta)\,\text{Log}\,\rho + \rho\,(\cos\theta + \theta\sin\theta) + \rho\cos\theta\,\text{Log}|z| - \rho\sin\theta\arg(-z) - \pi\rho|\sin\theta| + PI + O(1),$$

from which

$$(11)\qquad \psi(s) = O[\rho^{A-B-\frac{1}{2}-\rho\cos\theta}\exp\{\rho\cos\theta(1+\text{Log}|z|) + \rho\theta\sin\theta - \rho\sin\theta\arg(-z) - \pi\rho|\sin\theta|\}].$$

Now $|\theta| \leqq \frac{1}{2}\pi$, and we choose $|\arg(-z)| \leqq \frac{1}{2}\pi - \delta$, $\delta > 0$, in order to conclude that

$$\rho\theta\sin\theta - \rho\sin\theta\arg(-z) - \pi\rho|\sin\theta| \leqq -\delta\rho|\sin\theta|.$$

If we also choose $|z| < e^{-1}$, Log $|z| < -1$, we may conclude that, for $\theta \neq \pm\frac{1}{2}\pi$,

$$\rho \cos \theta \, (1 + \text{Log} \, |z|) < 0.$$

Since $\sin^2 \theta + \cos^2 \theta = 1$, the coefficient of ρ in our exponent is always negative.

Once more, for s on C_n, $\psi(s) = O[\exp(-c\rho)]$, $c > 0$, as $\rho \to \infty$. Again (8) holds true, and we may let $n \to \infty$ in each member of equation (5), page 99. Since $p = q$, the right member of (5) approaches an entire function, a series convergent for all finite z. The first term on the left approaches a Barnes integral known (Theorem 34) to represent an analytic function for $\text{Re}(z) < 0$. We may now drop the unnecessary restriction $|z| < e^{-1}$, which was merely a tool in our proof, since two analytic functions shown to be identical in some region are identical throughout their common region of analyticity.

THEOREM 36. *If* $\text{Re}(z) < 0$ *and if no* a_m *or* b_j *is zero or a negative integer,*

$$\frac{1}{2\pi i} \int_B \frac{(-z)^s \Gamma(-s) \prod_{m=1}^{q} \Gamma(a_m + s) \, ds}{\prod_{j=1}^{q} \Gamma(b_j + s)}$$

$$= \frac{\prod_{m=1}^{q} \Gamma(a_m)}{\prod_{j=1}^{q} \Gamma(b_j)} \, {}_qF_q\left[\begin{matrix} a_1, a_2, \cdots, a_q; \\ b_1, b_2, \cdots, b_q; \end{matrix} \quad z\right],$$

in which B is a Barnes path of integration, Figure 5, page 95.

56. A useful integral. Consider the integral

$$(1) \quad A(t) = \int_0^t x^{\alpha-1}(t-x)^{\beta-1} \, {}_pF_q\left[\begin{matrix} a_1, \cdots, a_p; \\ b_1, \cdots, b_q; \end{matrix} \quad cx^k(t-x)^s\right] dx$$

in which the parameters are subject to the conditions that

(a) $\text{Re}(\alpha) > 0$, $\text{Re}(\beta) > 0$;
(b) k and s are to be non-negative integers not both zero;
(c) No b_j is to be zero or a negative integer;
(d) $p \leqq q + 1$, unless some a_m is a nonpositive integer, in which case p may be any positive integer.

We evaluate $A(t)$ by term-by-term integration. Let $x = tv$. Then

$$A(t) = t^{\alpha+\beta-1} \int_0^1 v^{\alpha-1}(1-v)^{\beta-1} {}_pF_q\left[\begin{matrix} a_1, \cdots, a_p; \\ \\ b_1, \cdots, b_q; \end{matrix} \quad ct^{k+s}v^k(1-v)^s\right] dv$$

$$= t^{\alpha+\beta-1} \int_0^1 \sum_{n=0}^{\infty} \frac{\prod_{m=1}^{p} (a_m)_n t^{n(k+s)} c^n v^{\alpha-1+kn}(1-v)^{\beta-1+sn} dv}{n! \prod_{j=1}^{q} (b_j)_n},$$

so that

$$A(t) = t^{\alpha+\beta-1} \sum_{n=0}^{\infty} \frac{\prod_{m=1}^{p} (a_m)_n c^n t^{n(s+k)} B(\alpha + kn, \beta + sn)}{n! \prod_{j=1}^{q} (b_j)_n}, \tag{2}$$

in which $B(z_1, z_2)$ is the Beta function of Chapter 2. Now

$$B(\alpha + kn, \beta + sn) = \frac{\Gamma(\alpha + kn)\Gamma(\beta + sn)}{\Gamma(\alpha + \beta + kn + sn)}$$

$$= \frac{\Gamma(\alpha)\Gamma(\beta)}{\Gamma(\alpha + \beta)} \cdot \frac{(\alpha)_{kn}(\beta)_{sn}}{(\alpha + \beta)_{(k+s)n}}.$$

In Lemma 6, page 22, we found that

$$(\alpha)_{kn} = k^{nk} \prod_{i=1}^{k} \left(\frac{\alpha + i - 1}{k}\right)_n,$$

which permits us to write

(3) $B(\alpha + kn, \beta + sn)$

$$= \frac{B(\alpha, \beta) k^{nk} s^{sn} \prod_{i=1}^{k} \left(\frac{\alpha + i - 1}{k}\right)_n \prod_{u=1}^{s} \left(\frac{\beta + u - 1}{s}\right)_n}{(k + s)^{(k+s)n} \prod_{w=1}^{k+s} \left(\frac{\alpha + \beta + w - 1}{k + s}\right)_n}.$$

The use of (3) in the right member of (2) yields a hypergeometric function with $(p + k + s)$ numerator parameters and $(q + k + s)$ denominator parameters. The precise result follows.

THEOREM 37. *If* $\text{Re}(\alpha) > 0$, $\text{Re}(\beta) > 0$, *and if k and s are positive integers, then inside the region of convergence of the resultant series*

$$(4)\qquad \int_0^t x^{\alpha-1}(t-x)^{\beta-1}\,{}_pF_q\left[\begin{matrix} a_1,\cdots,a_p; \\ b_1,\cdots,b_q; \end{matrix}\; cx^k(t-x)^s\right]dx$$

$$= B(\alpha,\beta)t^{\alpha+\beta-1}\,{}_{p+k+s}F_{q+k+s}$$

$$\left[\begin{matrix} a_1,\cdots,a_p,\dfrac{\alpha}{k},\dfrac{\alpha+1}{k},\cdots,\dfrac{\alpha+k-1}{k},\dfrac{\beta}{s},\dfrac{\beta+1}{s},\cdots,\dfrac{\beta+s-1}{s}; \\ b_1,\cdots,b_q,\dfrac{\alpha+\beta}{k+s},\dfrac{\alpha+\beta+1}{k+s},\cdots,\dfrac{\alpha+\beta+k+s-1}{k+s}; \end{matrix}\; \frac{k^ks^sct^{k+s}}{(k+s)^{k+s}}\right].$$

If either, but not both, of k and s is zero, it is a simple matter to modify the steps in the derivation of (4) and thus arrive at the pertinent result.

THEOREM 38. *If* $\text{Re}(\alpha)>0$, $\text{Re}(\beta)>0$, *and if k is a positive integer, then inside the region of convergence of the resultant series*

$$(5)\qquad \int_0^t x^{\alpha-1}(t-x)^{\beta-1}\,{}_pF_q\left[\begin{matrix} a_1,\cdots,a_p; \\ b_1,\cdots,b_q; \end{matrix}\; cx^k\right]dx$$

$$= B(\alpha,\beta)t^{\alpha+\beta-1}\cdot$$

$${}_{p+k}F_{q+k}\left[\begin{matrix} a_1,\cdots,a_p,\dfrac{\alpha}{k},\dfrac{\alpha+1}{k},\cdots,\dfrac{\alpha+k-1}{k}; \\ b_1,\cdots,b_q,\dfrac{\alpha+\beta}{k},\dfrac{\alpha+\beta+1}{k},\cdots,\dfrac{\alpha+\beta+k-1}{k}; \end{matrix}\; ct^k\right].$$

The reader can easily write the corresponding theorem for $k=0$, $s>0$.

EXAMPLE: In equation (4) of Theorem 37 choose $p=0$ (no a's), $q=1$ (one b), $b_1=1$, $\alpha=1$, $\beta=1$, $k=1$, $s=1$, $c=-\frac{1}{4}$. The result is

$$(6)\qquad \int_0^t {}_0F_1\left[\begin{matrix} -; \\ 1; \end{matrix}\; -\tfrac{1}{4}x(t-x)\right]dx = B(1,1)t\,{}_2F_3\left[\begin{matrix} 1,1; \\ 1,\frac{2}{2},\frac{3}{2}; \end{matrix}\; -\frac{1\cdot1\cdot t^2}{4\cdot 2^2}\right]$$

$$= t\,{}_0F_1\left[\begin{matrix} -; \\ \frac{3}{2}; \end{matrix}\; -\frac{t^2}{16}\right].$$

The reader may already know, or will discover in the next chapter, that $J_0(z)$, the Bessel function of the first kind and of index zero, is given by

$$J_0(z) = {}_0F_1(-;1;-\tfrac{1}{4}z^2).$$

The ${}_0F_1$ in the second form of the right member of (6) is elementary. Indeed

$$\begin{aligned} t\,{}_0F_1\left(-;\frac{3}{2};-\frac{t^2}{16}\right) &= \sum_{n=0}^{\infty} \frac{(-1)^n t^{2n+1}}{2^{4n}(\frac{3}{2})_n n!} \\ &= \sum_{n=0}^{\infty} \frac{(-1)^n t^{2n+1}}{2^{2n}(2)_{2n}} = 2\sum_{n=0}^{\infty} \frac{(-1)^n t^{2n+1}}{2^{2n+1}(2n+1)!} \\ &= 2\sin \tfrac{1}{2}t. \end{aligned}$$

Therefore (6) may be rewritten in the form

$$(7) \qquad \int_0^t J_0\left(\sqrt{x(t-x)}\right) dx = 2\sin\tfrac{1}{2}t.$$

EXERCISES

1. Show that

$${}_0F_1\left[\begin{matrix} -; \\ a; \end{matrix}\; x\right] {}_0F_1\left[\begin{matrix} -; \\ b; \end{matrix}\; x\right] = {}_2F_3\left[\begin{matrix} \frac{1}{2}a+\frac{1}{2}b,\ \frac{1}{2}a+\frac{1}{2}b-\frac{1}{2}; \\ a,\ b,\ a+b-1; \end{matrix}\; 4x\right].$$

You may use the result in Ex. 6, page 69.

2. Show that

$$\int_0^t x^{\frac{1}{4}}(t-x)^{-\frac{1}{4}}[1-x^2(t-x)^2]^{-\frac{1}{4}}dx = \tfrac{1}{2}\pi t\,{}_2F_1\left[\begin{matrix} \frac{1}{4},\ \frac{3}{4}; \\ 1; \end{matrix}\; \frac{t^4}{16}\right].$$

3. With the aid of Theorem 8, page 21, show that

$$\frac{\Gamma(1+\frac{1}{2}a)}{\Gamma(1+a)} = \frac{\cos\frac{1}{2}\pi a\,\Gamma(1-a)}{\Gamma(1-\frac{1}{2}a)}$$

and that

$$\frac{\Gamma(1+a-b)}{\Gamma(1+\frac{1}{2}a-b)} = \frac{\sin\pi(b-\frac{1}{2}a)\Gamma(b-\frac{1}{2}a)}{\sin\pi(b-a)\Gamma(b-a)}.$$

Thus put Dixon's theorem, Theorem 33, page 92, in the form

$$\begin{aligned} &{}_3F_2\left[\begin{matrix} a,\ b,\ c; \\ 1+a-b,\ 1+a-c; \end{matrix}\; 1\right] \\ &= \frac{\cos\frac{1}{2}\pi a \sin\pi(b-\frac{1}{2}a)}{\sin\pi(b-a)} \cdot \frac{\Gamma(1-a)\Gamma(b-\frac{1}{2}a)\Gamma(1+a-c)\Gamma(1+\frac{1}{2}a-b-c)}{\Gamma(1-\frac{1}{2}a)\Gamma(b-a)\Gamma(1+\frac{1}{2}a-c)\Gamma(1+a-b-c)}. \end{aligned}$$

4. Use the result in Ex. 3 to show that if n is a non-negative integer,

$$ {}_3F_2\left[\begin{matrix}-2n, \alpha, 1-\beta-2n; \\ 1-\alpha-2n, \beta;\end{matrix}\; 1\right] = \frac{(2n)!(\alpha)_n(\beta-\alpha)_n}{n!(\alpha)_{2n}(\beta)_n}. $$

5. With the aid of the formula in Ex. 4 prove Ramanujan's theorem:

$$ {}_1F_1\left[\begin{matrix}\alpha; \\ \beta;\end{matrix}\; x\right] {}_1F_1\left[\begin{matrix}\alpha; \\ \beta;\end{matrix}\; -x\right] = {}_2F_3\left[\begin{matrix}\alpha, \beta-\alpha; \\ \beta, \frac{1}{2}\beta, \frac{1}{2}\beta+\frac{1}{2};\end{matrix}\; \frac{x^2}{4}\right]. $$

6. Let $\gamma_n = {}_3F_2(-n, 1-a-n, 1-b-n; a, b; 1)$. Use the result in Ex. 3 to show that $\gamma_{2n+1} = 0$ and

$$ \gamma_{2n} = \frac{(-1)^n(2n)!(a+b-1)_{3n}}{n!(a)_n(b)_n(a+b-1)_{2n}}. $$

7. With the aid of the result in Ex. 6 show that

$$ {}_0F_2(-; a, b; t)\; {}_0F_2(-; a, b; -t) $$

$$ = {}_3F_8\left[\begin{matrix}\frac{1}{3}(a+b-1), \frac{1}{3}(a+b), \frac{1}{3}(a+b+1); \\ a, b, \frac{1}{2}a, \frac{1}{2}a+\frac{1}{2}, \frac{1}{2}b, \frac{1}{2}b+\frac{1}{2}, \frac{1}{2}(a+b-1), \frac{1}{2}(a+b);\end{matrix}\; \frac{-27t^2}{64}\right]. $$

8. Prove that

$$ \sum_{k=0}^{n} \frac{(-1)^{n-k}(\gamma-b-c)_{n-k}(\gamma-b)_k(\gamma-c)_k x^{n-k}}{k!(n-k)!(\gamma)_k}\; {}_3F_2\left[\begin{matrix}-k, b, c; \\ 1-\gamma+b-k, 1-\gamma+c-k;\end{matrix}\; x\right] $$

$$ = \frac{(\gamma-b)_n(\gamma-c)_n(1-x)^n}{n!(\gamma)_n}\; {}_3F_2\left[\begin{matrix}-\frac{1}{2}n, -\frac{1}{2}n+\frac{1}{2}, 1-\gamma-n; \\ 1-\gamma+b-n, 1-\gamma+c-n;\end{matrix}\; \frac{-4x}{(1-x)^2}\right], $$

and note the special case $\gamma = b + c$, Whipple's theorem, page 88.

Exs. 9–11 below use the notation of the Laplace transform as in Ex. 16, page 71.

9. Show that

$$ L\left\{t^c\; {}_pF_q\left[\begin{matrix}a_1, \cdots, a_p; \\ b_1, \cdots, b_q;\end{matrix}\; zt\right]\right\} = \frac{\Gamma(1+c)}{s^{1+c}}\; {}_{p+1}F_q\left[\begin{matrix}1+c, a_1, \cdots, a_p; \\ b_1, \cdots, b_q;\end{matrix}\; \frac{z}{s}\right]. $$

10. Show that

$$ L^{-1}\left\{\frac{1}{s}\; {}_pF_{q+1}\left[\begin{matrix}a_1, \cdots, a_p; \\ s+1, b_1, \cdots, b_q;\end{matrix}\; z\right]\right\} = {}_pF_{q+1}\left[\begin{matrix}a_1, \cdots, a_p; \\ 1, b_1, \cdots, b_q;\end{matrix}\; z(1-e^{-t})\right]. $$

11. Show that

$$ L^{-1}\left\{\frac{s^k}{(s-z)^{k+1}}\right\} = {}_1F_1\left[\begin{matrix}k+1; \\ 1;\end{matrix}\; zt\right]. $$

12. Show that

$$\frac{d}{dz}\,{}_pF_q\left[\begin{matrix}a_1, \cdots, a_p; \\ \\ b_1, \cdots, b_q;\end{matrix}\quad z\right] = \frac{\prod_{m=1}^{p} a_m}{\prod_{j=1}^{q} b_j}\,{}_pF_q\left[\begin{matrix}a_1 + 1, \cdots, a_p + 1; \\ \\ b_1 + 1, \cdots, b_q + 1;\end{matrix}\quad z\right].$$

13. In Ex. 19, page 71, we found that the complete elliptic integral of the first kind is given by

$$K(k) = \tfrac{1}{2}\pi\,{}_2F_1(\tfrac{1}{2}, \tfrac{1}{2}; 1; k^2).$$

Show that

$$\int_0^t K\left(\sqrt{x(t - x)}\right) dx = \pi \operatorname{Arcsin}(\tfrac{1}{2}t).$$

CHAPTER 6

Bessel Functions

57. Remarks. No other special functions have received such detailed treatment in readily available treatises* as have the Bessel functions. Consequently we here present only a brief introduction to the subject, including those results which will be used in later chapters of this book.

For a discussion of orthogonality properties and of zeros of Bessel functions, see Churchill [2] and Watson [1]. An extremely simple result on zeros of Bessel functions appears in Ex. 13 at the end of this chapter.

58. Definition of $J_n(z)$. We already know that the ${}_0F_0$ is an exponential and that the ${}_1F_0$ is a binomial. It is natural to examine next the most general ${}_0F_1$, the only other ${}_pF_q$ with less than two parameters. The function we shall study is not precisely the ${}_0F_1$ but one that has an extra factor as in (1) below.

We define $J_n(z)$, for n not a negative integer,

$$(1) \qquad J_n(z) = \frac{(z/2)^n}{\Gamma(1+n)}\, {}_0F_1\left(-;\, 1+n;\, -\frac{z^2}{4}\right).$$

If n is a negative integer, we put

*The most striking example is Watson's exhaustive (804 pages) work; Watson [1]. An exposition sufficiently thorough for most readers will be found in Chapter 17 of Whittaker and Watson [1].

(2) $$J_n(z) = (-1)^n J_{-n}(z).$$

Equations (1) and (2) together define $J_n(z)$ for all finite z and n. The function $J_n(z)$ is called "the Bessel function of the first kind of index n."

Since

$$ {}_0F_1\left(-; 1+n; -\frac{z^2}{4}\right) = \sum_{k=0}^{\infty} \frac{(-z^2/4)^k}{(1+n)_k k!} = \sum_{k=0}^{\infty} \frac{\Gamma(1+n)(-1)^k z^{2k}}{k!\,\Gamma(1+n+k)2^{2k}}, $$

the relation (1) is equivalent to

(3) $$J_n(z) = \sum_{k=0}^{\infty} \frac{(-1)^k z^{2k+n}}{2^{2k+n} k!\,\Gamma(1+n+k)}.$$

Note also the immediate result

(4) $$J_n(-z) = (-1)^n J_n(z).$$

59. Bessel's differential equation. We know a differential equation satisfied by any ${}_0F_1$ by specializing the result in Section 46. The equation

(1) $$[\theta(\theta + b - 1) - y]u = 0; \qquad \theta = y\frac{d}{dy},$$

has $u = {}_0F_1(-; b; y)$ as one solution. Equation (1) can also be written

(2) $$y\frac{d^2u}{dy^2} + b\frac{du}{dy} - u = 0.$$

We now put $b = 1 + n$, $y = -z^2/4$ in (2) to obtain

(3) $$zu'' + (2n + 1)u' + zu = 0,$$

in which primes denote differentiations with respect to z. One solution of (3) is $u = {}_0F_1(-; 1 + n; -z^2/4)$. We seek an equation satisfied by $w = z^n u$. Hence in (3) we now put $u = z^{-n}w$ and arrive at the differential equation

(4) $$z^2w'' + zw' + (z^2 - n^2)w = 0,$$

of which one solution is $w = z^n {}_0F_1(-; 1 + n; -z^2/4)$.

Equation (4) is Bessel's differential equation. If n is not an integer, two linearly independent solutions of (4) are

$$w_1 = J_n(z) \tag{5}$$

and

$$w_2 = J_{-n}(z) \tag{6}$$

and they are valid for all finite z.

If n is zero or a positive integer, (5) is still a valid solution of Bessel's equation, but then (6) is not linearly independent of (5). For integral n, a second solution to accompany $J_n(z)$ is logarithmic in character and can be obtained by standard procedures.* The result is

$$w_3 = Y_n(z) = J_n(z) \log z + \sum_{k=0}^{n-1} \frac{(-1)^{k+1}(n-1)!z^{2k-n}}{2^{2k+1-n}k!(1-n)_k} + \tfrac{1}{2} \sum_{k=0}^{\infty} \frac{(-1)^{k+1}(H_k + H_{k+n})z^{2k+n}}{2^{2k+n}k!(k+n)!} \tag{7}$$

for $n > 1$. In (7) the common convention $H_0 = 0$ has been used. For $n = 0$, the first series on the right in (7) is to be omitted; for $n = 1$, the first series on the right in (7) is to be replaced by the single term $(-z^{-1})$.

60. Differential recurrence relations. In the equation

$$J_n(z) = \sum_{k=0}^{\infty} \frac{(-1)^k z^{2k+n}}{2^{2k+n}k!\,\Gamma(1+n+k)} \tag{1}$$

we may multiply both members by z^n and then differentiate throughout with respect to z to obtain

$$\frac{d}{dz}\,[z^n J_n(z)] = \sum_{k=0}^{\infty} \frac{(-1)^k z^{2k+2n-1}}{2^{2k+n-1}k!\,\Gamma(n+k)} \tag{2}$$

in which we have canceled the factor $(2k + 2n)$ in numerator and denominator, using $\Gamma(1 + n + k) = (n + k)\Gamma(n + k)$. Equation (2) can be rewritten as

$$\frac{d}{dz}\,[z^n J_n(z)] = z^n \sum_{k=0}^{\infty} \frac{(-1)^k z^{2k+n-1}}{2^{2k+n-1}k!\,\Gamma(1+n-1+k)},$$

and we thus see that the right member is $z^n J_{n-1}(z)$. We conclude that

$$\frac{d}{dz}\,[z^n J_n(z)] = z^n J_{n-1}(z). \tag{3}$$

*In Rainville [1], pertinent techniques are explained on pages 285–291 (for $n = 0$) and on pages 299–303 (for $n > 0$).

Equation (3) may also be put in the form

$$(4) \qquad zJ_n'(z) = zJ_{n-1}(z) - nJ_n(z),$$

which is called a *differential recurrence relation*, differential because of the differentiation involved, recurrence because of the presence of different indices n and $(n - 1)$.

Next we return to equation (1), insert the factor z^{-n} on each side, and again differentiate each member with respect to z to obtain

$$(5) \qquad \frac{d}{dz}[z^{-n}J_n(z)] = \sum_{k=1}^{\infty} \frac{(-1)^k z^{2k-1}}{2^{2k+n-1}(k-1)!\Gamma(1+n+k)}.$$

A shift of index from k to $(k + 1)$ yields

$$\frac{d}{dz}[z^{-n}J_n(z)] = \sum_{k=0}^{\infty} \frac{(-1)^{k+1} z^{2k+1}}{2^{2k+n+1}k!\Gamma(1+n+1+k)},$$

from which we obtain

$$(6) \qquad \frac{d}{dz}[z^{-n}J_n(z)] = -z^{-n}J_{n+1}(z).$$

Equation (6) can be expressed equally well as

$$(7) \qquad zJ_n'(z) = -zJ_{n+1}(z) + nJ_n(z).$$

In equations (7) and

$$(4) \qquad zJ_n'(z) = zJ_{n-1}(z) - nJ_n(z)$$

we have two differential recurrence relations. From them it follows also that

$$(8) \qquad 2J_n'(z) = J_{n-1}(z) - J_{n+1}(z).$$

61. A pure recurrence relation. Elimination of $J_n'(z)$ from the relations (4) and (7) of the preceding section gives us at once the pure recurrence relation

$$(1) \qquad 2nJ_n(z) = z[J_{n-1}(z) + J_{n+1}(z)].$$

It is also instructive to obtain (1) from the sole contiguous function relation possessed by the ${}_0F_1$ function.

Consider the set of Bessel functions $J_n(z)$ for non-negative integral index. If we write (1) in the form

$$(2) \qquad J_n(z) = \frac{2(n-1)}{z}J_{n-1}(z) - J_{n-2}(z),$$

we obtain each J_n of the set in terms of the two preceding it; J_2 from J_0 and J_1, J_3 from J_1 and J_2, etc. In this way we can, for integral n, write

$$(3)\qquad J_n(z) = A_n(z)J_0(z) + B_n(z)J_1(z).$$

The coefficients $A_n(z)$ and $B_n(z)$ are then polynomials in $1/z$. These are simple special cases of Lommel's polynomials $R_{m,\nu}(z)$ which may be encountered* by applying the same process to (2), with n replaced by $(\nu + m)$ to arrive at the result

$$(4)\qquad J_{\nu+m}(z) = R_{m,\nu}(z)J_\nu(z) - R_{m-1,\nu+1}(z)J_{\nu-1}(z).$$

The Lommel polynomial is a ${}_2F_3$ (see Watson [1:297]):

$$(5)\qquad R_{n,\nu}\left(\frac{1}{z}\right) = (\nu)_n(2z)^n\, {}_2F_3\left[\begin{matrix} -\tfrac{1}{2}n, -\tfrac{1}{2}n + \tfrac{1}{2}; \\ \nu, -n, 1 - \nu - n; \end{matrix} \; -\frac{1}{z^2}\right].$$

62. A generating function. Our approach (Section 58) to the function $J_n(z)$ was from the hypergeometric standpoint, which is natural here because this book is largely concerned with functions of hypergeometric character. Some authors approach $J_n(z)$ by first defining it, for integral n only, by means of a generating function† relation which we shall now obtain.

Lemma 12. *For* $n \geqq 1$,

$$(1)\qquad \sum_{k=0}^{n} A(k, n) = \sum_{k=0}^{[n/2]} A(k, n) + \sum_{k=0}^{[(n-1)/2]} A(n - k, n).$$

Proof: First note that for integral $n \geqq 1$,

$$(2)\qquad n = 1 + [\tfrac{1}{2}n] + [\tfrac{1}{2}(n - 1)],$$

in which [ ] is the usual greatest integer symbol. Equation (2) is easily verified separately for n even and for n odd.

Next note that

$$(3)\qquad \sum_{k=0}^{n} A(k, n) = \sum_{k=0}^{[n/2]} A(k, n) + \sum_{k=1+[n/2]}^{1+[n/2]+[(n-1)/2]} A(k, n).$$

In the last summation in (3) replace k by $(n - k)$; that is, k by $1 + [n/2] + [(n - 1)/2] - k$. Then

*See Watson [1:294].

†See Chapter 8 for some detail on the generating function concept.

$$\sum_{k=0}^{n} A(k, n) = \sum_{k=0}^{[n/2]} A(k, n) + \sum_{k=[(n-1)/2]}^{0} A(n - k, n),$$

from which Lemma 12 follows by reversing the order of the second summation on the right.

THEOREM 39. *For $t \neq 0$ and for all finite z,*

$$\exp\left[\frac{1}{2}z\left(t - \frac{1}{t}\right)\right] = \sum_{n=-\infty}^{\infty} J_n(z)t^n. \tag{4}$$

Proof: Let us collect powers of z in the summation

$$\sum_{n=-\infty}^{\infty} J_n(z)t^n = \sum_{n=-\infty}^{-1} J_n(z)t^n + \sum_{n=0}^{\infty} J_n(z)t^n$$

$$= \sum_{n=0}^{\infty} J_{-n-1}(z)t^{-n-1} + \sum_{n=0}^{\infty} J_n(z)t^n.$$

We defined $J_m(z)$ for negative integral m in Section 58. Using that definition we get

$$\sum_{n=-\infty}^{\infty} J_n(z)t^n = \sum_{n=0}^{\infty} (-1)^{n+1}J_{n+1}(z)t^{-n-1} + \sum_{n=0}^{\infty} J_n(z)t^n$$

$$= \sum_{n,k=0}^{\infty} \frac{(-1)^{n+k+1}t^{-n-1}z^{n+2k+1}}{2^{n+2k+1}k!(n+1+k)!} + \sum_{n,k=0}^{\infty} \frac{(-1)^k t^n z^{n+2k}}{2^{n+2k}k!(n+k)!}$$

$$= \sum_{n=0}^{\infty}\sum_{k=0}^{[n/2]} \frac{(-1)^{n-k+1}t^{-n+2k-1}z^{n+1}}{2^{n+1}k!(n+1-k)!} + \sum_{n=0}^{\infty}\sum_{k=0}^{[n/2]} \frac{(-1)^k t^{n-2k}z^n}{2^n k!(n-k)!}$$

$$= \sum_{n=1}^{\infty}\sum_{k=0}^{[(n-1)/2]} \frac{(-1)^{n-k}t^{k-(n-k)}}{k!(n-k)!} \cdot \frac{z^n}{2^n} + 1 + \sum_{n=1}^{\infty}\sum_{k=0}^{[n/2]} \frac{(-1)^k t^{n-k-k}}{k!(n-k)!} \cdot \frac{z^n}{2^n}.$$

We now use Lemma 12 to conclude that

$$\sum_{n=-\infty}^{\infty} J_n(z)t^n = 1 + \sum_{n=1}^{\infty}\sum_{k=0}^{n} \frac{(-1)^k t^{n-k-k}}{k!(n-k)!} \cdot \frac{z^n}{2^n}$$

$$= \sum_{n=0}^{\infty}\sum_{k=0}^{n} \frac{t^{n-k}(-t^{-1})^k}{k!(n-k)!} \cdot \frac{z^n}{2^n}$$

$$= \sum_{n=0}^{\infty} \frac{\left(t - \frac{1}{t}\right)^n z^n}{n!2^n} = \exp\left[\frac{1}{2}z\left(t - \frac{1}{t}\right)\right].$$

See also Ex. 23 at the end of this chapter.

63. Bessel's integral. Theorem 39 of the preceding section may be interpreted as giving the Laurent expansion, valid near the essential singularity $t = 0$, for the function $\exp\,[\frac{1}{2}z(t - t^{-1})]$. The Laurent series coefficient is known. Indeed,

$$(1) \qquad J_n(z) = \frac{1}{2\pi i}\int^{(0+)} u^{-n-1} \exp\left[\frac{1}{2}z(u - u^{-1})\right] du,$$

in which the contour $(0+)$ is a simple closed path encircling the origin $u = 0$ in the positive direction.

In (1) let us choose the particular path

$$u = e^{i\theta} = \cos\theta + i\sin\theta,$$

θ running from $(-\pi)$ to π. Then $u^{-1} = \cos\theta - i\sin\theta$, and (1) yields

$$J_n(z) = \frac{1}{2\pi}\int_{-\pi}^{\pi} \exp[-ni\theta + iz\sin\theta]\, d\theta$$

$$= \frac{1}{2\pi}\int_{-\pi}^{\pi} \cos(n\theta - z\sin\theta)\, d\theta - \frac{i}{2\pi}\int_{-\pi}^{\pi} \sin(n\theta - z\sin\theta)\, d\theta.$$

In the last two integrals the former has an even function of θ as integrand, the latter an odd function of θ as integrand. Hence

$$J_n(z) = \frac{1}{\pi}\int_0^{\pi} \cos(n\theta - z\sin\theta)\, d\theta,$$

which is Bessel's integral for $J_n(z)$.

THEOREM 40. *For integral n,*

$$(2) \qquad J_n(z) = \frac{1}{\pi}\int_0^{\pi} \cos(n\theta - z\sin\theta)\, d\theta.$$

Bessel's integral representation of $J_n(z)$ can be* extended to non-integral n. The result, called *Schläfli's integral*, is

$$(3) \quad J_n(z) = \frac{1}{\pi}\int_0^{\pi} \cos(n\theta - z\sin\theta)\, d\theta - \frac{\sin n\pi}{\pi}\int_0^{\infty} \exp(-n\theta - z\sinh\theta)\, d\theta,$$

valid for $\mathrm{Re}(z) > 0$. Equation (3) will not be used in our work.

64. Index half an odd integer. Let us put into hypergeometric form the elementary expansion

*See Whittaker and Watson [1:362].

(1) $$\sin z = \sum_{k=0}^{\infty} \frac{(-1)^k z^{2k+1}}{(2k+1)!}.$$

Since $(2k+1)! = (2)_{2k}$, equation (1) yields

$$\sin z = \sum_{k=0}^{\infty} \frac{(-1)^k z^{2k+1}}{2^{2k} k! \left(\frac{3}{2}\right)_k},$$

or

(2) $$\sin z = z\, {}_0F_1(-;\tfrac{3}{2}; -\tfrac{1}{4}z^2).$$

Now

$$J_{\frac{1}{2}}(z) = \frac{(z/2)^{\frac{1}{2}}}{\Gamma(\frac{3}{2})}\, {}_0F_1(-;\tfrac{3}{2}; -\tfrac{1}{4}z^2)$$

and $\Gamma(\frac{3}{2}) = \frac{1}{2}\sqrt{\pi}$. Hence

(3) $$J_{\frac{1}{2}}(z) = \left(\frac{2}{\pi z}\right)^{\frac{1}{2}} \sin z.$$

In much the same manner the elementary expansion

$$\cos z = \sum_{n=0}^{\infty} \frac{(-1)^n z^{2n}}{(2n)!},$$

or

(4) $$\cos z = {}_0F_1(-;\tfrac{1}{2}; -\tfrac{1}{4}z^2)$$

leads us to the relation

(5) $$J_{-\frac{1}{2}}(z) = \left(\frac{2}{\pi z}\right)^{\frac{1}{2}} \cos z.$$

In Section 61 we derived the pure recurrence relation

(6) $$J_n(z) = 2(n-1)z^{-1}J_{n-1}(z) - J_{n-2}(z).$$

In (6) replace n by $(n + \frac{1}{2})$ to obtain

(7) $$J_{n+\frac{1}{2}}(z) = (2n-1)z^{-1}J_{n-\frac{1}{2}}(z) - J_{n-\frac{3}{2}}(z).$$

Let n be a positive integer and iterate (7) to see that

(8) $$J_{n+\frac{1}{2}}(z) = P_1(z^{-1})J_{\frac{1}{2}}(z) + P_2(z^{-1})J_{-\frac{1}{2}}(z)$$

in which P_1 and P_2 are polynomials in their arguments.

From equations (3), (5), and (8) it follows that for integral n

$$J_{n+\frac{1}{2}}(z) = A(z) \cos z + B(z) \sin z$$

in which $A(z)$ and $B(z)$ are polynomials in $z^{-\frac{1}{2}}$.

Bessel functions of index half an odd integer are often called *spherical* Bessel functions. They, as well as most other Bessel functions, are encountered in various physical problems. Spherical Bessel functions led to the definition and study of Bessel polynomials which we discuss to some extent later in this book.

65. Modified Bessel functions. Many physical problems lead to the study of Bessel functions of pure imaginary argument. This in turn leads to the definition of such functions as

$$(1) \qquad I_n(z) = i^{-n}J_n(iz) = \frac{(z/2)^n}{\Gamma(1+n)}\, {}_0F_1\left(-;1+n;\frac{z^2}{4}\right),$$

n not a negative integer. The function $I_n(z)$ is called a *modified* Bessel function of the first kind of index n. A study of $J_n(z)$ for complex z includes corresponding properties of $I_n(z)$ by simple changes of variables. The function I_n is related to J_n in much the same way that the hyperbolic functions are related to the trigonometric functions. Some elementary properties of $I_n(z)$ will be found in the exercises below.

66. Neumann polynomials. From Theorem 39, page 113, we obtain, for $w \neq 0$ and for all finite z,

$$(1) \qquad \exp[\tfrac{1}{2}z(w - w^{-1})] = \sum_{n=-\infty}^{\infty} J_n(z)w^n,$$

which can equally well (Ex. 2, page 120) be written

$$(2) \qquad \exp[\tfrac{1}{2}z(w - w^{-1})] = J_0(z) + \sum_{n=1}^{\infty} J_n(z)[w^n + (-1)^n w^{-n}],$$

because $J_{-n}(z) = (-1)^n J_n(z)$. In (2) put $w = t + \sqrt{t^2+1}$ and note that $(-w^{-1}) = t - \sqrt{t^2+1}$. The result is

$$(3) \qquad e^{zt} = J_0(z) + \sum_{n=1}^{\infty} J_n(z)[(t + \sqrt{t^2+1})^n + (t - \sqrt{t^2+1})^n].$$

Let us define $f_n(t)$ by

$$(4) \qquad f_n(t) = (t + \sqrt{t^2+1})^n + (t - \sqrt{t^2+1})^n, \qquad n \geqq 0.$$

Then $f_n(t)$ is a polynomial in t and (3) now appears as

$$(5) \qquad e^{zt} = \tfrac{1}{2}f_0(t)J_0(z) + \sum_{n=1}^{\infty} f_n(t)J_n(z).$$

The Laplace transform (Churchill [1]) of a polynomial in t is a polynomial in s^{-1}. Let $2O_n(s)$ be the Laplace transform of our $f_n(t)$:

$$(6) \qquad 2O_n(s) = L\{f_n(t)\} = \int_0^\infty e^{-st} f_n(t)\,dt.$$

Then from (5), since $L\{e^{zt}\} = (s-z)^{-1}$, we obtain

$$(7) \qquad \frac{1}{s-z} = O_0(s)J_0(z) + 2\sum_{n=1}^\infty O_n(s)J_n(z).$$

The polynomials $O_n(s)$ are called *Neumann polynomials*.

Let us assume that the series in (7) is sufficiently well behaved (proved below) that the manipulations to be performed are legitimate. Differentiation of (7) yields

$$(8) \qquad -(s-z)^{-2} = O_0'(s)J_0(z) + 2\sum_{n=1}^\infty O_n'(s)J_n(z)$$

and

$$(9) \qquad (s-z)^{-2} = O_0(s)J_0'(z) + 2\sum_{n=1}^\infty O_n(s)J_n'(z).$$

Now $2J_n'(z) = J_{n-1}(z) - J_{n+1}(z)$ and $J_0'(z) = -J_1(z)$, so that (9) may be written as

$$\begin{aligned}(s-z)^{-2} &= -O_0(s)J_1(z) + \sum_{n=1}^\infty O_n(s)J_{n-1}(z) - \sum_{n=1}^\infty O_n(s)J_{n+1}(z)\\ &= -O_0(s)J_1(z) + \sum_{n=0}^\infty O_{n+1}(s)J_n(z) - \sum_{n=2}^\infty O_{n-1}(s)J_n(z),\end{aligned}$$

or

$$(10) \qquad (s-z)^{-2} = O_1(s)J_0(z) + \sum_{n=1}^\infty [O_{n+1}(s) - O_{n-1}(s)]J_n(z).$$

From (8) and (10) it follows that

$$[O_0'(s) + O_1(s)]J_0(z) + \sum_{n=1}^\infty [2O_n'(s) + O_{n+1}(s) - O_{n-1}(s)]J_n(z) = 0.$$

Since for each n the function $z^{-n}J_n(z)$ is nonzero at $z = 0$, it follows that an expansion of the form

$$\sum_{n=0}^\infty a_n J_n(z)$$

is unique.

Hence $O_1(s) = -O_0'(s)$ and

(11) $$O_{n+1}(s) = O_{n-1}(s) - 2O_n'(s), \qquad n \geqq 1.$$

We know that $O_0(s) = s^{-1}$ and now that $O_1(s) = s^{-2}$. The Neumann polynomials may now be described as follows:

(12) $$O_0(s) = s^{-1}, \qquad O_1(s) = s^{-2},$$

(13) $$O_n(s) = O_{n-2}(s) - 2O'_{n-1}(s), \qquad n \geqq 2.$$

The $O_n(s)$ are uniquely determined by the description (12) and (13).

THEOREM 41. *The Neumann polynomials defined by* (12) *and* (13) *above are given by* $O_0(s) = s^{-1}$ *and*

(14) $$O_n(s) = \frac{n}{4}\sum_{k=0}^{[n/2]} \frac{(n-1-k)!(2/s)^{n+1-2k}}{k!}, \qquad n \geqq 1.$$

Proof: From (14), $O_1(s) = \frac{1}{4}(2/s)^2 = s^{-2}$. Also, for $n \geqq 2$,

$$\begin{aligned} O_{n-2}(s) &= \frac{n-2}{4}\sum_{k=0}^{\left[\frac{n-2}{2}\right]} \frac{(n-3-k)!(2/s)^{n-1-2k}}{k!} \\ &= \frac{n-2}{4}\sum_{k=1}^{[n/2]} \frac{(n-2-k)!(2/s)^{n+1-2k}}{(k-1)!} \end{aligned}$$

and

$$\begin{aligned} O'_{n-1}(s) &= \frac{n-1}{4}\sum_{k=0}^{\left[\frac{n-1}{2}\right]} \frac{(n-2k)(n-2-k)!(-2/s^2)(2/s)^{n-1-2k}}{k!} \\ &= -\frac{n-1}{8}\sum_{k=0}^{[n/2]} \frac{(n-2k)(n-2-k)!(2/s)^{n+1-2k}}{k!}. \end{aligned}$$

Therefore, for the $O_n(s)$ of (14),

$$\begin{aligned} &O_{n-2}(s) - 2O'_{n-1}(s) \\ &= \frac{1}{4}\sum_{k=0}^{[n/2]} \frac{[(n-2)k + (n-1)(n-2k)](n-2-k)!(2/s)^{n+1-2k}}{k!} \\ &= O_n(s), \end{aligned}$$

as desired.

By Theorem 41 the dominating term in $O_n(s)$ is $2^{n-1}n!\, s^{-n-1}$. The dominating term in $J_n(z)$ is $(\frac{1}{2}z)^n/n!$. Therefore, as $n \to \infty$,

$$O_n(s)J_n(z) = \frac{2^{n-1}n!}{s^{n+1}} \cdot \frac{z^n}{2^n n!}(1 + \epsilon_n)$$

in which $\epsilon_n \to 0$. For $|z| \leqq r$, choose $|s| \geqq R$ where $R > r$. Thus for n sufficiently large,

$$\left|O_n(s)J_n(z)\right| < c\left(\frac{r}{R}\right)^n, \quad c \text{ a constant.}$$

Then the expansion (7) is absolutely and uniformly convergent, and the manipulations performed on it are justified.

For the moment let F denote the right member of (7). Because of (11) the right members of (8) and (9) have zero as their sum. Hence

$$\frac{\partial F}{\partial s} + \frac{\partial F}{\partial z} = 0$$

from which it follows that the right member of (7) is a function of the single argument $(s - z)$. But at $z = 0$, that right member is $O_0(s) = s^{-1}$. Hence, once again, the left member of (7) is $(s - z)^{-1}$.

67. Neumann series. On the basis of the expansion

$$\text{(1)} \qquad \frac{1}{s - z} = O_0(s)J_0(z) + 2\sum_{n=1}^{\infty} O_n(s)J_n(z)$$

and the Cauchy integral formula

$$\text{(2)} \qquad f(z) = \frac{1}{2\pi i}\int_C \frac{f(s)\,ds}{s - z},$$

where C is $|s| = r$ described in the positive direction, we obtain at once

$$f(z) = \frac{1}{2\pi i}J_0(z)\int_C \frac{f(s)\,ds}{s} + \frac{1}{\pi i}\sum_{n=1}^{\infty} J_n(z)\int_C f(s)O_n(s)\,ds,$$

or

$$\text{(3)} \qquad f(z) = \sum_{n=0}^{\infty} a_nJ_n(z),$$

in which $a_0 = f(0)$ and

$$\text{(4)} \qquad a_n = \frac{1}{\pi i}\int_C f(s)O_n(s)\,ds, \quad n \geqq 1.$$

That is, if $f(z)$ is analytic in $|z| \leqq r$, then $f(z)$ can be expanded into the Neumann series (3), with coefficients as described, and the expansion is valid for $|z| < r$.

Some of the expansions in the exercises may be obtained by this method, if it seems desirable.

EXERCISES

1. By collecting powers of x in the summation on the left, show that

$$\sum_{n=0}^{\infty} J_{2n+1}(x) = \tfrac{1}{2}\int_0^x J_0(y)\,dy.$$

2. Put the equation of Theorem 39, page 113, into the form

$$\text{(A)} \qquad \exp[\tfrac{1}{2}z(t - t^{-1})] = J_0(z) + \sum_{n=1}^{\infty} J_n(z)[t^n + (-1)^n t^{-n}].$$

Use equation (A) with $t = i$ to conclude that

$$\cos z = J_0(z) + 2\sum_{k=1}^{\infty} (-1)^k J_{2k}(z),$$

$$\sin z = 2\sum_{k=0}^{\infty} (-1)^k J_{2k+1}(z).$$

3. Use $t = e^{i\theta}$ in equation (A) of Ex. 2 to obtain the results

$$\cos (z \sin \theta) = J_0(z) + 2\sum_{k=1}^{\infty} J_{2k}(z) \cos 2k\theta,$$

$$\sin (z \sin \theta) = 2\sum_{k=0}^{\infty} J_{2k+1}(z) \sin(2k + 1)\theta.$$

4. Use Bessel's integral, page 114, to obtain for integral n the relations

$$\text{(B)} \qquad [1 + (-1)^n]J_n(z) = \frac{2}{\pi}\int_0^\pi \cos n\theta \cos(z \sin \theta)\,d\theta,$$

$$\text{(C)} \qquad [1 - (-1)^n]J_n(z) = \frac{2}{\pi}\int_0^\pi \sin n\theta \sin(z \sin \theta)\,d\theta.$$

With the aid of (B) and (C) show that for integral k,

$$J_{2k}(z) = \frac{1}{\pi}\int_0^\pi \cos 2k\theta \cos(z \sin \theta)\,d\theta,$$

$$J_{2k+1}(z) = \frac{1}{\pi}\int_0^\pi \sin(2k + 1)\theta \sin(z \sin \theta)\,d\theta,$$

$$\int_0^\pi \cos(2k + 1)\theta \cos(z \sin \theta)\,d\theta = 0,$$

$$\int_0^\pi \sin 2k\theta \sin(z \sin \theta)\,d\theta = 0.$$

5. Expand $\cos(z \sin \theta)$ and $\sin(z \sin \theta)$ in Fourier series over the interval $-\pi < \theta < \pi$. Thus use Ex. 4 to obtain in another way the expansions in Ex. 3.

6. In the product of $\exp[\tfrac{1}{2}x(t - t^{-1})]$ by $\exp[-\tfrac{1}{2}x(t - t^{-1})]$, obtain the coefficient of t^0 and thus show that

$$J_0^2(x) + 2\sum_{n=1}^{\infty} J_n^2(x) = 1.$$

For real x conclude that $|J_0(x)| \leqq 1$ and $|J_n(x)| \leqq 2^{-\frac{1}{2}}$ for $n \geqq 1$.

7. Use Bessel's integral to show that $|J_n(x)| \leqq 1$ for real x and integral n.

8. By iteration of equation (8), page 111, show that

$$2^m \frac{d^m}{dz^m} J_n(z) = \sum_{k=0}^{m} (-1)^{m-k} C_{m,k} J_{n+m-2k}(z),$$

where $C_{m,k}$ is the binomial coefficient.

9. Use the result in Ex. 1, page 105, to obtain the product of two Bessel functions of equal argument.

Ans. $$J_n(z)J_m(z) = \frac{(z/2)^{n+m}}{\Gamma(1+n)\Gamma(1+m)} {}_2F_3\left[\begin{matrix} \tfrac{1}{2}(n+m+1), \tfrac{1}{2}(n+m+2); & \\ & -z^2 \\ 1+n, 1+m, 1+n+m; & \end{matrix}\right].$$

10. Start with the power series for $J_n(z)$ and use the form (2), page 18, of the Beta function to arrive at the equation

$$J_n(z) = \frac{2(\frac{1}{2}z)^n}{\Gamma(\frac{1}{2})\Gamma(n+\frac{1}{2})} \int_0^{\frac{1}{2}\pi} \sin^{2n}\varphi \cos(z\cos\varphi)\, d\varphi,$$

for $\mathrm{Re}(n) > -\frac{1}{2}$.

11. Use the property

$$\frac{d}{dx}\, {}_0F_1(-;a;u) = \frac{1}{a}\frac{du}{dx}\, {}_0F_1(-;a+1;u)$$

to obtain the differential recurrence relation (6) of Section 60.

12. Expand

$${}_0F_1\left[\begin{matrix} -; & \\ & \dfrac{2xt-t^2}{4} \\ 1+\alpha; & \end{matrix}\right]$$

in a series of powers of x and thus arrive at the result

$$\left(\frac{t-2x}{t}\right)^{-\frac{1}{2}\alpha} J_\alpha\left(\sqrt{t^2-2xt}\right) = \sum_{n=0}^{\infty} \frac{J_{\alpha+n}(t)x^n}{n!}.$$

13. Use the relations (3) and (6) of Section 60 to prove that: For real x, between any two consecutive zeros of $x^{-n}J_n(x)$, there lies one and only one zero of $x^{-n}J_{n+1}(x)$.

14. For the function $I_n(z)$ of Section 65 obtain the following properties by using the methods, but not the results, of this chapter:

$$\begin{aligned} zI_n'(z) &= zI_{n-1}(z) - nI_n(z), \\ zI_n'(z) &= zI_{n+1}(z) + nI_n(z), \\ 2I_n'(z) &= I_{n-1}(z) + I_{n+1}(z), \\ 2nI_n(z) &= z[I_{n-1}(z) - I_{n+1}(z)]. \end{aligned}$$

15. Show that $I_n(z)$ is one solution of the equation

$$z^2w'' + zw' - (z^2+n^2)w = 0.$$

16. Show that, for $\mathrm{Re}(n) > -\frac{1}{2}$,

$$I_n(z) = \frac{2(\frac{1}{2}z)^n}{\Gamma(\frac{1}{2})\Gamma(n+\frac{1}{2})} \int_0^{\frac{1}{2}\pi} \sin^{2n}\varphi \cosh(z\cos\varphi)\, d\varphi.$$

17. For negative integral n define $I_n(z) = (-1)^n I_{-n}(z)$, thus completing the definition in Section 65. Show that $I_n(-z) = (-1)^n I_n(z)$ and that

$$\exp[\tfrac{1}{2}z(t + t^{-1})] = \sum_{n=-\infty}^{\infty} I_n(z)t^n.$$

18. Use the integral evaluated in Section 56 to show that

$$\int_0^t [\sqrt{x(t-x)}]^n J_n(\sqrt{x(t-x)})\, dx = 2^{-n}\sqrt{\pi}\, t^{n+\frac{1}{2}} J_{n+\frac{1}{2}}(\tfrac{1}{2}t).$$

19. By the method of Ex. 18 show that

$$\int_0^1 \sqrt{1-x}\,\sin(\alpha\sqrt{x})\, dx = \pi\alpha^{-1}J_2(\alpha),$$

and, in general, that

$$\int_0^1 (1-x)^{c-1}x^{\frac{1}{2}n}J_n(\alpha\sqrt{x})\, dx = \Gamma(c)\left(\frac{2}{\alpha}\right)^c J_{n+c}(\alpha).$$

20. Show that

$$\int_0^t \exp[-2x(t-x)]I_0[2x(t-x)]\, dx = \int_0^t \exp(-\beta^2)\, d\beta.$$

21. Show that

$$\int_0^t [x(t-x)]^{-\frac{1}{2}} \exp[4x(t-x)]\, dx = \pi \exp(\tfrac{1}{2}t^2)\, I_0(\tfrac{1}{2}t^2).$$

22. Obtain Neumann's expansion

$$\left(\frac{1}{2}z\right)^n = \sum_{k=0}^{\infty} \frac{(n+2k)(n+k-1)!J_{n+2k}(z)}{k!}, \quad n \geqq 1.$$

23. Prove Theorem 39, page 113, by forming the product of the series for $\exp(\frac{1}{2}zt)$ and the series for $\exp(-\frac{1}{2}zt^{-1})$.

CHAPTER 7

The Confluent Hypergeometric Function

68. Basic properties of the $_1F_1$. The functions $_0F_0$ (the exponential) and $_1F_0$ (the binomial) are elementary. We have devoted some time to the study of the $_0F_1$, a Bessel function, and to the $_2F_1$, the ordinary hypergeometric function. Except for terminating series, we are interested in the $_pF_q$ mainly when $p \leqq q + 1$ so that the series has a region of convergence. To complete the introduction to special properties of the $_pF_q$ when $q = 0, 1$, we need only to consider the $_1F_1$.

The series

$$(1) \qquad {}_1F_1(a; b; z) = \sum_{n=0}^{\infty} \frac{(a)_n z^n}{(b)_n n!},$$

in which $b \neq$ zero or a negative integer is convergent for all finite z. This function is also known as the Pochhammer-Barnes confluent hypergeometric function. An equation satisfied by the $_1F_1$ can be obtained by confluence of singularities from a Fuchsian* equation with three singular points. Other common notations for the $_1F_1$ are

$$(2) \qquad \Phi(a; b; z) = M(a, b, z) = {}_1F_1(a; b; z).$$

There are certain properties of the $_1F_1$ which follow from the fact that it is a generalized hypergeometric function; these properties

*Fuchsian equations and the concept of confluence are treated in Chapters 5 and 7 of Rainville [2].

are not peculiar to the ${}_1F_1$. By specializing results obtained in Chapter 5, we obtain the following facts.

The function $w = {}_1F_1(a; b; z)$ is a solution of the differential equation

$$[\theta(\theta + b - 1) - z(\theta + a)]\, w = 0; \quad \theta = z\frac{d}{dz}, \tag{3}$$

an equation which may also be written

$$zw'' + (b - z)w' - aw = 0. \tag{4}$$

If b is nonintegral, the general solution of (3) or (4) is

$$w = A\, {}_1F_1(a; b; z) + Bz^{1-b}\, {}_1F_1(a + 1 - b; 2 - b; z), \tag{5}$$

with A and B as arbitrary constants. If b is integral, the general solution may involve $\log z$ in the usual way, since $z = 0$ is a regular singular point of the differential equation (3) or (4).

There is a canonical set of three relations between the ${}_1F_1$ and pairs of its contiguous functions. They may be written in the form

$$(a-b+1)\, {}_1F_1(a; b; z) = a\, {}_1F_1(a+1; b; z) - (b-1)\, {}_1F_1(a; b-1; z), \tag{6}$$

$$b(a+z)\, {}_1F_1(a; b; z) = ab\, {}_1F_1(a+1; b; z) - (a-b)z\, {}_1F_1(a; b+1; z), \tag{7}$$

$$b\, {}_1F_1(a; b; z) = b\, {}_1F_1(a-1; b; z) + z\, {}_1F_1(a; b+1; z). \tag{8}$$

If $\mathrm{Re}(b) > \mathrm{Re}(a) > 0$,

$${}_1F_1(a; b; z) = \frac{\Gamma(b)}{\Gamma(a)\Gamma(b - a)}\int_0^1 e^{zt}t^{a-1}(1 - t)^{b-a-1}\, dt. \tag{9}$$

If neither a nor b is a nonpositive integer, if $\mathrm{Re}(z) < 0$, and if the path of integration is one of Barnes' type, page 95,

$${}_1F_1(a; b; z) = \frac{\Gamma(b)}{2\pi i\Gamma(a)}\int_B \frac{\Gamma(a + s)\Gamma(-s)(-z)^s\, ds}{\Gamma(b + s)}. \tag{10}$$

69. Kummer's first formula. We next obtain results which are characteristic of the ${}_1F_1$ in the sense that they are not held in common by all ${}_pF_q$'s.

Consider the product

$$\begin{aligned} e^{-z}{}_1F_1(a; b; z) &= \left(\sum_{n=0}^{\infty} \frac{(-1)^n z^n}{n!}\right)\left(\sum_{n=0}^{\infty} \frac{(a)_n z^n}{(b)_n n!}\right) \\ &= \sum_{n=0}^{\infty}\sum_{k=0}^{n} \frac{(-1)^{n-k}(a)_k z^n}{(b)_k k!(n - k)!}. \end{aligned}$$

Since $1/(n-k)! = (-1)^k(-n)_k/n!$, we may write

$$e^{-z}\,{}_1F_1(a;b;z) = \sum_{n=0}^{\infty}\sum_{k=0}^{n}\frac{(-n)_k(a)_k}{(b)_k k!}\cdot\frac{(-1)^n z^n}{n!}$$

$$= \sum_{n=0}^{\infty}{}_2F_1(-n,a;b;1)\frac{(-1)^n z^n}{n!}.$$

But we already know (page 69) that

$$(1) \qquad {}_2F_1(-n,a;b;1) = \frac{(b-a)_n}{(b)_n}.$$

Then

$$e^{-z}\,{}_1F_1(a;b;z) = \sum_{n=0}^{\infty}\frac{(b-a)_n(-z)^n}{(b)_n n!}.$$

THEOREM 42. *If b is neither zero nor a negative integer,*

$$(2) \qquad {}_1F_1(a;b;z) = e^z\,{}_1F_1(b-a;b;-z).$$

This is Kummer's first formula.

70. Kummer's second formula. Examination of Kummer's first formula soon arouses interest in the special case when the two ${}_1F_1$ functions have the same parameters. This happens when $b - a = a$, $b = 2a$. We then obtain

$${}_1F_1(a;2a;z) = e^z\,{}_1F_1(a;2a;-z),$$

or

$$(1) \qquad e^{-\frac{1}{2}z}\,{}_1F_1(a;2a;z) = e^{\frac{1}{2}z}{}_1F_1(a;2a;-z).$$

More pleasantly, (1) may be expressed by saying that the function

$$e^{-z}\,{}_1F_1(a;2a;2z)$$

is an even function of z. Since, by straightforward multiplication,

$$e^{-z}\,{}_1F_1(a;2a;2z) = \sum_{n=0}^{\infty}\sum_{k=0}^{n}\frac{(a)_k(-1)^{n-k}2^k z^n}{(2a)_k k!(n-k)!}$$

$$= \sum_{n=0}^{\infty}\sum_{k=0}^{n}\frac{(a)_k(-n)_k 2^k}{(2a)_k k!}\cdot\frac{(-z)^n}{n!}$$

$$= \sum_{n=0}^{\infty}{}_2F_1(-n,a;2a;2)\frac{(-z)^n}{n!},$$

we may conclude, from the fact that the left member is an even function of z, that for k a non-negative integer

$$(2) \qquad {}_2F_1(-2k - 1, a; 2a; 2) = 0,$$

and also that

$$(3) \qquad e^{-z}\, {}_1F_1(a; 2a; 2z) = \sum_{k=0}^{\infty} {}_2F_1(-2k, a; 2a; 2) \frac{z^{2k}}{(2k)!}.$$

Next let us determine a differential equation satisfied by the function $w = e^{-z}\, {}_1F_1(a; 2a; 2z)$. We know that $y = {}_1F_1(a; b; x)$ is a solution of the equation

$$(4) \qquad x\frac{d^2y}{dx^2} + (b - x)\frac{dy}{dx} - ay = 0.$$

In (4) put $b = 2a$, $x = 2z$, and $y = e^z w$. The result is

$$(5) \qquad zw'' + 2aw' - zw = 0,$$

of which one solution must be $w = e^{-z}\, {}_1F_1(a; 2a; 2z)$.

In equation (5) change the independent variable to $\sigma = \frac{1}{4}z^2$ and thus arrive at the equation

$$(6) \qquad \sigma^2\frac{d^2w}{d\sigma^2} + (a + \tfrac{1}{2})\sigma\frac{dw}{d\sigma} - \sigma w = 0,$$

or

$$(7) \qquad [\theta(\theta + a + \tfrac{1}{2} - 1) - \sigma]\, w = 0; \quad \theta = \sigma\frac{d}{d\sigma}.$$

Equation (7) is a differential equation for the ${}_0F_1$ function with denominator parameter $(a + \frac{1}{2})$ and argument $\sigma = \frac{1}{4}z^2$. Hence, if $a + \frac{1}{2}$ is nonintegral (that is, if $2a$ is not an odd integer), the general solution of (7) is

$$(8) \quad w = A\,{}_0F_1(-; a + \tfrac{1}{2}; \tfrac{1}{4}z^2) + B(z^2)^{\frac{1}{2}-a}\, {}_0F_1(-; \tfrac{3}{2} - a; \tfrac{1}{4}z^2).$$

But (7) is also satisfied by $w_1 = e^{-z}\, {}_1F_1(a; 2a; 2z)$. Therefore there exist constants A and B such that the right member of (8) becomes $e^{-z}\, {}_1F_1(a; 2a; 2z)$. In the usual manner it is easy to see that $B = 0$ and $A = 1$.

THEOREM 43. *If $2a$ is not an odd integer < 0,*

$$(9) \qquad e^{-z}\, {}_1F_1(a; 2a; 2z) = {}_0F_1(-; a + \tfrac{1}{2}; \tfrac{1}{4}z^2).$$

If $2a$ is an odd positive integer, the second term on the right in (8) is replaced by a solution involving $\log z$, and the same argument again shows that $B = 0$, $A = 1$, so that (9) holds. Equation (9) is known as Kummer's second formula.

We may use (9) and (3) to conclude that

$$\sum_{k=0}^{\infty} {}_2F_1(-2k, a; 2a; 2)\frac{z^{2k}}{(2k)!} = \sum_{k=0}^{\infty} \frac{z^{2k}}{2^{2k}k!(a+\frac{1}{2})_k},$$

from which it follows that

$$(10) \qquad {}_2F_1(-2k, a; 2a; 2) = \frac{(\frac{1}{2})_k}{(a+\frac{1}{2})_k}.$$

Theorem 43 may be interpreted as a relation between a particular ${}_1F_1$ and the modified Bessel function of the first kind of index $(a - \frac{1}{2})$.

Many other properties of the ${}_1F_1$ will be found in Chapter 6 of volume one of the Bateman Manuscript Project volumes, Erdélyi [1].

The Whittaker functions* $W_{k,m}$ are expressible as linear combinations of ${}_1F_1$'s. Two subsidiary solutions of the basic Whittaker equation

$$(11) \qquad \frac{d^2W}{dz^2} + \left[-\frac{1}{4} + \frac{k}{z} + \frac{\frac{1}{4} - m^2}{z^2}\right] W = 0$$

are, if $2m$ is not an integer,

$$M_{k,m}(z) = z^{m+\frac{1}{2}}e^{-\frac{1}{2}z}\, {}_1F_1(\tfrac{1}{2} + m - k; 2m + 1; z)$$

and $M_{k,-m}(z)$. See also Chapter 8 of Rainville [2] for more detail on equation (11).

EXERCISES

1. The function

$$\operatorname{erf}(x) = \frac{2}{\sqrt{\pi}}\int_0^x \exp(-t^2)\, dt$$

was defined on page 36. Show that

$$\operatorname{erf}(x) = \frac{2x}{\sqrt{\pi}}\, {}_1F_1\left(\frac{1}{2}; \frac{3}{2}; -x^2\right).$$

2. The incomplete Gamma function may be defined by the equation

$$\gamma(\alpha, x) = \int_0^x e^{-t}t^{\alpha-1}\, dt, \quad \operatorname{Re}(\alpha) > 0.$$

Show that

$$\gamma(\alpha, x) = \alpha^{-1}x^{\alpha}\, {}_1F_1(\alpha; \alpha + 1; -x).$$

*See Whittaker and Watson [1] and Whittaker [1].

3. Prove that

$$(b)\ \frac{d^k}{dz^k}\left[e^{-z}\,{}_1F_1(a;b;z)\right] = (-1)^k(b-a)_k e^{-z}\,{}_1F_1(a;b+k;z).$$

You may find it helpful to use Kummer's first formula, Theorem 42.

4. Show that

$${}_1F_1(a;b;z) = \frac{1}{\Gamma(a)}\int_0^\infty e^{-t}t^{a-1}\,{}_0F_1(-;b;zt)\,dt.$$

5. Show, with the aid of the result in Ex. 4, that

$$\int_0^\infty \exp(-t^2)t^{2a-n-1}J_n(zt)\,dt = \frac{\Gamma(a)z^n}{2^{n+1}\Gamma(n+1)}\,{}_1F_1\left(a;n+1;\ -\frac{z^2}{4}\right).$$

6. If k and n are non-negative integers, show that

$$F\left[\begin{matrix}-k,\ \alpha+n; \\ \alpha;\end{matrix}\ 1\right] = 0, \qquad \text{for } k > n,$$

$$= \frac{(-n)_k}{(\alpha)_k}, \qquad \text{for } 0 \leqq k \leqq n.$$

CHAPTER 8

Generating Functions

71. The generating function concept. Consider a function $F(x,t)$ which has a formal (it need not converge) power series expansion in t:

(1) $$F(x, t) = \sum_{n=0}^{\infty} f_n(x)t^n.$$

The coefficient of t^n in (1) is, in general, a function of x. We say that the expansion (1) of $F(x,t)$ has *generated* the set $f_n(x)$ and that $F(x,t)$ is a *generating function* for the $f_n(x)$. If for some set of values of x, usually a region in the complex x-plane, the function $F(x,t)$ is analytic at $t = 0$, the series in (1) converges in some region around $t = 0$. Convergence is not necessary for the relation (1) to define the $f_n(x)$ and to be useful in obtaining properties of those functions.

Before proceeding to a discussion of some of the uses of generating functions, we wish to extend the foregoing definition slightly. Let c_n; $n = 0, 1, 2, \cdots$, be a specified sequence independent of x and t. We say that $G(x,t)$ is a generating function of the set $g_n(x)$ if

(2) $$G(x, t) = \sum_{n=0}^{\infty} c_n g_n(x)t^n.$$

If the c_n and $g_n(x)$ in (2) are assigned, and we can determine the sum function $G(x,t)$ as a finite sum of products of a finite number of known special functions of one argument, we say that the generating function $G(x,t)$ is known.

The question arises as to what is a "known special function." It is, of course, a matter of opinion or convention. We consider as known any function which has received individual attention in at least one research publication. In this terminology we follow the late Harry Bateman (1882–1946). Bateman, who probably knew more about special functions than anyone else, is said to have known of about a thousand of them.

The necessity for some terminology such as that defined above can be appreciated after examination of certain publications purporting to obtain new generating functions for classical polynomials.

Generating functions will play a large role in our study of polynomial sets. For example, we shall define the Legendre polynomials $P_n(x)$ by

$$(3) \qquad (1 - 2xt + t^2)^{-\frac{1}{2}} = \sum_{n=0}^{\infty} P_n(x)t^n,$$

and the Hermite polynomials $H_n(x)$ by

$$(4) \qquad \exp(2xt - t^2) = \sum_{n=0}^{\infty} \frac{H_n(x)t^n}{n!}.$$

We shall find (page 201) that the Laguerre polynomials $L_n^{(\alpha)}(x)$ possess the generating relation

$$(5) \qquad e^t {}_0F_1(-; 1 + \alpha; -xt) = \sum_{n=0}^{\infty} \frac{L_n^{(\alpha)}(x)t^n}{(1 + \alpha)_n}.$$

One of our major problems will be the search for generating functions for known polynomial sets. Certain purely manipulative techniques will be found to accomplish much in this direction, but more systematic attacks are highly desirable. Unfortunately the box score to date reveals that no known systematic theory has produced results comparable to those attained by manipulative skill. Since the latter usually requires long practice and training, it is hoped that in the future the tide will swing toward theoretical developments capable of producing practical new results. Some start in that direction has been made by Sheffer [1] (see Chapter 13) and Boas and Buck [1], the latter to be touched upon in this chapter. See also Weisner [1].

We shall find that if polynomials $f_n(x)$ are generated by

$$(1) \qquad F(x, t) = \sum_{n=0}^{\infty} f_n(x)t^n,$$

certain properties of $f_n(x)$ are readily deduced from known properties of $F(x,t)$. This idea will be used frequently in the study of specific polynomials in Chapters 10, 11, 12, 16, 17, and 18. In the present chapter we seek properties held in common by many polynomial sets.

72. Generating functions of the form $G(2xt - t^2)$. Each of the generating functions in (3) and (4) of the preceding section is a function of the single combination $(2xt - t^2)$. By studying the generating relation

$$(1) \qquad G(2xt - t^2) = \sum_{n=0}^{\infty} g_n(x)t^n,$$

in which $G(u)$ has a formal power-series expansion, we arrive at properties held in common by $P_n(x)$ and $H_n(x)/n!$, where $P_n(x)$ is the Legendre polynomial and $H_n(x)$ the Hermite polynomial. Let

$$(2) \qquad F = G(2xt - t^2).$$

Then

$$(3) \qquad \frac{\partial F}{\partial x} = 2tG', \qquad \frac{\partial F}{\partial t} = (2x - 2t)G',$$

in which the argument of G' is omitted because it remains $(2xt - t^2)$ throughout. From equations (3) we find that the F of (2) satisfies the partial differential equation

$$(4) \qquad (x - t)\frac{\partial F}{\partial x} - t\frac{\partial F}{\partial t} = 0.$$

Since

$$F = G(2xt - t^2) = \sum_{n=0}^{\infty} g_n(x)t^n,$$

it follows from (4) that

$$\sum_{n=0}^{\infty} xg_n'(x)t^n - \sum_{n=0}^{\infty} g_n'(x)t^{n+1} - \sum_{n=0}^{\infty} ng_n(x)t^n = 0,$$

or

$$(5) \qquad \sum_{n=0}^{\infty} xg_n'(x)t^n - \sum_{n=0}^{\infty} ng_n(x)t^n = \sum_{n=1}^{\infty} g_{n-1}'(x)t^n.$$

In (5), equate coefficients of t^n to obtain the following result.

THEOREM 44. *From*

$$G(2xt - t^2) = \sum_{n=0}^{\infty} g_n(x)t^n$$

it follows that $g_0'(x) = 0$, *and for* $n \geqq 1$,

(6) $$xg_n'(x) - ng_n(x) = g'_{n-1}(x).$$

The differential recurrence relation (6) is common to all sets $g_n(x)$ possessing a generating function of the form used in (1). For the choice $G(u) = (1 - u)^{-\frac{1}{2}}$, the $g_n(x)$ become the Legendre polynomials $P_n(x)$, as stated in (3) of Section 71. Hence the $P_n(x)$ satisfy the relation

(7) $$xP_n'(x) - nP_n(x) = P'_{n-1}(x).$$

For the choice $G(u) = \exp(u)$, the $g_n(x)$ become $H_n(x)/n!$ by (4) of Section 71. Hence the Hermite polynomials satisfy the relation

$$\frac{xH_n'(x)}{n!} - \frac{nH_n(x)}{n!} = \frac{H'_{n-1}(x)}{(n-1)!},$$

or

(8) $$xH_n'(x) - nH_n(x) = nH'_{n-1}(x).$$

73. Sets generated by $e^t\psi(xt)$. The generating function in (5) of Section 71 suggests that we consider sets $\sigma_n(x)$ defined by

(1) $$e^t\psi(xt) = \sum_{n=0}^{\infty} \sigma_n(x)t^n.$$

For a short discussion of these polynomials, see also Rainville [4].

Let

(2) $$F = e^t\psi(xt).$$

Then

(3) $$\frac{\partial F}{\partial x} = te^t\psi',$$

(4) $$\frac{\partial F}{\partial t} = e^t\psi + xe^t\psi'.$$

We eliminate ψ and ψ' from the three equations (2), (3), (4), and thus obtain

(5) $$x\frac{\partial F}{\partial x} - t\frac{\partial F}{\partial t} = -tF.$$

Since

$$F = e^t\psi(xt) = \sum_{n=0}^{\infty} \sigma_n(x)t^n,$$

equation (5) yields

$$\sum_{n=0}^{\infty} x\sigma_n'(x)t^n - \sum_{n=0}^{\infty} n\sigma_n(x)t^n = -\sum_{n=0}^{\infty} \sigma_n(x)t^{n+1}$$

$$= -\sum_{n=1}^{\infty} \sigma_{n-1}(x)t^n,$$

from which the next theorem follows.

THEOREM 45. *From* $e^t\psi(xt) = \sum_{n=0}^{\infty} \sigma_n(x)t^n$, *it follows that* $\sigma_0'(x) = 0$, *and for* $n \geqq 1$,

$$x\sigma_n'(x) - n\sigma_n(x) = -\sigma_{n-1}(x). \tag{6}$$

Next let us assume that the function ψ in (1) has the formal power-series expansion

$$\psi(u) = \sum_{n=0}^{\infty} \gamma_n u^n. \tag{7}$$

Then (1) yields

$$\sum_{n=0}^{\infty} \sigma_n(x)t^n = \left(\sum_{n=0}^{\infty} \frac{t^n}{n!}\right)\left(\sum_{n=0}^{\infty} \gamma_n x^n t^n\right)$$

$$= \sum_{n=0}^{\infty} \sum_{k=0}^{n} \frac{\gamma_k x^k t^n}{(n-k)!},$$

so that

$$\sigma_n(x) = \sum_{k=0}^{n} \frac{\gamma_k x^k}{(n-k)!}. \tag{8}$$

Now consider the sum

$$\sum_{n=0}^{\infty} (c)_n \sigma_n(x)t^n = \sum_{n=0}^{\infty} \sum_{k=0}^{n} \frac{(c)_n \gamma_k x^k t^n}{(n-k)!}$$

$$= \sum_{n,k=0}^{\infty} \frac{(c)_{n+k} \gamma_k x^k t^{n+k}}{n!}$$

$$= \sum_{k=0}^{\infty} \sum_{n=0}^{\infty} \frac{(c+k)_n t^n}{n!} \cdot \frac{(c)_k \gamma_k (xt)^k}{1}$$

$$= \sum_{k=0}^{\infty} \frac{(c)_k \gamma_k (xt)^k}{(1-t)^{c+k}}.$$

THEOREM 46. *From*

$$e^t\psi(xt) = \sum_{n=0}^{\infty} \sigma_n(x)t^n, \qquad \psi(u) = \sum_{n=0}^{\infty} \gamma_n u^n,$$

it follows that for arbitrary c

$$(9) \qquad (1-t)^{-c} F\left(\frac{xt}{1-t}\right) = \sum_{n=0}^{\infty} (c)_n \sigma_n(x) t^n,$$

in which

$$(10) \qquad F(u) = \sum_{n=0}^{\infty} (c)_n \gamma_n u^n.$$

The role of Theorem 46 is as follows: If a set $\sigma_n(x)$ has a generating function of the form $e^t\psi(xt)$, Theorem 46 yields for $\sigma_n(x)$ another generating function of the form exhibited in (9). For instance, if $\psi(u)$ is a specified ${}_pF_q$, the theorem gives for $\sigma_n(x)$ a class (c is arbitrary) of generating functions involving a ${}_{p+1}F_q$. Furthermore, if c is chosen equal to a denominator parameter of the original ${}_pF_q$, the second generating function becomes one involving a ${}_pF_{q-1}$.

Let us now apply Theorems 45 and 46 to Laguerre polynomials. As stated in Section 71, we shall show in Chapter 12 that the Laguerre polynomials possess the generating relation

$$(11) \qquad e^t\, {}_0F_1(-;1+\alpha;-xt) = \sum_{n=0}^{\infty} \frac{L_n^{(\alpha)}(x)t^n}{(1+\alpha)_n}.$$

We use Theorem 45 of this section to conclude that $L_0^{(\alpha)}(x)$ is a constant, and for $n \geqq 1$,

$$\frac{x}{(1+\alpha)_n}\frac{d}{dx}L_n^{(\alpha)}(x) - \frac{nL_n^{(\alpha)}(x)}{(1+\alpha)_n} = -\frac{L_{n-1}^{(\alpha)}(x)}{(1+\alpha)_{n-1}},$$

or

$$(12) \qquad x\frac{d}{dx}L_n^{(\alpha)}(x) = nL_n^{(\alpha)}(x) - (\alpha+n)L_{n-1}^{(\alpha)}(x).$$

In applying Theorem 46 to the Laguerre polynomials, note that $\sigma_n(x) = L_n^{(\alpha)}(x)/(1+\alpha)_n$ and that

$$\psi(u) = {}_0F_1(-;1+\alpha;-u) = \sum_{n=0}^{\infty} \frac{(-1)^n u^n}{n!(1+\alpha)_n}.$$

Then $\gamma_n = (-1)^n/[n!(1+\alpha)_n]$, and

$$F(u) = \sum_{n=0}^{\infty} \frac{(-1)^n (c)_n u^n}{n!(1+\alpha)_n} = {}_1F_1(c;1+\alpha;-u).$$

Therefore Theorem 46 yields

$$(13)\qquad (1-t)^{-c}\,{}_1F_1\left(c;1+\alpha;\ \frac{-xt}{1-t}\right)=\sum_{n=0}^{\infty}\frac{(c)_n L_n^{(\alpha)}(x)t^n}{(1+\alpha)_n},$$

a class of generating relations for $L_n^{(\alpha)}(x)$.

In (13) the choice $c=1+\alpha$ is appealing. With that choice we obtain

$$(14)\qquad (1-t)^{-1-\alpha}\exp\left(\frac{-xt}{1-t}\right)=\sum_{n=0}^{\infty}L_n^{(\alpha)}(x)t^n.$$

74. The generating functions $A(t)\exp[-xt/(1-t)]$. Equation (14) suggests that we consider sets $y_n(x)$ generated by

$$(1)\qquad A(t)\exp\left(\frac{-xt}{1-t}\right)=\sum_{n=0}^{\infty}y_n(x)t^n.$$

From

$$(2)\qquad F=A(t)\exp\left(\frac{-xt}{1-t}\right)$$

it follows that

$$(3)\qquad (1-t)\frac{\partial F}{\partial x}=-tF.$$

Hence

$$\sum_{n=0}^{\infty}y_n'(x)t^n-\sum_{n=0}^{\infty}y_n'(x)t^{n+1}=-\sum_{n=0}^{\infty}y_n(x)t^{n+1},$$

which readily yields $y_0'(x)=0$. and for $n\geqq 1$,

$$(4)\qquad y_n'(x)=y_{n-1}'(x)-y_{n-1}(x).$$

Now (3) can be rewritten as

$$(5)\qquad \frac{\partial F}{\partial x}=-\frac{t}{1-t}F$$

so that we obtain

$$\begin{aligned}\sum_{n=0}^{\infty}y_n'(x)t^n&=-\left(\sum_{n=0}^{\infty}t^{n+1}\right)\left(\sum_{n=0}^{\infty}y_n(x)t^n\right)\\&=-\sum_{n=0}^{\infty}\sum_{k=0}^{n}y_k(x)t^{n+1}\\&=-\sum_{n=1}^{\infty}\sum_{k=0}^{n-1}y_k(x)t^n.\end{aligned}$$

Hence, for $n \geqq 1$,

$$y_n'(x) = -\sum_{k=0}^{n-1} y_k(x). \tag{6}$$

Of course (6) can also be obtained from (4) by iteration and summation.

THEOREM 47. *From*

$$A(t)\exp\left(\frac{-xt}{1-t}\right) = \sum_{n=0}^{\infty} y_n(x)t^n$$

if follows that $y_0'(x) = 0$, *and for* $n \geqq 1$,

$$y_n'(x) = y_{n-1}'(x) - y_{n-1}(x), \tag{4}$$

$$y_n'(x) = -\sum_{k=0}^{n-1} y_k(x). \tag{6}$$

Since, by (14) of the preceding section, the choice $A(t) = (1-t)^{-1-\alpha}$ yields $y_n(x) = L_n^{(\alpha)}(x)$, we have shown that the Laguerre polynomials satisfy, for $n \geqq 1$,

$$\frac{d}{dx}L_n^{(\alpha)}(x) = \frac{d}{dx}L_{n-1}^{(\alpha)}(x) - L_{n-1}^{(\alpha)}(x) \tag{7}$$

and

$$\frac{d}{dx}L_n^{(\alpha)}(x) = -\sum_{k=0}^{n-1} L_k^{(\alpha)}(x). \tag{8}$$

In equations (7) above and (12) of Section 73 we have two differential recurrence relations for $L_n^{(\alpha)}(x)$. These polynomials are completely determined by the two relations once $L_0^{(\alpha)}(x)$, a constant, is specified. The value of $L_0^{(\alpha)}(x)$ is easily found by putting $t = 0$ in a generating relation. Indeed, $L_0^{(\alpha)}(x) = 1$. Thus we see that the Laguerre polynomials are essentially determined by the fact that they have both a generating function of the form

$$e^t\psi(xt)$$

and a generating function of the form

$$A(t)\exp\left(\frac{-xt}{1-t}\right)$$

without specification of the functions ψ and A. For more detail on Laguerre polynomials see Chapter 12.

75. Another class of generating functions. Later we shall encounter many polynomial sets each of which has a generating function of the form next to be considered. Let $\psi(u)$ have a formal power-series expansion

$$\psi(u) = \sum_{n=0}^{\infty} \gamma_n u^n, \qquad \gamma_0 \neq 0. \tag{1}$$

Define the polynomials $f_n(x)$ by

$$(1-t)^{-c}\psi\left(\frac{-4xt}{(1-t)^2}\right) = \sum_{n=0}^{\infty} f_n(x)t^n. \tag{2}$$

THEOREM 48. *The polynomials $f_n(x)$ defined by* (1) *and* (2) *have the following properties:*

$$f_n(x) = \frac{(c)_n}{n!}\sum_{k=0}^{n} \frac{(-n)_k(c+n)_k\gamma_k x^k}{(\frac{1}{2}c)_k(\frac{1}{2}c+\frac{1}{2})_k}, \tag{3}$$

$$x^n = \frac{(c)_{2n}}{2^{2n}\gamma_n}\sum_{k=0}^{n} \frac{(-1)^k(c+2k)f_k(x)}{(n-k)!(c)_{n+k+1}}, \tag{4}$$

$$xf_n'(x) - nf_n(x) = -(c+n-1)f_{n-1}(x) - xf_{n-1}'(x), \qquad n \geqq 1, \tag{5}$$

$$xf_n'(x) - nf_n(x) = -c\sum_{k=0}^{n-1} f_k(x) - 2x\sum_{k=0}^{n-1} f_k'(x), \qquad n \geqq 1, \tag{6}$$

$$xf_n'(x) - nf_n(x) = \sum_{k=0}^{n-1} (-1)^{n-k}(c+2k)f_k(x), \qquad n \geqq 1. \tag{7}$$

For $c = 1$, equations (3) and (5) appear in Sister Celine's work, Fasenmyer [1].

Proof: To obtain (3), consider

$$\begin{aligned}\sum_{n=0}^{\infty} f_n(x)t^n &= \sum_{k=0}^{\infty} \frac{(-4)^k\gamma_k x^k t^k}{(1-t)^{c+2k}}\\ &= \sum_{n,k=0}^{\infty} \frac{(-4)^k(c)_{n+2k}\gamma_k x^k t^{n+k}}{(c)_{2k}n!}\\ &= \sum_{n=0}^{\infty}\sum_{k=0}^{n} \frac{(-1)^k(c)_{n+k}\gamma_k x^k t^n}{(n-k)!(\frac{1}{2}c)_k(\frac{1}{2}c+\frac{1}{2})_k},\end{aligned}$$

from which (3) follows by equating coefficients of t^n.

Next, in (2) put

$$\frac{-4t}{(1-t)^2} = v.$$

Then

$$t = 1 - \frac{2}{1 + \sqrt{1 - v}} = \frac{-v}{(1 + \sqrt{1 - v})^2}$$

and (2) becomes

$$\psi(xv) = \left(\frac{2}{1 + \sqrt{1 - v}}\right)^c \sum_{k=0}^{\infty} \frac{f_k(x)(-1)^k v^k}{(1 + \sqrt{1 - v})^{2k}}$$

or

$$\psi(xv) = \sum_{k=0}^{\infty} \frac{(-1)^k f_k(x) v^k}{2^{2k}} \left(\frac{2}{1 + \sqrt{1 - v}}\right)^{c+2k}.$$

In Ex. 10, page 70, we found that

$$(8) \qquad \left(\frac{2}{1 + \sqrt{1 - v}}\right)^{2\gamma - 1} = {}_2F_1\left[\begin{matrix} \gamma, \gamma - \frac{1}{2}; \\ 2\gamma; \end{matrix} \; v\right].$$

The use of (8) with $2\gamma = c + 2k + 1$ leads to

$$\psi(xv) = \sum_{k=0}^{\infty} {}_2F_1\left[\begin{matrix} \frac{1}{2}(1 + c + 2k), \frac{1}{2}(c + 2k); \\ 1 + c + 2k; \end{matrix} \; v\right] \frac{(-1)^k f_k(x) v^k}{2^{2k}}$$

$$= \sum_{n,k=0}^{\infty} \frac{(c + 2k)_{2n}(-1)^k f_k(x) v^{n+k}}{2^{2n}(1 + c + 2k)_n n! 2^{2k}}$$

$$= \sum_{n,k=0}^{\infty} \frac{(c)_{2n+2k}(c + 2k)(-1)^k f_k(x) v^{n+k}}{2^{2n+2k}(c)_{n+1+2k} n!}.$$

Therefore

$$\sum_{n=0}^{\infty} \gamma_n x^n v^n = \sum_{n=0}^{\infty} \sum_{k=0}^{n} \frac{(c)_{2n}(c + 2k)(-1)^k f_k(x) v^n}{2^{2n}(c)_{n+1+k}(n - k)!},$$

which yields equation (4).

In order to derive (5), (6), and (7), put

$$(9) \qquad F = (1 - t)^{-c} \psi\left(\frac{-4xt}{(1 - t)^2}\right).$$

Then

$$(10) \qquad \frac{\partial F}{\partial x} = -4t(1 - t)^{-c-2} \psi',$$

$$\frac{\partial F}{\partial t} = c(1-t)^{-c-1}\psi - 4x(1+t)(1-t)^{-c-3}\psi'. \tag{11}$$

Therefore F satisfies the partial differential equation

$$x(1+t)\frac{\partial F}{\partial x} - t(1-t)\frac{\partial F}{\partial t} = -ctF. \tag{12}$$

Equation (12) can be put in the forms

$$x\frac{\partial F}{\partial x} - t\frac{\partial F}{\partial t} = -ctF - t^2\frac{\partial F}{\partial t} - xt\frac{\partial F}{\partial x}, \tag{13}$$

$$x\frac{\partial F}{\partial x} - t\frac{\partial F}{\partial t} = \frac{-ct}{1-t}F - \frac{2xt}{1-t}\frac{\partial F}{\partial x}, \tag{14}$$

$$x\frac{\partial F}{\partial x} - t\frac{\partial F}{\partial t} = \frac{-ct}{1+t}F - \frac{2t^2}{1+t}\frac{\partial F}{\partial t}. \tag{15}$$

Since

$$F = \sum_{n=0}^{\infty} f_n(x)t^n,$$

equation (13) yields

$$\begin{aligned}
\sum_{n=0}^{\infty} [xf_n'(x) - nf_n(x)]t^n \\
&= -c\sum_{n=0}^{\infty} f_n(x)t^{n+1} - \sum_{n=0}^{\infty} nf_n(x)t^{n+1} - \sum_{n=0}^{\infty} xf_n'(x)t^{n+1} \\
&= -\sum_{n=1}^{\infty} (c+n-1)f_{n-1}(x)t^n - \sum_{n=1}^{\infty} xf_{n-1}'(x)t^n,
\end{aligned}$$

which leads to (5).

Equation (14) yields

$$\begin{aligned}
\sum_{n=0}^{\infty} [xf_n'(x) - nf_n(x)]t^n \\
&= -c\left(\sum_{n=0}^{\infty} t^{n+1}\right)\left(\sum_{n=0}^{\infty} f_n(x)t^n\right) - 2x\left(\sum_{n=0}^{\infty} t^{n+1}\right)\left(\sum_{n=0}^{\infty} f_n'(x)t^n\right) \\
&= -c\sum_{n=0}^{\infty}\sum_{k=0}^{n} f_k(x)t^{n+1} - 2x\sum_{n=0}^{\infty}\sum_{k=0}^{n} f_k'(x)t^{n+1} \\
&= -c\sum_{n=1}^{\infty}\sum_{k=0}^{n-1} f_k(x)t^n - 2x\sum_{n=1}^{\infty}\sum_{k=0}^{n-1} f_k'(x)t^n,
\end{aligned}$$

which leads to (6).

From (15) we obtain

$$\sum_{n=0}^{\infty} [xf_n'(x) - nf_n(x)]t^n$$

$$= -c\left(\sum_{n=0}^{\infty} (-1)^n t^{n+1}\right)\left(\sum_{n=0}^{\infty} f_n(x)t^n\right) - 2\left(\sum_{n=0}^{\infty} (-1)^n t^{n+1}\right)\left(\sum_{n=0}^{\infty} nf_n(x)t^n\right)$$

$$= -\sum_{n=0}^{\infty}\sum_{k=0}^{n} (-1)^{n-k}(c + 2k)f_k(x)t^{n+1}$$

$$= \sum_{n=1}^{\infty}\sum_{k=0}^{n-1} (-1)^{n-k}(c + 2k)f_k(x)t^n,$$

which gives (7).

Equation (7) was obtained by Dickinson [1] by a somewhat different method.

76. Boas and Buck generating functions. In 1956 Boas and Buck [1] studied a large class of generating functions of polynomial sets. Some of their work appeared also in their earlier mimeographed reports which are not generally available. A rough statement of one of the main results in Boas and Buck [1] is that a necessary and sufficient condition for the polynomials $p_n(x)$ to have a generating function of the form

$$A(t)\psi(xH(t)) = \sum_{n=0}^{\infty} p_n(x)t^n \tag{1}$$

is that sequences of numbers α_k and β_k exist such that, for $n \geqq 1$,

$$xp_n'(x) - np_n(x) = -\sum_{k=0}^{n-1} \alpha_k p_{n-1-k}(x) - x\sum_{k=0}^{n-1} \beta_k p_{n-1-k}'(x). \tag{2}$$

We now present, with minor variations in notation, that part of Boas' and Buck's work which we wish to have available for later chapters.

Let

$$\psi(t) = \sum_{n=0}^{\infty} \gamma_n t^n, \qquad \gamma_0 \neq 0, \tag{3}$$

$$A(t) = \sum_{n=0}^{\infty} a_n t^n, \qquad a_0 \neq 0, \tag{4}$$

$$H(t) = \sum_{n=0}^{\infty} h_n t^{n+1}, \qquad h_0 \neq 0. \tag{5}$$

THEOREM 49. *If $p_n(x)$ is defined by (1), with (3), (4), and (5) holding, $p_n(x)$ is a polynomial in x and $p_n(x)$ is of degree precisely n if and only if $\gamma_n \neq 0$.*

Proof: Put

$$p_n(x) = \sum_{k=0}^{\infty} s(k, n)x^k. \tag{6}$$

Then

$$A(t)\psi(xH(t)) = \sum_{n,k=0}^{\infty} s(k, n)x^k t^n,$$

so that m differentiations with respect to x, followed by our putting $x = 0$, yield

$$A(t)[H(t)]^m \psi^{(m)}(0) = \sum_{n=0}^{\infty} m!\, s(m, n)t^n. \tag{7}$$

Because of (3), (4), and (5),

$$A(t)[H(t)]^m \psi^{(m)}(0) = a_0 h_0{}^m m! \gamma_m t^m + \sum_{n=m+1}^{\infty} C(m, n)t^n, \tag{8}$$

in which the precise nature of $C(m,n)$ is not important to us.

Comparison of (7) and (8) leads to

$$s(m, n) = 0 \qquad \text{for } n < m, \tag{9}$$

$$s(m, m) = a_0 h_0{}^m \gamma_m. \tag{10}$$

The condition (9) shows that $p_n(x)$ is a polynomial of degree $\leqq n$. The condition (10), with m replaced by n, shows that $p_n(x)$ is of degree precisely n if and only if $\gamma_n \neq 0$, since $a_0 h_0 \neq 0$ by (4) and (5).

THEOREM 50. *For the polynomials $p_n(x)$ defined by (1), with (3), (4), and (5) holding ,and $\gamma_n \neq 0$, there exist sequences of numbers α_k and β_k such that, for $n \geqq 1$,*

$$xp_n'(x) - np_n(x) = -\sum_{k=0}^{n-1} \alpha_k p_{n-1-k}(x) - x\sum_{k=0}^{n-1} \beta_k p'_{n-1-k}(x). \tag{2}$$

Indeed,

$$\frac{tA'(t)}{A(t)} = \sum_{n=0}^{\infty} \alpha_n t^{n+1}, \tag{11}$$

$$\frac{tH'(t)}{H(t)} = 1 + \sum_{n=0}^{\infty} \beta_n t^{n+1}. \tag{12}$$

Proof: Put

$$F = A(t)\psi[xH(t)]. \tag{13}$$

Then

$$\frac{\partial F}{\partial x} = H(t)A(t)\psi', \tag{14}$$

$$\frac{\partial F}{\partial t} = A'(t)\psi + xH'(t)A(t)\psi . \tag{15}$$

As usual, we eliminate ψ and ψ' with the aid of equations (13), (14), and (15). The result may be written in the form

$$\frac{xtH'(t)}{H(t)} \cdot \frac{\partial F}{\partial x} - t\frac{\partial F}{\partial t} = -\frac{tA'(t)}{A(t)} \cdot F. \tag{16}$$

If we define α_n and β_n by (11) and (12) and recall that

$$F = \sum_{n=0}^{\infty} p_n(x)t^n,$$

equation (16) leads us to

$$\left[1 + \sum_{n=0}^{\infty} \beta_n t^{n+1}\right]\left[\sum_{n=0}^{\infty} xp_n'(x)t^n\right] - \sum_{n=0}^{\infty} np_n(x)t^n$$
$$= -\left[\sum_{n=0}^{\infty} \alpha_n t^{n+1}\right]\left[\sum_{n=0}^{\infty} p_n(x)t^n\right],$$

or

$$\begin{aligned}\sum_{n=0}^{\infty} [xp_n'(x) - np_n(x)]t^n &= -\sum_{n=0}^{\infty}\sum_{k=0}^{n} \alpha_k p_{n-k}(x)t^{n+1} - x\sum_{n=0}^{\infty}\sum_{k=0}^{n} \beta_k p'_{n-k}(x)t^{n+1} \\ &= -\sum_{n=1}^{\infty}\sum_{k=0}^{n-1} \alpha_k p_{n-1-k}(x)t^n - x\sum_{n=1}^{\infty}\sum_{k=0}^{n-1} \beta_k p'_{n-1-k}(x)t^n,\end{aligned} \tag{17}$$

from which

$$xp_n'(x) - np_n(x) = -\sum_{k=0}^{n-1} \alpha_k p_{n-1-k}(x) - x\sum_{k=0}^{n-1} \beta_k p'_{n-1-k}(x) \tag{2}$$

follows at once. It is important that the α_k and β_k in (2) are independent of n.

EXAMPLE: Consider the polynomials $f_n(x)$ of Section 75 in which

$$(1 - t)^{-c}\psi\left(\frac{-4xt}{(1 - t)^2}\right) = \sum_{n=0}^{\infty} f_n(x)t^n. \tag{18}$$

The $f_n(x)$ fit into the Boas and Buck theory with

$$A(t) = (1-t)^{-c}, \qquad H(t) = \frac{-4t}{(1-t)^2},$$

$$\frac{tA'(t)}{A(t)} = \sum_{n=0}^{\infty} ct^{n+1}, \qquad \frac{tH'(t)}{H(t)} = 1 + \sum_{n=0}^{\infty} 2t^{n+1}.$$

Hence $\alpha_n = c$, $\beta_n = 2$, and the relation (2) becomes

$$(19) \qquad xf_n'(x) - nf_n(x) = -c\sum_{k=0}^{n-1} f_{n-1-k}(x) - 2x\sum_{k=0}^{n-1} f'_{n-1-k}(x),$$

which is equation (6) of Theorem 48, page 137, with the right member written in reverse order. Any one of equations (5), (6), and (7) of Theorem 48 can be obtained from any other. For the Boas and Buck generating function, the results corresponding to (5) and (7) of Theorem 48 are complicated and are therefore omitted.

The Boas and Buck work applies to the polynomials considered in Sections 73, 74, 75 but not to those of Section 72.

77. An extension. Consider the generating relation

$$(1) \qquad A(t)\psi\big(xH(t) + g(t)\big) = \sum_{n=0}^{\infty} f_n(x)t^n$$

in which

$$(2) \qquad \psi(t) = \sum_{n=0}^{\infty} \gamma_n t^n, \qquad \gamma_0 \neq 0,$$

$$(3) \qquad A(t) = \sum_{n=0}^{\infty} a_n t^n, \qquad a_0 \neq 0,$$

$$(4) \qquad H(t) = \sum_{n=0}^{\infty} h_n t^{n+1}, \qquad h_0 \neq 0,$$

and

$$(5) \qquad g(t) = \sum_{n=0}^{\infty} g_n t^{n+2}.$$

Note that $g(t)$ is permitted to be identically zero. It is not necessary to require that $g'(0) = 0$, but this involves no loss of generality, as can be seen by employing a translation in the x-plane.

THEOREM 51. *If $f_n(x)$ is defined by* (1) *with* (2), (3), (4), *and* (5) *holding, $f_n(x)$ is a polynomial in x, and $f_n(x)$ is of degree precisely n if and only if $\gamma_n \neq 0$.*

Proof: We parallel the proof of Theorem 49. Put

$$f_n(x) = \sum_{k=0}^{\infty} s(k, n)x^k. \tag{6}$$

Then

$$A(t)\psi(xH(t) + g(t)) = \sum_{n,k=0}^{\infty} s(k, n)x^k t^n,$$

from which we obtain

$$A(t)[H(t)]^m\psi^{(m)}(g(t)) = \sum_{n=0}^{\infty} m!s(m, n)t^n. \tag{7}$$

Because of (2), (3), (4), and (5),

$$A(t)[H(t)]^m\psi^{(m)}(g(t)) = a_0h_0{}^m m!\gamma_m t^m + \sum_{n=m+1}^{\infty} C(m, n)t^n, \tag{8}$$

in which the nature of $C(m,n)$ is, fortunately, unimportant to us.

Comparison of (7) and (8) leads to

$$s(m,n) = 0 \qquad \text{for } n < m, \tag{9}$$

$$s(m,m) = a_0h_0{}^m\gamma_m, \tag{10}$$

from which the conclusions in Theorem 51 follow.

THEOREM 52. *For the polynomials $f_n(x)$ defined by* (1), *with* (2), (3), (4), *and* (5) *holding, and $\gamma_n \neq 0$, there exist sequences of numbers α_k, β_k and δ_k such that, for $n \geqq 1$,*

$$xf_n'(x) - nf_n(x) = -\sum_{k=0}^{n-1} \alpha_k f_{n-1-k}(x) - \sum_{k=0}^{n-1} (\beta_k x + \delta_k)f'_{n-1-k}(x). \tag{11}$$

Indeed,

$$\frac{tA'(t)}{A(t)} = \sum_{n=0}^{\infty} \alpha_n t^{n+1}, \tag{12}$$

$$\frac{tH'(t)}{H(t)} = 1 + \sum_{n=0}^{\infty} \beta_n t^{n+1}, \tag{13}$$

$$\frac{tg'(t)}{H(t)} = \sum_{n=0}^{\infty} \delta_n t^{n+1}. \tag{14}$$

Proof: Put

$$F = A(t)\psi(xH(t) + g(t)). \tag{15}$$

Then

(16) $$\frac{\partial F}{\partial x} = A(t)H(t)\psi',$$

(17) $$\frac{\partial F}{\partial t} = A'(t)\psi + A(t)[xH'(t) + g'(t)]\psi'.$$

Eliminate ψ and ψ' from (15), (16), and (17) to obtain

(18) $$\left[\frac{xtH'(t)}{H(t)} + \frac{tg'(t)}{H(t)}\right]\frac{\partial F}{\partial x} - t\frac{\partial F}{\partial t} = -\frac{tA'(t)}{A(t)}F.$$

Since

$$F = \sum_{n=0}^{\infty} f_n(x)t^n,$$

it follows from (18) with the aid of (12), (13), and (14) that

$$\left(1 + \sum_{n=0}^{\infty} \beta_n t^{n+1}\right)\left(\sum_{n=0}^{\infty} xf_n'(x)t^n\right) + \left(\sum_{n=0}^{\infty} \delta_n t^{n+1}\right)\left(\sum_{n=0}^{\infty} f_n'(x)t^n\right)$$

$$- \sum_{n=0}^{\infty} nf_n(x)t^n = -\left(\sum_{n=0}^{\infty} \alpha_n t^{n+1}\right)\left(\sum_{n=0}^{\infty} f_n(x)t^n\right).$$

Therefore,

$$\sum_{n=0}^{\infty} [xf_n'(x) - nf_n(x)]t^n$$

$$= - \sum_{n=0}^{\infty}\sum_{k=0}^{n} [(x\beta_k + \delta_k)f'_{n-k}(x) + \alpha_k f_{n-k}(x)]t^{n+1},$$

from which (11) follows after a shift from n to $(n - 1)$ on the right.

The polynomials $g_n(x)$ of Section 72 fit into the above scheme with $\alpha_n = 0$, $\beta_n = 0$, $\delta_0 = -1$, and $\delta_n = 0$ for $n \geqq 1$.

EXERCISES

1. From $e^t\psi(xt) = \sum_{n=0}^{\infty} \sigma_n(x)t^n$, show that

$$\sigma_n(xy) = \sum_{k=0}^{n} \frac{y^k(1 - y)^{n-k}\sigma_k(x)}{(n - k)!},$$

and in particular that

$$2^n\sigma_n\left(\frac{1}{2}x\right) = \sum_{k=0}^{n} \frac{\sigma_k(x)}{(n - k)!}.$$

2. Consider the set (called Appell polynomials) $\alpha_n(x)$ generated by

$$e^{xt}A(t) = \sum_{n=0}^{\infty} \alpha_n(x)t^n.$$

Show that $\alpha'_0(x) = 0$, and that for $n \geqq 1$, $\alpha_n'(x) = \alpha_{n-1}(x)$.

3. Apply Theorem 50, page 141, to the polynomials $\sigma_n(x)$ of Section 73 and thus obtain Theorem 45.

4. The polynomials $\sigma_n(x)$ of Ex. 3 and Section 73 are defined by

$$e^t\psi(xt) = \sum_{n=0}^{\infty} \sigma_n(x)t^n, \tag{A}$$

but by equation (9), page 134, they also satisfy

$$(1-t)^{-c} F\left(\frac{xt}{1-t}\right) = \sum_{n=0}^{\infty} (c)_n\sigma_n(x)t^n, \tag{B}$$

for a certain function F. By applying Theorem 50, page 141, to (B), conclude that the $\sigma_n(x)$ of (A) satisfy the relation

$$(c)_n[x\sigma_n'(x) - n\sigma_n(x)] = -\sum_{k=0}^{n-1} (c)_k[c\sigma_k(x) + x\sigma_k'(x)],$$

for arbitrary c.

5. Apply Theorem 50, page 141, to the polynomials $y_n(x)$ defined by (1), page 135. You do not, of course, get Theorem 47, since that theorem depended upon the specific character of the exponential.

6. Apply Theorem 50 to the Laguerre polynomials through the generating relation (14), page 135, to get

$$xDL_n^{(\alpha)}(x) - nL_n^{(\alpha)}(x) = -\sum_{k=0}^{n-1} [(1+\alpha)L_k^{(\alpha)}(x) + xDL_k^{(\alpha)}(x)],$$

in which $D = d/dx$. Use the above relation in conjunction with equation (8), page 136, to derive the differential equation

$$xD^2L_n^{(\alpha)}(x) + (1+\alpha-x)DL_n^{(\alpha)}(x) + nL_n^{(\alpha)}(x) = 0$$

for the Laguerre polynomials.

7. The Humbert polynomials $h_n(x)$ are defined by

$$(1 - 3xt + t^3)^{-\nu} = \sum_{n=0}^{\infty} h_n(x)t^n.$$

Use Theorem 52, page 144, to conclude that

$$xh_n'(x) - nh_n(x) = h_{n-2}'(x).$$

8. For the $y_n(x)$ of Section 74 show that

$$F = A(t)\exp\left(\frac{-xt}{1-t}\right)$$

satisfies the equation

$$x\frac{\partial F}{\partial x} - t\frac{\partial F}{\partial t} = -t^2\frac{\partial F}{\partial t} - \frac{(1-t)tA'(t)}{A(t)}F$$

and draw what conclusions you can about $y_n(x)$.

9. For polynomials $a_n(x)$ defined by

$$(1-t)^{-c}A\left(\frac{-xt}{1-t}\right) = \sum_{n=0}^{\infty} a_n(x)t^n$$

obtain what results you can parallel to those of Theorem 48, page 137.

CHAPTER 9

Orthogonal Polynomials

78. Simple sets of polynomials. A set of polynomials $\{\varphi_n(x)\}$; $n = 0, 1, 2, \cdots$, is called a *simple set* if $\varphi_n(x)$ is of degree precisely n in x so that the set contains one polynomial of each degree. One immediate result of the definition of a simple set of polynomials is that any polynomial can be expressed linearly in terms of the elements of that simple set.

THEOREM 53. *If* $\{\varphi_n(x)\}$ *is a simple set of polynomials and if* $P(x)$ *is a polynomial of degree* m, *there exist constants* c_k *such that*

$$P(x) = \sum_{k=0}^{m} c_k \varphi_k(x). \tag{1}$$

The c_k are functions of k and of any parameters involved in $P(x)$.

Proof: Let the highest degree term in $P(x)$ be $a_m x^m$, and the highest degree term in $\varphi_m(x)$ be $b_m x^m$. Note that $b_m \neq 0$. Form the polynomial

$$P(x) - c_m \varphi_m(x) \tag{2}$$

in which $c_m = a_m/b_m$. The polynominal (2) is of degree at most $(m-1)$. On this polynomial use the same procedure as was used on $P(x)$, thus reducing the degree again. Iteration of the process yields (1).

79. Orthogonality. Consider a simple set of real polynomials $\varphi_n(x)$. If there exists an interval $a < x < b$ and a function $w(x) > 0$ on that interval, and if

$$(1)\qquad \int_a^b w(x)\varphi_n(x)\varphi_m(x)\,dx = 0, \qquad m \neq n,$$

we say that the polynomials $\varphi_n(x)$ are *orthogonal* with respect to the weight function $w(x)$ over the interval $a < x < b$. Because we have taken $w(x) > 0$ and $\varphi_n(x)$ real, it follows that

$$\int_a^b w(x)\varphi_n^2(x)\,dx \neq 0.$$

With due attention to convergence, either or both endpoints of the interval of orthogonality may be taken to be infinite. The concept of orthogonality used here has been extended in many directions, but the simple version above is all we use. A large number of the sets of polynomials encountered later in the book are orthogonal sets. The limits of integration in (1) are important but the form in which the interval of orthogonality is stated (open or closed) is not vital.

80. An equivalent condition for orthogonality. The following theorem is of use in our study of polynomial sets.

THEOREM 54. *If the $\varphi_n(x)$ form a simple set of real polynomials and $w(x) > 0$ on $a < x < b$, a necessary and sufficient condition that the set $\varphi_n(x)$ be orthogonal with respect to $w(x)$ over the interval $a < x < b$ is that*

$$(1)\qquad \int_a^b w(x)x^k\varphi_n(x)\,dx = 0, \qquad k = 0, 1, 2, \cdots, (n-1).$$

Proof: Suppose (1) is satisfied. Since x^k forms a simple set, there exist constants $b(k,m)$ such that

$$(2)\qquad \varphi_m(x) = \sum_{k=0}^{m} b(k, m)x^k.$$

For the moment, let $m < n$. Then

$$\int_a^b w(x)\varphi_n(x)\varphi_m(x)\,dx = \sum_{k=0}^{m} b(k, m)\int_a^b w(x)x^k\varphi_n(x)\,dx = 0,$$

since m, and therefore each k, is less than n. If $m > n$, interchange m and n in the above argument. We have shown that if (1) is satisfied, it follows that

$$(3)\qquad \int_a^b w(x)\varphi_n(x)\varphi_m(x)\,dx = 0, \qquad m \neq n.$$

Now suppose (3) is satisfied. The $\varphi_n(x)$ form a simple set, so there exist constants $a(m,k)$ such that

$$x^k = \sum_{m=0}^{k} a(m, k)\varphi_m(x). \tag{4}$$

For any k in the range $0 \leqq k < n$

$$\int_a^b w(x)x^k\varphi_n(x)\,dx = \sum_{m=0}^{k} a(m, k)\int_a^b w(x)\varphi_m(x)\varphi_n(x)\,dx = 0,$$

since $m \leqq k < n$ so that $m \neq n$. Therefore (1) follows from (3), and the proof of Theorem 54 is complete.

From Theorem 54 we obtain at once that the orthogonal set $\varphi_n(x)$ has the property that

$$\int_a^b w(x)\varphi_n(x)P(x)\,dx = 0, \tag{5}$$

for every polynomial $P(x)$ of degree $<n$.

It is useful to note that since

$$\int_a^b w(x)\varphi_n^2(x)\,dx \neq 0,$$

it follows that also

$$\int_a^b w(x)x^n\varphi_n(x)\,dx \neq 0. \tag{6}$$

81. Zeros of orthogonal polynomials. Certain elementary information about the location of the zeros of any set of real orthogonal polynomials is easily obtained.

THEOREM 55. *If the simple set of real polynomials $\varphi_n(x)$ is orthogonal with respect to $w(x) > 0$ over the interval $a < x < b$, the zeros of $\varphi_n(x)$ are distinct and all lie in the open interval $a < x < b$.*

Proof: Since, for $n > 0$,

$$\int_a^b w(x)\varphi_n(x)\,dx = 0,$$

the integrand must change sign at least once in the open interval $a < x < b$. Since $w(x) > 0$, $\varphi_n(x)$ must change sign at least once in $a < x < b$. Let the polynomial $\varphi_n(x)$ change sign at precisely the points $\alpha_1, \alpha_2, \cdots, \alpha_s$ in $a < x < b$. The α's are the zeros of odd multiplicity of $\varphi_n(x)$ in $a < x < b$. Since $\varphi_n(x)$ is a polynomial, $s \leqq n$. Now form the polynomial

$$\psi(x) = \prod_{i=1}^{s} (x - \alpha_i).$$

If $s < n$,

$$\int_a^b w(x)\varphi_n(x)\psi(x)\,dx = 0, \tag{1}$$

since $\psi(x)$ is a polynomial of degree less than n. But the integrand in (1) cannot change sign in $a < x < b$ because $\varphi_n(x)$ and $\psi(x)$ change sign at precisely the same points and $w(x) > 0$. Therefore $s < n$ is impossible, and we must have $s = n$. Thus $\varphi_n(x)$ has n roots of odd multiplicity in $a < x < b$. Since $\varphi_n(x)$ is a polynomial of degree n, it has exactly n roots, multiplicity counted, so that its roots are distinct and all lie in $a < x < b$.

82. Expansion of polynomials. Let $f(x)$ and $h(x)$ be any two functions for which the integrals to be involved exist, and let an interval $a < x < b$ and a weight function $w(x) > 0$ on that interval be stipulated. We define the symbol (f, h) by

$$(f, h) = \int_a^b w(x)f(x)h(x)\,dx. \tag{1}$$

The symbol (f, h) has the properties

$$\begin{aligned}(f, h) &= (h, f),\\ (f_1 + f_2, h) &= (f_1, h) + (f_2, h),\\ (cf, h) &= c(f, h), \text{ for constant } c,\\ (fg, h) &= (f, gh).\end{aligned}$$

For a simple set of real polynomials $\varphi_n(x)$ orthogonal with respect to $w(x)$ on the interval $a < x < b$, we already know that

$$(\varphi_n, \varphi_m) = 0, \qquad m \neq n, \tag{2}$$

and

$$(\varphi_n, \varphi_n) \neq 0. \tag{3}$$

For convenience let us also define a sequence of numbers g_n by

$$g_n = (\varphi_n, \varphi_n) = \int_a^b w(x)\varphi_n^2(x)\,dx \neq 0. \tag{4}$$

Theorem 53, page 147, becomes particularly pleasant when the $\varphi_n(x)$ form an orthogonal set, for we can then obtain a simple formula for the coefficients in the expansion.

THEOREM 56. *Let $\varphi_n(x)$ be a simple set of real polynomials orthogonal with respect to $w(x) > 0$ over the interval $a < x < b$, and let $P(x)$ be a polynomial of degree m. Then*

$$P(x) = \sum_{k=0}^{m} C_k \varphi_k(x), \tag{5}$$

in which $C_k = g_k^{-1}(P, \varphi_k)$; that is,

$$C_k = \frac{\int_a^b w(x) P(x) \varphi_k(x)\, dx}{\int_a^b w(x) \varphi_k^2(x)\, dx}. \tag{6}$$

Proof: We know the expansion (5) exists (Theorem 53). From (5) we obtain, for $0 \leqq n \leqq m$,

$$\int_a^b w(x) P(x) \varphi_n(x)\, dx = \sum_{k=0}^{m} C_k \int_a^b w(x) \varphi_k(x) \varphi_n(x)\, dx,$$

or

$$(P, \varphi_n) = \sum_{k=0}^{m} C_k (\varphi_k, \varphi_n), \tag{7}$$

from which, by (2) and (4),

$$(P, \varphi_n) = C_n(\varphi_n, \varphi_n) = C_n g_n.$$

Thus $C_n = g_n^{-1}(P, \varphi_n)$ which is equivalent to (6).

The statement in equation (5) of Section 80 can now be expressed by writing

$$(P, \varphi_n) = 0 \tag{8}$$

for every polynomial P of degree $<n$.

83. The three-term recurrence relation. Every orthogonal set of polynomials possesses a three-term recurrence relation of a simple nature.

THEOREM 57. *If $\varphi_n(x)$ is a simple set of real polynomials orthogonal with respect to $w(x) > 0$ on $a < x < b$, there exist sequences of numbers A_n, B_n, C_n such that for $n \geqq 1$,*

$$x\varphi_n(x) = A_n \varphi_{n+1}(x) + B_n \varphi_n(x) + C_n \varphi_{n-1}(x) \tag{1}$$

in which $A_n \neq 0$ and $C_n \neq 0$.

Proof: Since $x\varphi_n(x)$ is a polynomial of degree $(n+1)$, we know by Theorem 56 that

$$x\varphi_n(x) = \sum_{k=0}^{n+1} a(k, n)\varphi_k(x),$$

in which

$$a(k,n) = g_k^{-1}(x\varphi_n, \varphi_k) = g_k^{-1}(\varphi_n, x\varphi_k).$$

By (8) of Section 82 we see that $(\varphi_n, x\varphi_k) = 0$ for $k < (n-1)$. Therefore the relation (1) of Theorem 57 exists.

If A_n were zero for any n, the right member of (1) would be of degree $\leqq n$ and the left member of degree $(n+1)$. Hence $A_n \neq 0$. We still must show that $C_n \neq 0$.

It is now convenient to introduce a symbol π_m to denote a polynomial of degree $\leqq m$. The symbol π_m is not to stand for a specific polynomial but merely to stipulate that the degree of the polynomial does not exceed the subscript used. When π_m occurs more than once in a discussion, there is no implication that the polynomials indicated are related in any manner other than the fact that none of them is of degree $> m$.

Let h_n denote the leading coefficient in our $\varphi_n(x)$. Then

$$\varphi_n(x) = h_n x^n + \pi_{n-1} \tag{2}$$

and $h_n \neq 0$ because $\varphi_n(x)$ is of degree precisely n. Now

$$x\varphi_{n-1}(x) = \frac{h_{n-1}}{h_n}\varphi_n(x) + \pi_{n-1}, \tag{3}$$

as can be seen by examining the leading coefficient on each side.

In equation (1) of Theorem 57 we know, by Theorem 56, that

$$C_n = g_{n-1}^{-1}(x\varphi_n, \varphi_{n-1}) = g_{n-1}^{-1}(\varphi_n, x\varphi_{n-1})$$

so that by (3),

$$C_n = g_{n-1}^{-1}\left[\left(\varphi_n, \frac{h_{n-1}}{h_n}\varphi_n\right) + (\varphi_n, \pi_{n-1})\right] = \frac{h_{n-1}}{h_n g_{n-1}}(\varphi_n, \varphi_n)$$

from which

$$C_n = \frac{g_n h_{n-1}}{g_{n-1} h_n}. \tag{4}$$

Thus $C_n \neq 0$ for $n \geqq 1$, and the proof of Theorem 57 is complete. That B_n in Theorem 57 can be zero, even for all n, will be seen in specific examples in later chapters.

By comparison of leading coefficients in equation (1) we find that

$$A_n = \frac{h_n}{h_{n+1}}. \tag{5}$$

Now (1) can be written

$$x\varphi_n(x) = \frac{h_n}{h_{n+1}}\varphi_{n+1}(x) + B_n\varphi_n(x) + \frac{g_n h_{n-1}}{g_{n-1}h_n}\varphi_{n-1}(x),$$

or in even more promising form as

$$\frac{x\varphi_n(x)}{g_n} = \frac{h_n}{g_n h_{n+1}}\varphi_{n+1}(x) + \frac{B_n}{g_n}\varphi_n(x) + \frac{h_{n-1}}{g_{n-1}h_n}\varphi_{n-1}(x), \tag{6}$$

in which the coefficients of $\varphi_{n+1}(x)$ and $\varphi_{n-1}(x)$ are the same except for a shift of index.

In our treatment of specific polynomial sets in later chapters we shall obtain the pure recurrence relations explicitly and thus have no direct need for Theorem 57. That theorem is useful in the general discussions, as in Section 84, and is a powerful tool for showing that a polynomial set is not an orthogonal set. If a set of polynomials $\psi_n(x)$ does not possess a three-term recurrence relation of the form in Theorem 57, the set $\psi_n(x)$ is not an orthogonal set.

A widely known theorem of Favard [1] states essentially that any real polynomial set which satisfies a pure recurrence relation of the type in Theorem 57 is orthogonal with respect to some weight function over some interval with Stieltjes (not necessarily just Riemann) integration used. No method of finding the weight function and interval is given. Recently Dickinson, Pollak and Wannier [1] have, for a special subclass of such polynomials, constructively shown orthogonality over a denumerable set of points.

84. The Christoffel-Darboux formula. In equation (6) of the preceding section put

$$t_n = \frac{h_n}{g_n h_{n+1}}. \tag{1}$$

The three-term recurrence relation may then be written

$$g_n^{-1}x\varphi_n(x) = t_n\varphi_{n+1}(x) + g_n^{-1}B_n\varphi_n(x) + t_{n-1}\varphi_{n-1}(x). \tag{2}$$

Then

$$g_n^{-1}x\varphi_n(x)\varphi_n(y) = t_n\varphi_{n+1}(x)\varphi_n(y) + t_{n-1}\varphi_n(y)\varphi_{n-1}(x) + g_n^{-1}B_n\varphi_n(x)\varphi_n(y)$$

and

$$g_n^{-1}y\varphi_n(x)\varphi_n(y) = t_n\varphi_{n+1}(y)\varphi_n(x) + t_{n-1}\varphi_n(x)\varphi_{n-1}(y) + g_n^{-1}B_n\varphi_n(x)\varphi_n(y)$$

from which it follows that

$$\begin{aligned} g_n^{-1}(y - x)\varphi_n(x)\varphi_n(y) &= t_n[\varphi_{n+1}(y)\varphi_n(x) - \varphi_{n+1}(x)\varphi_n(y)] \\ &\quad - t_{n-1}[\varphi_n(y)\varphi_{n-1}(x) - \varphi_n(x)\varphi_{n-1}(y)]. \end{aligned}$$

Next put

$$j_n(x, y) = t_n[\varphi_{n+1}(y)\varphi_n(x) - \varphi_{n+1}(x)\varphi_n(y)]. \tag{3}$$

Then

$$g_n^{-1}(y - x)\varphi_n(x)\varphi_n(y) = j_n(x, y) - j_{n-1}(x, y)$$

so that

$$\sum_{k=1}^{n} g_k^{-1}(y - x)\varphi_k(x)\varphi_k(y) = j_n(x, y) - j_0(x, y). \tag{4}$$

Now, by (3),

$$j_0(x, y) = t_0[\varphi_1(y)\varphi_0(x) - \varphi_1(x)\varphi_0(y)]$$

and $\varphi_0(x) = \varphi_0(y) = h_0$, a constant. Also $t_0 = h_0g_0^{-1}h_1^{-1}$. Let $\varphi_1(x) = h_1x + c$, in which c is constant. Then

$$\begin{aligned} j_0(x, y) &= h_0g_0^{-1}h_1^{-1}[(h_1y + c)h_0 - (h_1x + c)h_0] \\ &= h_0^2g_0^{-1}(y - x) = g_0^{-1}(y - x)\varphi_0(x)\varphi_0(y). \end{aligned}$$

We then transfer the $j_0(x, y)$ term from the right to the left in equation (4) and obtain

$$\sum_{k=0}^{n} g_k^{-1}(y - x)\varphi_k(x)\varphi_k(y) = j_n(x, y). \tag{5}$$

Using (1) and (3) in (5), we arrive at the desired result, the Christoffel-Darboux formula, equation (6) of the following theorem.

THEOREM 58. *Let $\varphi_n(x)$ be a simple set of real polynomials orthogonal with respect to $w(x) > 0$ on $a < x < b$. Let h_n be the leading coefficient in $\varphi_n(x)$ so that*

$$\varphi_n(x) = h_nx^n + \pi_{n-1}$$

and let

$$g_k = (\varphi_k, \varphi_k) = \int_a^b w(x)\varphi_k^2(x)\,dx.$$

Then

$$\sum_{k=0}^{n} g_k^{-1}\varphi_k(x)\varphi_k(y) = \frac{h_n}{g_nh_{n+1}} \cdot \frac{\varphi_{n+1}(y)\varphi_n(x) - \varphi_{n+1}(x)\varphi_n(y)}{y - x}. \tag{6}$$

85. Normalization; Bessel's inequality. For theoretical discussions it is convenient to replace the orthogonal polynomials $\varphi_n(x)$ by polynomials

$$\psi_n(x) = g_n^{-\frac{1}{2}}\varphi_n(x) \tag{1}$$

which are called *orthonormal* polynomials. Note that

$$(\psi_n, \psi_n) = (g_n^{-\frac{1}{2}}\varphi_n, g_n^{-\frac{1}{2}}\varphi_n) = g_n^{-1}(\varphi_n, \varphi_n) = 1.$$

The use of (1) increases the neatness of many formulas. When specific polynomials are being used, the normalization process is of little help. We shall concentrate on specific polynomials in most of the later chapters.

Let $\psi_n(x)$ be an orthonormal polynomial set over the interval (a, b) with weight function $w(x) > 0$. Let

$$s_n(x) = \sum_{k=0}^{n} c_k\psi_k(x), \tag{2}$$

where

$$c_k = (f, \psi_k) = \int_a^b w(y)f(y)\psi_k(y)\,dy, \tag{3}$$

$f(y)$ as yet unrestricted except that the integrals involved exist. Consider

$$\int_a^b w(x)[f(x) - s_n(x)]^2\,dx \tag{4}$$

$$= \int_a^b w(x)f^2(x)\,dx - 2\int_a^b w(x)f(x)s_n(x)\,dx + \int_a^b w(x)s_n^2(x)\,dx.$$

Now

$$\int_a^b w(x)f(x)s_n(x)\,dx = \sum_{k=0}^{n} c_k\int_a^b w(x)f(x)\psi_k(x)\,dx = \sum_{k=0}^{n} c_k^2$$

and, since the $\psi_n(x)$ form an orthonormal set,

$$\int_a^b w(x)s_n^2(x)\,dx = \int_a^b w(x)\left[\sum_{k=0}^{n} c_k\psi_k(x)\right]^2 dx$$

$$= \sum_{k=0}^{n} c_k^2\int_a^b w(x)\psi_k^2(x)\,dx = \sum_{k=0}^{n} c_k^2.$$

It follows that equation (4) becomes

$$\int_a^b w(x)[f(x) - s_n(x)]^2dx = \int_a^b w(x)f^2(x)\,dx - \sum_{k=0}^{n} c_k^2. \tag{5}$$

Since the left member of (5) is never negative,

$$\sum_{k=0}^{n} c_k^2 \leqq \int_a^b w(x) f^2(x)\, dx, \tag{6}$$

which is Bessel's inequality. The right member of (6) is independent of n. Hence $\sum_{n=0}^{\infty} c_n^2$ converges, and it follows that $c_n \to 0$ as $n \to \infty$.

THEOREM 59. *If the polynomials $\psi_n(x)$ form an orthonormal set on the interval (a, b) with respect to the weight function $w(x) > 0$ and if $\int_a^b w(x) f^2(x)\, dx$ exists,*

$$\lim_{n \to \infty} \int_a^b w(x) f(x) \psi_n(x)\, dx = 0. \tag{7}$$

If we wish to state the result corresponding to (7) for orthogonal polynomials $\varphi_n(x)$, not necessarily normalized, we need merely to define

$$g_n = \int_a^b w(x) \varphi_n^2(x)\, dx$$

and then replace (7) by

$$\lim_{n \to \infty} g_n^{-\frac{1}{2}} \int_a^b w(x) f(x) \varphi_n(x)\, dx = 0. \tag{8}$$

For much additional material on general orthogonal polynomials the reader should consult the following: Szegö [1], Chapter 10 of Erdélyi [2], Jackson [1], and Shohat [1].

CHAPTER 10

Legendre Polynomials

86. A generating function. We define the Legendre polynomials $P_n(x)$ by the generating relation

$$(1) \qquad (1 - 2xt + t^2)^{-\frac{1}{2}} = \sum_{n=0}^{\infty} P_n(x)t^n,$$

in which $(1 - 2xt + t^2)^{-\frac{1}{2}}$ denotes the particular branch which $\rightarrow 1$ as $t \rightarrow 0$. We shall first show that $P_n(x)$ is a polynomial of degree precisely n.

Since $(1 - z)^{-\alpha} = {}_1F_0(\alpha; -; z)$, we may write

$$\begin{aligned}(1 - 2xt + t^2)^{-\frac{1}{2}} &= \sum_{n=0}^{\infty} \frac{(\frac{1}{2})_n(2xt - t^2)^n}{n!} \\ &= \sum_{n=0}^{\infty}\sum_{k=0}^{n} \frac{(\frac{1}{2})_n(-1)^k(2x)^{n-k}t^{n+k}}{k!(n-k)!} \\ &= \sum_{n=0}^{\infty}\sum_{k=0}^{[n/2]} \frac{(-1)^k(\frac{1}{2})_{n-k}(2x)^{n-2k}t^n}{k!(n-2k)!},\end{aligned}$$

by equation (13), page 58. We thus obtain

$$(2) \qquad P_n(x) = \sum_{k=0}^{[n/2]} \frac{(-1)^k(\frac{1}{2})_{n-k}(2x)^{n-2k}}{k!(n-2k)!},$$

from which it follows that $P_n(x)$ is a polynomial of degree precisely n in x. Equation (2) also yields

$$P_n(x) = \frac{2^n(\frac{1}{2})_n x^n}{n!} + \pi_{n-2} \tag{3}$$

in which π_{n-2} is a polynomial of degree $(n - 2)$ in x.

If in (1) we replace x by $(-x)$ and t by $(-t)$, the left member does not change. Hence

$$P_n(-x) = (-1)^n P_n(x), \tag{4}$$

so that $P_n(x)$ is an odd function of x for n odd, an even function of x for n even. Equation (4) follows just as easily from (2).

In equation (1) put $x = 1$ to obtain

$$(1 - t)^{-1} = \sum_{n=0}^{\infty} P_n(1)t^n,$$

from which

$$P_n(1) = 1, \tag{5}$$

which combines with (4) to give

$$P_n(-1) = (-1)^n. \tag{6}$$

From (1) with $x = 0$, we get

$$(1 + t^2)^{-\frac{1}{2}} = \sum_{n=0}^{\infty} P_n(0)t^n.$$

But

$$(1 + t^2)^{-\frac{1}{2}} = \sum_{n=0}^{\infty} \frac{(-1)^n(\frac{1}{2})_n t^{2n}}{n!}.$$

Hence

$$P_{2n+1}(0) = 0, \qquad P_{2n}(0) = \frac{(-1)^n(\frac{1}{2})_n}{n!}, \tag{7}$$

results just as easily obtained directly from (2).

Equation (2) yields

$$P_n'(x) = \sum_{k=0}^{[(n-1)/2]} \frac{(-1)^k 2(\frac{1}{2})_{n-k}(2x)^{n-1-2k}}{k!(n-1-2k)!} \tag{8}$$

and from (8) it follows that

$$P_{2n}'(0) = 0, \qquad P_{2n+1}'(0) = \frac{(-1)^n 2(\frac{1}{2})_{n+1}}{n!} = \frac{(-1)^n(\frac{3}{2})_n}{n!}. \tag{9}$$

87. Differential recurrence relations. We already know from Section 72 that the generating relation

(1) $$(1 - 2xt + t^2)^{-\frac{1}{2}} = \sum_{n=0}^{\infty} P_n(x)t^n$$

implies the differential recurrence relation

(2) $$xP_n'(x) = nP_n(x) + P_{n-1}'(x).$$

From (1) it follows by the usual method (differentiation) that

(3) $$(1 - 2xt + t^2)^{-\frac{3}{2}} = \sum_{n=1}^{\infty} P_n'(x)t^{n-1},$$

(4) $$(x - t)(1 - 2xt + t^2)^{-\frac{3}{2}} = \sum_{n=1}^{\infty} nP_n(x)t^{n-1}.$$

Since $1 - t^2 - 2t(x - t) = (1 - 2xt + t^2)$, we may multiply the left member of (3) by $(1 - t^2)$, the left member of (4) by $2t$, subtract and obtain the left member of (1). In this way we find that

$$\sum_{n=1}^{\infty} P_n'(x)t^{n-1} - \sum_{n=1}^{\infty} P_n'(x)t^{n+1} - \sum_{n=1}^{\infty} 2nP_n(x)t^n = \sum_{n=0}^{\infty} P_n(x)t^n,$$

or

$$\sum_{n=0}^{\infty} P_{n+1}'(x)t^n - \sum_{n=2}^{\infty} P_{n-1}'(x)t^n = \sum_{n=0}^{\infty} (2n + 1)P_n(x)t^n.$$

We thus obtain another differential recurrence relation

(5) $$(2n + 1)P_n(x) = P_{n+1}'(x) - P_{n-1}'(x).$$

Equations (2) and (5) are independent differential recurrence relations. From (2) and (5) other relations may be obtained, each useful in various ways. By combining (2) and (5), we find that

(6) $$xP_n'(x) = P_{n+1}'(x) - (n + 1)P_n(x).$$

Next in (6) shift index from n to $(n - 1)$ and substitute the resulting expression for $P_{n-1}'(x)$ into (2) to obtain

(7) $$(x^2 - 1)\, P_n'(x) = nxP_n(x) - nP_{n-1}(x).$$

88. The pure recurrence relation. The relation (7) above permits us to eliminate derivatives from other recurrence relations. Equation (2) of Section 87 yields

(1) $$x(x^2 - 1)P_n'(x) = n(x^2 - 1)P_n(x) + (x^2 - 1)P_{n-1}'(x),$$

and we may now substitute for $(x^2-1)P_n'(x)$ and for $(x^2-1)P_{n-1}'(x)$ from (7) of Section 87 to arrive at the identity

$$x[nxP_n(x) - nP_{n-1}(x)] = n(x^2 - 1)P_n(x) + (n - 1)xP_{n-1}(x) - (n - 1)P_{n-2}(x).$$

Collect terms in the above equation to obtain the pure recurrence relation

(2) $$nP_n(x) = (2n - 1)xP_{n-1}(x) - (n - 1)P_{n-2}(x), \qquad n \geqq 2.$$

Equation (2), with index shifted, is of the character of the pure recurrence relation for an orthogonal set of polynomials. See Theorem 57, page 151, and note that here we have an example in which $B_n = 0$. We shall show in Section 99 that the $P_n(x)$ form an orthogonal set.

Equation (2) furnishes a fairly easy method for computing successive Legendre polynomials. From the relation (2) of Section 86 we easily find that

$$P_0(x) = 1, \qquad P_1(x) = x.$$

Then the pure recurrence relation may be used to obtain

$$P_2(x) = \frac{3}{2}x^2 - \frac{1}{2}, \qquad P_3(x) = \frac{5}{2}x^3 - \frac{3}{2}x,$$

$$P_4(x) = \frac{35}{8}x^4 - \frac{15}{4}x^2 + \frac{3}{8}, \qquad P_5(x) = \frac{63}{8}x^5 - \frac{35}{4}x^3 + \frac{15}{8}x,$$

$$P_6(x) = \frac{1}{16}(231x^6 - 315x^4 + 105x^2 - 5),$$

etc.

89. Legendre's differential equation. We have already obtained the relations

(1) $$xP_n'(x) = nP_n(x) + P_{n-1}'(x),$$

(2) $$xP_n'(x) = P_{n+1}'(x) - (n + 1)P_n(x).$$

We now wish to eliminate the differences in subscript to find a relation involving only $P_n(x)$ and its derivatives.

In (2) replace n by $(n - 1)$ to get

(3) $$xP_{n-1}'(x) = P_n'(x) - nP_{n-1}(x),$$

from which also, by differentiation, we have

(4) $$xP_{n-1}''(x) = P_n''(x) - (n + 1)P_{n-1}'(x).$$

Both $P_{n-1}'(x)$ and $P_{n-1}''(x)$ can be obtained from (1) and put into (4) to yield

$$x[xP_n''(x)+P_n'(x)-nP_n'(x)]=P_n''(x)-(n+1)[xP_n'(x)-nP_n(x)].$$

A rearrangement of terms in the above equation gives us Legendre's differential equation for $P_n(x)$,

$$(5) \qquad (1-x^2)P_n''(x)-2xP_n'(x)+n(n+1)P_n(x)=0.$$

We shall return to this differential equation later when we take up the matter of orthogonality of the set of Legendre polynomials. Equation (5) is one of many natural starting points for the study of Legendre functions, which are solutions of (5) for nonintegral n.

90. The Rodrigues formula. In Section 86 we established that

$$(1) \qquad P_n(x) = \sum_{k=0}^{[n/2]} \frac{(-1)^k(\tfrac{1}{2})_{n-k}(2x)^{n-2k}}{k!(n-2k)!}.$$

We also know that $(2m)! = 2^{2m}(\tfrac{1}{2})_m m!$ and therefore that

$$(2) \qquad 2^{2n-2k}(\tfrac{1}{2})_{n-k} = \frac{(2n-2k)!}{(n-k)!}.$$

Employing (2) on the right in equation (1), we obtain

$$(3) \qquad P_n(x) = \sum_{k=0}^{[n/2]} \frac{(-1)^k(2n-2k)!x^{n-2k}}{2^n k!(n-k)!(n-2k)!}.$$

If $D = \dfrac{d}{dx}$, we know that

$$(4) \qquad D^s x^m = \frac{m!x^{m-s}}{(m-s)!}.$$

The expression $(2n-2k)!\,x^{n-2k}/(n-2k)!$ in equation (3) suggests the use of (4). Indeed, by (4),

$$D^n x^{2n-2k} = \frac{(2n-2k)!x^{n-2k}}{(n-2k)!},$$

so that (3) may be rewritten as

$$(5) \qquad P_n(x) = \sum_{k=0}^{[n/2]} \frac{(-1)^k D^n x^{2n-2k}}{2^n k!(n-k)!}.$$

Of course the $k!$ and $(n-k)!$ in (5) remind us of the binomial coefficient

$$C_{n,k} = \frac{n!}{k!(n-k)!}.$$

Since n is independent of k, equation (5) can now be put in the form

(6) $$P_n(x) = \frac{D^n}{2^n n!} \sum_{k=0}^{[n/2]} (-1)^k C_{n,k} x^{2n-2k}.$$

For $[n/2] < k \leqq n$, $0 \leqq 2n - 2k < n$, so that for those values of k, $D^n x^{2n-2k} = 0$. Hence the summation on the right in equation (6) can be extended to the range $k = 0$ to $k = n$. Thus we have

$$P_n(x) = \frac{D^n}{2^n n!} \sum_{k=0}^{n} (-1)^k C_{n,k} x^{2n-2k},$$

or

(7) $$P_n(x) = \frac{1}{2^n n!} D^n (x^2 - 1)^n,$$

which is called Rodrigues' formula. We shall use (7) to obtain another formula for $P_n(x)$.

Leibnitz' rule for the nth derivative of a product is

(8) $$D^n(uv) = \sum_{k=0}^{n} C_{n,k} (D^k u)(D^{n-k} v),$$

in which $D = d/dx$ and u and v are to be functions of x. The validity of (8) is easily shown by induction.

Since, by Rodrigues' formula,

$$P_n(x) = \frac{1}{2^n n!} D^n[(x - 1)^n (x + 1)^n],$$

the application of (8) with $u = (x - 1)^n$, $v = (x + 1)^n$, leads to

$$P_n(x) = \frac{1}{2^n n!} \sum_{k=0}^{n} C_{n,k} \frac{n!(x - 1)^{n-k}}{(n - k)!} \frac{n!(x + 1)^k}{k!},$$

or the beautifully symmetric result

(9) $$P_n(x) = \sum_{k=0}^{n} C_{n,k}^2 \left(\frac{x - 1}{2}\right)^{n-k} \left(\frac{x + 1}{2}\right)^k.$$

91. Bateman's generating function. Equation (9), when put in the form

(1) $$P_n(x) = \sum_{k=0}^{n} \frac{(n!)^2[\frac{1}{2}(x - 1)]^{n-k}[\frac{1}{2}(x + 1)]^k}{[(n - k)!]^2 (k!)^2},$$

reminds us of the Cauchy product of two power series,

(2) $$\left(\sum_{n=0}^{\infty} a_n t^n\right)\left(\sum_{n=0}^{\infty} b_n t^n\right) = \sum_{n=0}^{\infty} \sum_{k=0}^{n} a_k b_{n-k} t^n.$$

Hence we multiply each member of equation (1) by $t^n/(n!)^2$ and sum from $n = 0$ to ∞ to obtain

$$\sum_{n=0}^{\infty} \frac{P_n(x)t^n}{(n!)^2} = \sum_{n=0}^{\infty} \sum_{k=0}^{n} \frac{[\frac{1}{2}(x-1)]^{n-k}[\frac{1}{2}(x+1)]^k t^n}{[(n-k)!]^2(k!)^2}$$

$$= {}_0F_1(-;1;\tfrac{1}{2}t(x-1))\ {}_0F_1(-;1;\tfrac{1}{2}t(x+1)).$$

Thus from the Rodrigues formula we have obtained a second generating function:*

$$(3) \qquad {}_0F_1(-;1;\tfrac{1}{2}t(x-1))\ {}_0F_1(-;1;\tfrac{1}{2}t(x+1)) = \sum_{n=0}^{\infty} \frac{P_n(x)t^n}{(n!)^2}.$$

92. Additional generating functions. The generating function $(1 - 2xt + t^2)^{-\frac{1}{2}}$ used to define the Legendre polynomials can be expanded in powers of t in new ways, thus yielding additional results. For instance,

$$(1 - 2xt + t^2)^{-\frac{1}{2}} = [(1 - xt)^2 - t^2(x^2 - 1)]^{-\frac{1}{2}}$$

$$= (1 - xt)^{-1}\left[1 - \frac{t^2(x^2-1)}{(1-xt)^2}\right]^{-\frac{1}{2}}.$$

Therefore

$$(1 - xt)^{-1}\ {}_1F_0\left(\tfrac{1}{2}; -; \frac{t^2(x^2-1)}{(1-xt)^2}\right) = \sum_{n=0}^{\infty} P_n(x)t^n.$$

Now

$$(1 - xt)^{-1}\ {}_1F_0\left(\tfrac{1}{2}; -; \frac{t^2(x^2-1)}{(1-xt)^2}\right)$$

$$= \sum_{k=0}^{\infty} \frac{(\frac{1}{2})_k t^{2k}(x^2-1)^k}{k!(1-xt)^{2k+1}}$$

$$= \sum_{k=0}^{\infty} \sum_{n=0}^{\infty} \frac{(\frac{1}{2})_k (2k+1)_n (x^2-1)^k t^{2k+n} x^n}{k!n!}$$

$$= \sum_{n=0}^{\infty} \sum_{k=0}^{\infty} \frac{(\frac{1}{2})_k (n+2k)!(x^2-1)^k x^n t^{n+2k}}{k!n!(2k)!}$$

$$= \sum_{n=0}^{\infty} \sum_{k=0}^{[n/2]} \frac{(\frac{1}{2})_k n!(x^2-1)^k x^{n-2k} t^n}{k!(2k)!(n-2k)!}$$

$$= \sum_{n=0}^{\infty} \sum_{k=0}^{[n/2]} \frac{n!(x^2-1)^k x^{n-2k} t^n}{2^{2k}(k!)^2(n-2k)!}.$$

---

*This is a special case of a result published by Harry Bateman in 1905. See Bateman [1].

Hence we obtain a new form for $P_n(x)$:

$$(1) \qquad P_n(x) = \sum_{k=0}^{[n/2]} \frac{n!(x^2-1)^k x^{n-2k}}{2^{2k}(k!)^2(n-2k)!}.$$

Let us employ (1) to discover new generating functions for $P_n(x)$. Consider, for arbitrary c, the sum

$$\begin{aligned}
\sum_{n=0}^{\infty} \frac{(c)_n P_n(x) t^n}{n!} &= \sum_{n=0}^{\infty}\sum_{k=0}^{[n/2]} \frac{(c)_n (x^2-1)^k x^{n-2k} t^n}{2^{2k}(k!)^2(n-2k)!} \\
&= \sum_{n=0}^{\infty}\sum_{k=0}^{\infty} \frac{(c)_{n+2k}(x^2-1)^k x^n t^{n+2k}}{2^{2k}(k!)^2 n!} \\
&= \sum_{k=0}^{\infty}\sum_{n=0}^{\infty} \frac{(c+2k)_n (xt)^n}{n!} \cdot \frac{(c)_{2k}(x^2-1)^k t^{2k}}{2^{2k}(k!)^2} \\
&= \sum_{k=0}^{\infty} {}_1F_0(c+2k; -; xt) \frac{(\frac{1}{2}c)_k(\frac{1}{2}c+\frac{1}{2})_k (x^2-1)^k t^{2k}}{(k!)^2} \\
&= \sum_{k=0}^{\infty} \frac{(\frac{1}{2}c)_k(\frac{1}{2}c+\frac{1}{2})_k (x^2-1)^k t^{2k}}{(k!)^2 (1-xt)^{c+2k}} \\
&= (1-xt)^{-c}\, {}_2F_1\left[\begin{matrix} \frac{1}{2}c, \frac{1}{2}c+\frac{1}{2}; \\ 1; \end{matrix} \quad \frac{t^2(x^2-1)}{(1-xt)^2}\right].
\end{aligned}$$

We have thus discovered* the family of generating functions:

$$(2) \qquad (1-xt)^{-c}\, {}_2F_1\left[\begin{matrix} \frac{1}{2}c, \frac{1}{2}c+\frac{1}{2}; \\ 1; \end{matrix} \quad \frac{t^2(x^2-1)}{(1-xt)^2}\right] = \sum_{n=0}^{\infty} \frac{(c)_n P_n(x) t^n}{n!},$$

in which c may be any complex number. If c is unity, (2) degenerates into the generating relation used to define $P_n(x)$ at the start of this chapter. If c is taken to be zero or a negative integer, both members of (2) terminate, and only a finite set of Legendre polynomials is then generated by (2).

With the aid of Ex. 11, page 70, it is a simple matter to transform the left member of (2) into the form shown in (3) below. Let

$$\rho = (1 - 2xt + t^2)^{\frac{1}{2}}$$

then an equivalent form for (2) is

*Special cases of (2) have been known for a long time, but the general formula may have been first published by Brafman in 1951. See Brafman [1].

$$(3) \qquad \rho^{-c}\,{}_2F_1\left[\begin{matrix} c, 1-c; \\ \\ 1; \end{matrix} \quad \tfrac{1}{2}\left(1 - \frac{1-xt}{\rho}\right)\right] = \sum_{n=0}^{\infty} \frac{(c)_n P_n(x) t^n}{n!}.$$

Let us now return to (1) and consider the sum

$$\sum_{n=0}^{\infty} \frac{P_n(x)t^n}{n!} = \sum_{n=0}^{\infty} \sum_{k=0}^{[n/2]} \frac{(x^2-1)^k x^{n-2k} t^n}{2^{2k}(k!)^2(n-2k)!}$$

$$= \sum_{n,k=0}^{\infty} \frac{(x^2-1)^k x^n t^{n+2k}}{2^{2k}(k!)^2 n!}$$

$$= e^{xt}\,{}_0F_1\big(-;1;\tfrac{1}{4}t^2(x^2-1)\big).$$

We thus find another generating relation,

$$(4) \qquad e^{xt}\,{}_0F_1\big(-;1;\tfrac{1}{4}t^2(x^2-1)\big) = \sum_{n=0}^{\infty} \frac{P_n(x)t^n}{n!},$$

which can equally well be written in terms of a Bessel function as

$$(5) \qquad e^{xt}J_0(t\sqrt{1-x^2}) = \sum_{n=0}^{\infty} \frac{P_n(x)t^n}{n!}.$$

The relation (5) was being used at the beginning of this century. We have not been able to determine when or by whom it was first discovered.

93. Hypergeometric forms of $P_n(x)$. Return once more to the original definition of $P_n(x)$:

$$(1) \qquad (1-2xt+t^2)^{-\frac{1}{2}} = \sum_{n=0}^{\infty} P_n(x)t^n.$$

This time note that

$$(1-2xt+t^2)^{-\frac{1}{2}} = [(1-t)^2 - 2t(x-1)]^{-\frac{1}{2}}$$

$$= (1-t)^{-1}\left[1 - \frac{2t(x-1)}{(1-t)^2}\right]^{-\frac{1}{2}},$$

which permits us to write

$$\sum_{n=0}^{\infty} P_n(x)t^n = \sum_{k=0}^{\infty} \frac{(\frac{1}{2})_k 2^k t^k (x-1)^k}{k!(1-t)^{2k+1}}$$

$$= \sum_{n,k=0}^{\infty} \frac{(\frac{1}{2})_k (2k+1)_n 2^k (x-1)^k t^{n+k}}{k!\,n!}.$$

Thus we have

$$\sum_{n=0}^{\infty} P_n(x)t^n = \sum_{n=0}^{\infty}\sum_{k=0}^{\infty} \frac{(\frac{1}{2})_k 2^k (n+2k)!(x-1)^k t^{n+k}}{k!(2k)!n!}$$

$$= \sum_{n=0}^{\infty}\sum_{k=0}^{\infty} \frac{(n+2k)!(x-1)^k t^{n+k}}{k!2^k k! n!}$$

$$= \sum_{n=0}^{\infty}\sum_{k=0}^{n} \frac{(n+k)!(x-1)^k t^n}{2^k (k!)^2 (n-k)!}$$

$$= \sum_{n=0}^{\infty}\sum_{k=0}^{n} \frac{(-1)^k(-n)_k(n+1)_k(x-1)^k t^n}{2^k (k!)^2}.$$

Therefore

$$(2) \qquad P_n(x) = {}_2F_1\left[\begin{matrix} -n, n+1; \\ 1; \end{matrix} \quad \frac{1-x}{2}\right].$$

Since $P_n(-x) = (-1)^n P_n(x)$, it follows from (2) that also

$$(3) \qquad P_n(x) = (-1)^n \, {}_2F_1\left[\begin{matrix} -n, n+1; \\ 1; \end{matrix} \quad \frac{1+x}{2}\right].$$

Various formulas of Chapter 4 may now be applied to equation (2) to obtain other expressions for $P_n(x)$. It is interesting also to convert into hypergeometric form the results already derived in this chapter.

Equation (2) of Section 86, page 157, is

$$P_n(x) = \sum_{k=0}^{[n/2]} \frac{(-1)^k(\frac{1}{2})_{n-k}(2x)^{n-2k}}{k!(n-2k)!}.$$

Hence we may write

$$P_n(x) = \sum_{k=0}^{[n/2]} \frac{(\frac{1}{2})_n(-n)_{2k}(2x)^{n-2k}}{k!(\frac{1}{2}-n)_k n!}$$

$$= \frac{2^n(\frac{1}{2})_n x^n}{n!} \sum_{k=0}^{[n/2]} \frac{(-\frac{1}{2}n)_k(-\frac{1}{2}n+\frac{1}{2})_k x^{-2k}}{k!(\frac{1}{2}-n)_k},$$

or

$$(4) \qquad P_n(x) = \frac{(\frac{1}{2})_n(2x)^n}{n!} \, {}_2F_1\left[\begin{matrix} -\frac{1}{2}n, -\frac{1}{2}n+\frac{1}{2}; \\ \frac{1}{2}-n; \end{matrix} \quad \frac{1}{x^2}\right].$$

Equation (1) of Section 91, page 162, is

$$P_n(x) = \sum_{k=0}^{n} \frac{(n!)^2[\frac{1}{2}(x-1)]^{n-k}[\frac{1}{2}(x+1)]^k}{(k!)^2[(n-k)!]^2},$$

from which it follows that

$$P_n(x) = \sum_{k=0}^{n} \frac{(-n)_k(-n)_k[\frac{1}{2}(x-1)]^{n-k}[\frac{1}{2}(x+1)]^k}{(k!)^2}.$$

Therefore

$$(5) \qquad P_n(x) = \left(\frac{x-1}{2}\right)^n {}_2F_1\left[\begin{matrix}-n, -n; \\ 1;\end{matrix} \quad \frac{x+1}{x-1}\right],$$

or, by reversing the order of summation,

$$(6) \qquad P_n(x) = \left(\frac{x+1}{2}\right)^n {}_2F_1\left[\begin{matrix}-n, -n; \\ 1;\end{matrix} \quad \frac{x-1}{x+1}\right].$$

In Section 92, equation (1), page 164, is

$$P_n(x) = \sum_{k=0}^{[n/2]} \frac{n!(x^2-1)^k x^{n-2k}}{2^{2k}(k!)^2(n-2k)!}$$

from which

$$P_n(x) = \sum_{k=0}^{[n/2]} \frac{(-n)_{2k}(x^2-1)^k x^{n-2k}}{2^{2k}(k!)^2},$$

or

$$(7) \qquad P_n(x) = x^n \, {}_2F_1\left[\begin{matrix}-\frac{1}{2}n, -\frac{1}{2}n+\frac{1}{2}; \\ 1;\end{matrix} \quad \frac{x^2-1}{x^2}\right].$$

See also Exs. 14 and 15 at the end of this chapter.

94. Brafman's generating functions. Brafman [1] obtained a new* class of generating functions for Legendre polynomials as an incidental result of his work on Jacobi polynomials. In Chapter 16 we shall prove a theorem which contains as a special case the following result:

Let $\rho = (1 - 2xt + t^2)^{\frac{1}{2}}$ denote that branch for which $\rho \to 1$ as $t \to 0$. For arbitrary c,

---

*For negative integral c the generating relation (1) had been known for a long time.

$$(1)\qquad {}_2F_1\left[\begin{matrix} c, 1-c; & \dfrac{1-t-\rho}{2} \\ 1; & \end{matrix}\right] {}_2F_1\left[\begin{matrix} c, 1-c; & \dfrac{1+t-\rho}{2} \\ 1; & \end{matrix}\right] = \sum_{n=0}^{\infty} \frac{(c)_n(1-c)_n P_n(x) t^n}{(n!)^2}.$$

For proof of (1), put $\alpha = \beta = 0$ in the derivation of equation (2), page 272.

95. Special properties of $P_n(x)$. We have already shown that

$$(1)\qquad e^{xt} J_0(t\sqrt{1-x^2}) = \sum_{n=0}^{\infty} \frac{P_n(x) t^n}{n!}.$$

In (1) first put $x = \cos\alpha$, $t = v \sin\beta$; and second put $x = \cos\beta$, $t = v \sin\alpha$ to obtain the two relations

$$(2)\qquad \exp(v\cos\alpha\sin\beta) J_0(v\sin\beta\sin\alpha) = \sum_{n=0}^{\infty} \frac{P_n(\cos\alpha) v^n \sin^n\beta}{n!},$$

$$(3)\qquad \exp(v\cos\beta\sin\alpha) J_0(v\sin\alpha\sin\beta) = \sum_{n=0}^{\infty} \frac{P_n(\cos\beta) v^n \sin^n\alpha}{n!}.$$

Since $\sin(\beta - \alpha) = \sin\beta\cos\alpha - \cos\beta\sin\alpha$,

$$(4)\qquad \exp(v\cos\alpha\sin\beta) = \exp[v\sin(\beta-\alpha)] \exp(v\cos\beta\sin\alpha).$$

Now combine (2), (3), and (4) to arrive at the identity

$$\sum_{n=0}^{\infty} \frac{P_n(\cos\alpha) v^n \sin^n\beta}{n!} = \exp[v\sin(\beta-\alpha)] \sum_{n=0}^{\infty} \frac{P_n(\cos\beta) v^n \sin^n\alpha}{n!}$$

$$= \sum_{n=0}^{\infty}\sum_{k=0}^{n} \frac{\sin^{n-k}(\beta-\alpha)\sin^k\alpha\, P_k(\cos\beta) v^n}{k!(n-k)!},$$

from which it follows that

$$\sin^n\beta\, P_n(\cos\alpha) = \sum_{k=0}^{n} C_{n,k} \sin^{n-k}(\beta-\alpha)\sin^k\alpha\, P_k(\cos\beta),$$

in which $C_{n,k}$ is the binominal coefficient. This last equation can be written in the form (Rainville [5])

$$(5)\qquad P_n(\cos\alpha) = \left(\frac{\sin\alpha}{\sin\beta}\right)^n \sum_{k=0}^{n} C_{n,k} \left[\frac{\sin(\beta-\alpha)}{\sin\alpha}\right]^{n-k} P_k(\cos\beta).$$

Equation (5) relates $P_n(\cos\alpha)$ to a sum involving $P_k(\cos\beta)$ with α and β arbitrary. We make use of (5) later.

Let us return to the original definition of $P_n(x)$ and for convenience use $\rho = (1 - 2xt + t^2)^{\frac{1}{2}}$. We know that

$$(6) \qquad \sum_{n=0}^{\infty} P_n(x)t^n = \rho^{-1}.$$

In (6) replace x by $(x - t)/\rho$ and t by v/ρ to get

$$\sum_{n=0}^{\infty} P_n\left(\frac{x-t}{\rho}\right)\rho^{-n}v^n = \left[1 - \frac{2(x-t)v}{\rho^2} + \frac{v^2}{\rho^2}\right]^{-\frac{1}{2}}$$

$$= \rho[\rho^2 - 2(x-t)v + v^2]^{-\frac{1}{2}}.$$

We may now write

$$\sum_{n=0}^{\infty} P_n\left(\frac{x-t}{\rho}\right)\rho^{-n-1}v^n = [1 - 2xt + t^2 - 2xv + 2vt + v^2]^{-\frac{1}{2}}$$

$$= [1 - 2x(t+v) + (t+v)^2]^{-\frac{1}{2}},$$

which by (6) yields

$$\sum_{n=0}^{\infty} P_n\left(\frac{x-t}{\rho}\right)\rho^{-n-1}v^n = \sum_{n=0}^{\infty} P_n(x)(t+v)^n$$

$$= \sum_{n=0}^{\infty}\sum_{k=0}^{n} \frac{n!P_n(x)t^kv^{n-k}}{k!(n-k)!}$$

$$= \sum_{n,k=0}^{\infty} \frac{(n+k)!P_{n+k}(x)t^kv^n}{k!n!}.$$

Equating coefficients of v^n in the above, we find that

$$(7) \qquad \rho^{-n-1}P_n\left(\frac{x-t}{\rho}\right) = \sum_{k=0}^{\infty} \frac{(n+k)!P_{n+k}(x)t^k}{k!n!},$$

in which $\rho = (1 - 2xt + t^2)^{\frac{1}{2}}$. Equation (7) can be used to transform identities involving Legendre polynomials and sometimes leads in that way to additional results. See Bedient [1].

96. More generating functions. As an example of the use of equation (7) of the preceding section, we shall apply (7) to the generating relation

$$(1) \qquad e^{xt}\,{}_0F_1(-;1;\tfrac{1}{4}t^2(x^2-1)) = \sum_{n=0}^{\infty} \frac{P_n(x)t^n}{n!},$$

obtained in Section 92. In (1) replace x by $(x - t)/\rho$, t by $-ty/\rho$, and multiply each member by ρ^{-1}, where $\rho = (1 - 2xt + t^2)^{\frac{1}{2}}$, to obtain

$$\rho^{-1}\exp\left[\frac{-ty(x-t)}{\rho^2}\right] {}_0F_1\left[\begin{matrix}-; \\ 1;\end{matrix}\quad \frac{y^2t^2(x^2-1)}{4\rho^4}\right]$$

$$= \sum_{n=0}^{\infty} \frac{(-1)^n \rho^{-n-1} P_n\left(\dfrac{x-t}{\rho}\right) t^n y^n}{n!}$$

$$= \sum_{n=0}^{\infty}\sum_{k=0}^{\infty} \frac{(-1)^n (n+k)! P_{n+k}(x) t^{n+k} y^n}{k!(n!)^2},$$

in which we have used equation (7) of Section 95.

Collect powers of t on the right in the last summation to see that

$$\rho^{-1}\exp\left[\frac{-ty(x-t)}{\rho^2}\right] {}_0F_1\left[\begin{matrix}-; \\ 1;\end{matrix}\quad \frac{y^2t^2(x^2-1)}{4\rho^4}\right]$$

$$= \sum_{n=0}^{\infty}\sum_{k=0}^{n} \frac{(-1)^{n-k} n! y^{n-k} P_n(x) t^n}{k![(n-k)!]^2}$$

$$= \sum_{n=0}^{\infty}\sum_{k=0}^{n} \frac{(-1)^k n! y^k P_n(x) t^n}{(k!)^2 (n-k)!}$$

$$= \sum_{n=0}^{\infty} {}_1F_1(-n; 1; y) P_n(x) t^n.$$

In Chapter 12 we shall encounter the simple Laguerre polynomials

$$(2) \qquad L_n(x) = {}_1F_1(-n; 1; x).$$

Using the notation in (2) we may now write the generating relation*

$$(3) \quad (1 - 2xt + t^2)^{-\frac{1}{2}} \exp\left[\frac{ty(t-x)}{1 - 2xt + t^2}\right] \cdot$$

$${}_0F_1\left[\begin{matrix}-; \\ 1;\end{matrix}\quad \frac{y^2t^2(x^2-1)}{4(1-2xt+t^2)^2}\right] = \sum_{n=0}^{\infty} L_n(y) P_n(x) t^n,$$

which we shall call a bilateral generating function. The relation (3) may be used to generate either $L_n(y)$ or $P_n(x)$.

---

*Equation (3) was first obtained by Weisner [1] by a method different from that used here. We use the method introduced by Bedient [1].

In terms of a Bessel function, equation (3) may be written compactly as

$$(4)\qquad \rho^{-1}\exp[ty(t-x)\rho^{-2}]J_0(ty\sqrt{1-x^2}\rho^{-2}) = \sum_{n=0}^{\infty} L_n(y)P_n(x)t^n$$

in which $\rho = (1-2xt+t^2)^{\frac{1}{2}}$ and in which x, y, and t are independent of each other. The procedure used to get (3) or (4) can be used to obtain further generating functions as indicated in the exercises at the end of this chapter.

97. Laplace's first integral form. In Section 92 we obtained the expansion

$$(1)\qquad P_n(x) = \sum_{k=0}^{[n/2]} \frac{n!x^{n-2k}(x^2-1)^k}{2^{2k}(k!)^2(n-2k)!},$$

which may be written as

$$(2)\qquad P_n(x) = \sum_{k=0}^{[n/2]} \frac{n!(\frac{1}{2})_k x^{n-2k}(x^2-1)^k}{k!(2k)!(n-2k)!}.$$

Now

$$\frac{(\frac{1}{2})_k}{k!} = \frac{\Gamma(\frac{1}{2}+k)}{\Gamma(\frac{1}{2})\Gamma(1+k)} = \frac{\Gamma(\frac{1}{2})\Gamma(\frac{1}{2}+k)}{\pi\Gamma(k+1)} = \frac{1}{\pi}B(\tfrac{1}{2}, \tfrac{1}{2}+k)$$

$$= \frac{2}{\pi}\int_0^{\frac{1}{2}\pi}\cos^{2k}\varphi\,d\varphi = \frac{1}{\pi}\int_0^{\pi}\cos^{2k}\varphi\,d\varphi.$$

Hence (2) leads us to

$$(3)\qquad P_n(x) = \frac{1}{\pi}\sum_{k=0}^{[n/2]} \frac{n!x^{n-2k}(x^2-1)^k}{(2k)!(n-2k)!}\int_0^{\pi}\cos^{2k}\varphi\,d\varphi.$$

Since $\int_0^{\pi}\cos^m\varphi\,d\varphi = 0$ for odd m, we may replace $2k$ by k in the summation on the right in equation (3). Thus equation (3) leads to the relation

$$(4)\qquad P_n(x) = \frac{1}{\pi}\sum_{k=0}^{n} \frac{n!x^{n-k}(x^2-1)^{\frac{1}{2}k}}{k!(n-k)!}\int_0^{\pi}\cos^k\varphi\,d\varphi,$$

in which each term involving an odd k is zero. Equation (4) in turn gives us Laplace's first integral for $P_n(x)$,

$$(5)\qquad P_n(x) = \frac{1}{\pi}\int_0^{\pi}[x+(x^2-1)^{\frac{1}{2}}\cos\varphi]^n\,d\varphi.$$

98. Some bounds on $P_n(x)$. Equation (5) of the preceding section yields certain simple properties of $P_n(x)$. In the range $-1<x<1$,

$$|x + (x^2 - 1)^{\frac{1}{2}} \cos \varphi| = \sqrt{x^2 + (1 - x^2)\cos^2 \varphi}$$
$$= \sqrt{x^2 \sin^2 \varphi + \cos^2 \varphi},$$

so that

$$(1) \qquad |x + (x^2 - 1)^{\frac{1}{2}}\cos \varphi| = \sqrt{1 - (1 - x^2)\sin^2 \varphi}, \qquad -1 < x < 1.$$

From (1) it follows that, except at $\varphi = 0$ and $\varphi = \pi$,

$$|x + (x^2 - 1)^{\frac{1}{2}} \cos \varphi| < 1,$$

which leads to

$$|P_n(x)| \leqq \frac{1}{\pi}\int_0^\pi |x + (x^2 - 1)^{\frac{1}{2}}\cos \varphi|^n \, d\varphi < \frac{1}{\pi}\int_0^\pi 1 \cdot d\varphi.$$

THEOREM 60. *For* $-1 < x < 1$, $|P_n(x)| < 1$.

The integral form

$$(2) \qquad P_n(x) = \frac{1}{\pi}\int_0^\pi [x + (x^2 - 1)^{\frac{1}{2}}\cos \varphi]^n \, d\varphi$$

and (1) combine to yield, for $-1 < x < 1$,

$$|P_n(x)| \leqq \frac{1}{\pi}\int_0^\pi [1 - (1 - x^2) \sin^2 \varphi]^{\frac{1}{2}n} \, d\varphi,$$

or

$$|P_n(x)| \leqq \frac{2}{\pi}\int_0^{\frac{1}{2}\pi} [1 - (1 - x^2) \sin^2 \varphi]^{\frac{1}{2}n} \, d\varphi.$$

For $0 < \varphi < \frac{1}{2}\pi$, $\sin \varphi > 2\varphi/\pi$. Hence

$$1 - (1 - x^2) \sin^2 \varphi < 1 - \frac{4\varphi^2(1 - x^2)}{\pi^2} < \exp\left[- \frac{4\varphi^2(1 - x^2)}{\pi^2}\right],$$

in which we use the fact that $1 - y < \exp(-y)$ for $y > 0$. We may now write

$$|P_n(x)| < \frac{2}{\pi}\int_0^{\frac{1}{2}\pi} \exp\left[- \frac{2n\varphi^2(1 - x^2)}{\pi^2}\right] d\varphi$$
$$< \frac{2}{\pi}\int_0^\infty \exp\left[\frac{-2n\varphi^2(1 - x^2)}{\pi^2}\right] d\varphi,$$

then introduce a new variable of integration $\beta = (\varphi/\pi)[2n(1 - x^2)]^{\frac{1}{2}}$ to obtain

$$|P_n(x)| < \frac{2}{\pi} \cdot \frac{\pi}{[2n(1 - x^2)]^{\frac{1}{2}}} \int_0^\infty \exp(-\beta^2)\, d\beta = \frac{\sqrt{\pi}}{\sqrt{2n(1 - x^2)}}.$$

THEOREM 61. *If $-1 < x < 1$ and if n is any positive integer*

$$|P_n(x)| < \left[\frac{\pi}{2n(1 - x^2)}\right]^{\frac{1}{2}}. \tag{3}$$

99. Orthogonality. We know that the Legendre polynomial $P_n(x)$ satisfies the differential equation

$$(1 - x^2)P_n''(x) - 2xP_n'(x) + n(n + 1)P_n(x) = 0. \tag{1}$$

Equation (1) can equally well be written

$$[(1 - x^2)P_n'(x)]' + n(n + 1)P_n(x) = 0. \tag{2}$$

We are now interested in obtaining integrals involving the product $P_n(x)P_m(x)$ of two Legendre polynomials. Hence we combine (2) with

$$[(1 - x^2)P_m'(x)]' + m(m + 1)P_m(x) = 0 \tag{3}$$

to get

$$\begin{aligned} P_m(x)[(1 - x^2)P_n'(x)]' - P_n(x)[(1 - x^2)P_m'(x)]' \\ + [n(n + 1) - m(m + 1)]\, P_n(x)P_m(x) = 0. \end{aligned} \tag{4}$$

But

$$\begin{aligned} &[(1 - x^2)\,\{P_m(x)P_n'(x) - P_m'(x)P_n(x)\}]' \\ &\qquad = (1 - x^2)P_m'(x)P_n'(x) + P_m(x)\,[(1 - x^2)P_n'(x)]' \\ &\qquad\quad - (1 - x^2)P_m'(x)P_n'(x) - P_n(x)\,[(1 - x^2)P_m'(x)]', \end{aligned}$$

so that (4) becomes

$$\begin{aligned} &[(1 - x^2)\{P_m(x)P_n'(x) - P_m'(x)P_n(x)\}]' \\ &\qquad + (n^2 - m^2 + n - m)P_n(x)P_m(x) = 0, \end{aligned}$$

or

$$\begin{aligned} &(n - m)(n + m + 1)P_n(x)P_m(x) \\ &\qquad = [(1 - x^2)\{P_m'(x)P_n(x) - P_m(x)P_n'(x)\}]'. \end{aligned} \tag{5}$$

From (5) it follows that for any finite limits of integration

$$(6)\quad (n - m)(n + m + 1)\int_a^b P_n(x)P_m(x)\,dx = \Big[(1 - x^2)\{P_m'(x)P_n(x) - P_m(x)P_n'(x)\}\Big]_a^b.$$

Since $(1 - x^2)$ vanishes at $x = 1$ and $x = -1$, we conclude that

$$(n - m)(n + m + 1)\int_{-1}^1 P_n(x)P_m(x)\,dx = 0.$$

Now m and n are non-negative integers, so $n + m + 1 \neq 0$. If also $n \neq m$, $n - m \neq 0$, and we may conclude that

$$(7)\qquad \int_{-1}^1 P_n(x)P_m(x)\,dx = 0, \qquad m \neq n.$$

In the terminology of Chapter 9, equation (7) means that the set of polynomials $P_n(x)$ is orthogonal with respect to the weight function unity on the interval $-1 < x < 1$. The Legendre polynomials therefore possess the properties held by all orthogonal polynomials.

THEOREM 62. *The zeros of $P_n(x)$ are distinct, and all lie in the open interval $-1 < x < 1$.*

THEOREM 63. *For $k = 0, 1, 2, \cdots, (n - 1)$,*

$$(8)\qquad \int_{-1}^1 x^k P_n(x)\,dx = 0.$$

We already know (Section 88) the three-term recurrence relation

$$(9)\qquad nP_n(x) = (2n - 1)xP_{n-1}(x) - (n - 1)P_{n-2}(x),$$

whose existence follows from the orthogonality property.

Later we shall need the value of $\int_{-1}^1 P_n^2(x)\,dx.$

From

$$(1 - 2xt + t^2)^{-1} = \left[\sum_{m=0}^\infty P_m(x)t^m\right]^2$$

we obtain

$$(10)\qquad \int_{-1}^1 \frac{dx}{1 - 2xt + t^2} = \sum_{m=0}^\infty \sum_{k=0}^m \int_{-1}^1 P_k(x)P_{m-k}(x)\,dx\, t^m.$$

In (10) the integral on the left is elementary; on the right each integral vanishes except when $k = m - k$. Therefore the only nonzero terms in the series are those for which m is even, $m = 2n$, and $k = n$. Hence

$$(11)\qquad \left[-\frac{1}{2t}\,\mathrm{Log}(1 - 2xt + t^2)\right]_{-1}^{1} = \sum_{n=0}^{\infty}\int_{-1}^{1} P_n^2(x)\,dx\,t^{2n}.$$

But the left member of (11) is

$$\begin{aligned}-\frac{1}{2t}\,\mathrm{Log}\frac{(1-t)^2}{(1+t)^2} &= t^{-1}[\mathrm{Log}(1+t) - \mathrm{Log}(1-t)]\\ &= \sum_{m=0}^{\infty}\frac{(-1)^m t^m}{m+1} + \sum_{m=0}^{\infty}\frac{t^m}{m+1}\\ &= \sum_{n=0}^{\infty}\frac{2t^{2n}}{2n+1},\end{aligned}$$

since each term for odd m drops out. We thus obtain from (11) the desired result:

$$(12)\qquad \int_{-1}^{1} P_n^2(x)\,dx = \frac{2}{2n+1}.$$

Since, by equation (3), page 158,

$$x^n = \frac{n!}{2^n(\frac{1}{2})_n}P_n(x) + \pi_{n-2}$$

the result (12) is easily converted into the form

$$(13)\qquad \int_{-1}^{1} x^n P_n(x)\,dx = \frac{n!}{2^n(\frac{1}{2})_{n+1}}.$$

As an application of (6) let us evaluate

$$(14)\qquad \int_{0}^{1} P_n(x)P_m(x)\,dx.$$

By (6), page 174,

$$\begin{aligned}(15)\qquad (n-m)(n+m+1)&\int_0^1 P_n(x)P_m(x)\,dx\\ &= P_m(0)P_n'(0) - P_m'(0)P_n(0).\end{aligned}$$

If m and n are both odd or both even, the right member of (15) is zero by (7) and (9) of Section 86, page 158, and the integral (14) vanishes unless $m = n$. If $m = n$, it follows from (12) and the fact that $P_n^2(x)$ is an even function of x, that

$$\int_0^1 P_n^2(x)\,dx = \frac{1}{2n+1}.$$

We are left with the need to evaluate (14) when one of m and n is odd and the other even. Let $n = 2k$, $m = 2s + 1$. Then by (15),

$$(16)\quad (2k - 2s - 1)(2k + 2s + 2)\int_0^1 P_{2k}(x)P_{2s+1}(x)\,dx$$
$$= P_{2s+1}(0)P'_{2k}(0) - P'_{2s+1}(0)P_{2k}(0).$$

In Section 86, equations (7) and (9), page 158, we found that

$$P_{2s+1}(0) = 0, \qquad P_{2k}(0) = \frac{(-1)^k(\tfrac{1}{2})_k}{k!}, \qquad P'_{2s+1}(0) = \frac{(-1)^s(\tfrac{3}{2})_s}{s!}.$$

Therefore it follows from (16) that

$$(17)\quad \int_0^1 P_{2k}(x)P_{2s+1}(x)\,dx = \frac{(-1)^{k+s}(\tfrac{1}{2})_k(\tfrac{1}{2})_{s+1}}{(k+s+1)(2s+1-2k)k!s!}.$$

100. An expansion theorem. We now seek an expansion of the form

$$(1)\qquad f(x) = \sum_{n=0}^{\infty} a_n P_n(x), \qquad -1 < x < 1.$$

From (1) we obtain a_n in a purely formal manner. With that value for a_n we then proceed to prove that the series on the right in (1) actually converges to $f(x)$, providing $f(x)$ is sufficiently well behaved.

From (1) it follows formally that

$$(2)\qquad \int_{-1}^1 f(x)P_m(x)\,dx = \sum_{n=0}^{\infty} a_n \int_{-1}^1 P_m(x)P_n(x)\,dx.$$

All the integrals on the right in (2) vanish except for the single term for which $n = m$. Therefore

$$(3)\qquad \int_{-1}^1 f(x)P_m(x)\,dx = a_m\int_{-1}^1 P_m^2(x)\,dx = \frac{2a_m}{2m+1}.$$

In (3) replace m by n and x by y to obtain

$$(4)\qquad a_n = (n + \tfrac{1}{2})\int_{-1}^1 f(y)P_n(y)\,dy.$$

Now that we know what a_n to use, we proceed to prove the desired result.*

THEOREM 64. *If on* $-1 \leqq x \leqq 1$ $f(x)$ *is continuous except for a finite number of finite discontinuities, if on* $-1 \leqq x \leqq 1$ $f'(x)$ *exists where* $f(x)$ *is continuous and the right hand and left hand derivatives of* $f(x)$ *exist at the discontinuities, and if*

$$a_n = (n + \tfrac{1}{2}) \int_{-1}^{1} f(y) P_n(y)\, dy, \tag{4}$$

then

$$\sum_{n=0}^{\infty} a_n P_n(x) = f(x), \qquad -1 < x < 1, \tag{5}$$

at the points of continuity of $f(x)$.

The series on the left in (5) converges to the mean value $\frac{1}{2}[f(x + 0) + f(x - 0)]$ at the points of discontinuity of $f(x)$, but we omit proof of that portion of the theorem.

Proof: Consider the left member of (5) with the a_n from (4):

$$\begin{aligned} \sum_{n=0}^{\infty} a_n P_n(x) &= \sum_{n=0}^{\infty} (n + \tfrac{1}{2}) \int_{-1}^{1} f(y) P_n(y) P_n(x)\, dy \\ &= \operatorname*{Lim}_{n \to \infty} \sum_{k=0}^{n} (k + \tfrac{1}{2}) \int_{-1}^{1} f(y) P_k(y) P_k(x)\, dy \\ &= \operatorname*{Lim}_{n \to \infty} \int_{-1}^{1} f(y) \sum_{k=0}^{n} (k + \tfrac{1}{2}) P_k(x) P_k(y)\, dy. \end{aligned}$$

The series on the left in (5) will converge to the sum $f(x)$ if and only if

$$\operatorname*{Lim}_{n \to \infty} \int_{-1}^{1} f(y) K_n(x, y)\, dy = f(x), \tag{6}$$

in which

$$K_n(x, y) = \sum_{k=0}^{n} (k + \tfrac{1}{2}) P_k(x) P_k(y). \tag{7}$$

To sum the series on the right in (7), we turn to the Christoffel-Darboux formula of Theorem 58, page 154,

$$\sum_{k=0}^{n} g_k^{-1} \varphi_k(x) \varphi_k(y) = \frac{h_n}{g_n h_{n+1}} \cdot \frac{\varphi_{n+1}(y)\varphi_n(x) - \varphi_{n+1}(x)\varphi_n(y)}{y - x}. \tag{8}$$

*Theorem 64 is a most elementary form of an expansion theorem. The aim here is to exhibit the underlying ideas with as little complication as possible.

For $\varphi_k(x) = P_k(x)$,

$$g_k = \int_{-1}^{1} P_k^2(x)\, dx = \frac{2}{2k+1} = \frac{1}{k+\frac{1}{2}},$$

and the leading coefficient in $P_n(x)$ is $h_n = 2^n(\frac{1}{2})_n/n!$. Hence $h_n/(g_n h_{n+1}) = \frac{1}{2}(n+1)$, and we conclude from (8) that

$$(9) \qquad K_n(x, y) = \frac{n+1}{2} \cdot \frac{P_{n+1}(x)P_n(y) - P_n(x)P_{n+1}(y)}{x - y}.$$

The condition (6) may be rewritten as

$$(10) \qquad \lim_{n\to\infty} \left[f(x) - \int_{-1}^{1} f(y)K_n(x, y)\, dy \right] = 0.$$

Since

$$\int_{-1}^{1} P_k(y)\, dy = 0 \qquad \text{for } k > 0,$$

we may write

$$\int_{-1}^{1} K_n(x, y)\, dy = \int_{-1}^{1} \sum_{k=0}^{n} (k + \tfrac{1}{2}) P_k(x) P_k(y)\, dy$$

$$= \int_{-1}^{1} \left[\tfrac{1}{2} + \sum_{k=1}^{n} (k + \tfrac{1}{2}) P_k(x) P_k(y) \right] dy$$

$$= \int_{-1}^{1} \tfrac{1}{2} dy = 1.$$

Now the condition (10) may be put in the form

$$\lim_{n\to\infty} \int_{-1}^{1} [f(x) - f(y)] K_n(x, y)\, dy = 0,$$

or because of (9),

$$(11) \quad \lim_{n\to\infty} \frac{n+1}{2} \int_{-1}^{1} \frac{f(x)-f(y)}{x-y} [P_{n+1}(x)P_n(y) - P_n(x)P_{n+1}(y)]\, dy = 0.$$

By Theorem 59, page 156, it follows that

$$(12) \qquad \lim_{n\to\infty} (n + \tfrac{1}{2})^{\frac{1}{2}} \int_{-1}^{1} g(y) P_n(y)\, dy = 0$$

for any $g(y)$ such that $\int_{-1}^{1} g^2(y)\, dy$ exists.

At a point of continuity of $f(y)$ we also assumed that $f'(y)$ exists. Hence, with

$$g(y) = \frac{f(x) - f(y)}{x - y},$$

$\int_{-1}^{1} g^2(y)\, dy$ does exists. It follows from (12) that both

$$\text{(13)} \qquad \lim_{n \to \infty} (n + \tfrac{1}{2})^{\frac{1}{2}} \int_{-1}^{1} \frac{f(x) - f(y)}{x - y} P_n(y)\, dy = 0,$$

and

$$\text{(14)} \qquad \lim_{n \to \infty} \left(n + \frac{3}{2}\right)^{\frac{1}{2}} \int_{-1}^{1} \frac{f(x) - f(y)}{x - y} P_{n+1}(y)\, dy = 0.$$

Then (11) will be satisfied at a point of continuity of $f(x)$ in $-1 < x < 1$ if we can show that each of

$$\frac{n+1}{2}\left(n + \frac{1}{2}\right)^{-\frac{1}{2}} P_{n+1}(x), \qquad \frac{n+1}{2}\left(n + \frac{3}{2}\right)^{-\frac{1}{2}} P_n(x)$$

is bounded as $n \to \infty$.

We use Theorem 61, page 173, to see that, on $-1 < x < 1$,

$$\left|\frac{n+1}{2}\left(n + \frac{1}{2}\right)^{-\frac{1}{2}} P_{n+1}(x)\right| < \left[\frac{(n+1)^2}{4(n + \frac{1}{2})} \cdot \frac{\pi}{2(n+1)(1 - x^2)}\right]^{\frac{1}{2}}$$

$$< \left[\frac{\pi(n+1)}{4(2n+1)(1 - x^2)}\right]^{\frac{1}{2}} < \left[\frac{\pi}{4(1 - x^2)}\right]^{\frac{1}{2}}$$

and

$$\left|\frac{n+1}{2}\left(n + \frac{3}{2}\right)^{-\frac{1}{2}} P_n(x)\right| < \left[\frac{(n+1)^2}{4(n + \frac{3}{2})} \cdot \frac{\pi}{2n(1 - x^2)}\right]^{\frac{1}{2}}$$

$$< \left[\frac{\pi(n^2 + 2n + 1)}{4(2n^2 + 3n)(1 - x^2)}\right]^{\frac{1}{2}} < \left[\frac{\pi}{4(1 - x^2)}\right]^{\frac{1}{2}}.$$

We have thus shown the validity of equation (5) of Theorem 64 at points of continuity of $f(x)$.

101. Expansion of x^n. We already know from Theorem 53, page 147, that any polynomial can be expanded in a series of Legendre polynomials merely because the $P_n(x)$ form a simple set. The orthogonality of the set $P_n(x)$ plays a role only in the determination of coefficients; Theorem 64 of Section 100 has no bearing on expansion of polynomials.

The expansion of x^n in a series of Legendre polynomials is useful. Consider

$$\sum_{n=0}^{\infty} P_n(x)t^n = (1 - 2xt + t^2)^{-\frac{1}{2}} = (1 + t^2)^{-\frac{1}{2}}\left[1 - \frac{2xt}{1 + t^2}\right]^{-\frac{1}{2}}$$

$$= \sum_{n=0}^{\infty} \frac{(\frac{1}{2})_n (2x)^n t^n}{n!(1 + t^2)^{n+\frac{1}{2}}},$$

from which we get

$$\sum_{n=0}^{\infty} \frac{(\frac{1}{2})_n (2x)^n t^n}{n!(1 + t^2)^n} = (1 + t^2)^{\frac{1}{2}} \sum_{n=0}^{\infty} P_n(x)t^n. \tag{1}$$

Now put

$$t = \frac{2v}{1 + \sqrt{1 - 4v^2}}$$

from which

$$1 + t^2 = \frac{2}{1 + \sqrt{1 - 4v^2}}, \qquad \frac{t}{1 + t^2} = v.$$

Equation (1) becomes

$$\sum_{n=0}^{\infty} \frac{(\frac{1}{2})_n (2x)^n v^n}{n!} = \sum_{n=0}^{\infty} P_n(x) v^n \left(\frac{2}{1 + \sqrt{1 - 4v^2}}\right)^{n+\frac{1}{2}}. \tag{2}$$

By Ex. 10, page 70,

$$\left(\frac{2}{1 + \sqrt{1 - 4v^2}}\right)^{n+\frac{1}{2}} = {}_2F_1\left[\begin{matrix} \frac{1}{2}(n + \frac{1}{2}), \frac{1}{2}(n + \frac{3}{2}); \\ n + \frac{3}{2}; \end{matrix} \; 4v^2\right]$$

$$= \sum_{k=0}^{\infty} \frac{(n + \frac{1}{2})_{2k} v^{2k}}{k!(n + \frac{3}{2})_k}$$

$$= \sum_{k=0}^{\infty} \frac{(\frac{1}{2})_{n+2k} (\frac{3}{2})_n v^{2k}}{(\frac{1}{2})_n k! (\frac{3}{2})_{n+k}}$$

or

$$\left(\frac{2}{1 + \sqrt{1 - 4v^2}}\right)^{n+\frac{1}{2}} = \sum_{k=0}^{\infty} \frac{(2n + 1)(\frac{1}{2})_{n+2k} v^{2k}}{k!(\frac{3}{2})_{n+k}}. \tag{3}$$

We may now conclude from (2) and (3) that

$$\sum_{n=0}^{\infty} \frac{(\frac{1}{2})_n (2x)^n v^n}{n!} = \sum_{n,k=0}^{\infty} \frac{(2n+1)(\frac{1}{2})_{n+2k} P_n(x) v^{n+2k}}{k!(\frac{3}{2})_{n+k}}$$

$$= \sum_{n=0}^{\infty} \sum_{k=0}^{[n/2]} \frac{(2n-4k+1)(\frac{1}{2})_n P_{n-2k}(x) v^n}{k!(\frac{3}{2})_{n-k}}.$$

Comparison of coefficients of v^n in the preceding equation yields the following result.

THEOREM 65. *For non-negative integral n,*

$$x^n = \frac{n!}{2^n} \sum_{k=0}^{[n/2]} \frac{(2n-4k+1) P_{n-2k}(x)}{k!(\frac{3}{2})_{n-k}}. \tag{4}$$

In later chapters we shall use Theorem 65 to find expansions of various polynomials in series of Legendre polynomials. See also Ex. 17 at the end of this chapter.

102. Expansion of analytic functions. We use Theorem 65 of the preceding section to find explicit expressions for the coefficients in the expansion of analytic functions in series of Legendre polynomials. The theory of such expansions is treated in several places. See Whittaker and Watson [1; 321–323] and Szegö [1]. For a general discussion of expansion of analytic functions in series of polynomials, see Boas and Buck [2].

If we have

$$f(x) = \sum_{n=0}^{\infty} \frac{a_n x^n}{n!}, \tag{1}$$

application of Theorem 65 yields

$$f(x) = \sum_{n=0}^{\infty} \sum_{k=0}^{[n/2]} \frac{a_n (2n-4k+1) P_{n-2k}(x)}{2^n k!(\frac{3}{2})_{n-k}}$$

$$= \sum_{n=0}^{\infty} \sum_{k=0}^{\infty} \frac{a_{n+2k} (2n+1) P_n(x)}{2^{n+2k} k!(\frac{3}{2})_{n+k}},$$

which is the desired expansion.

THEOREM 66. *If* $|x|$ *is sufficiently small and if*

$$f(x) = \sum_{n=0}^{\infty} \frac{a_n x^n}{n!}, \tag{1}$$

then

$$f(x) = \sum_{n=0}^{\infty} b_n P_n(x), \tag{2}$$

in which

$$b_n = \sum_{k=0}^{\infty} \frac{(2n+1)a_{n+2k}}{2^{n+2k}k!(\tfrac{3}{2})_{n+k}}. \tag{3}$$

The region of convergence of (2) is the interior of an ellipse with center at $x = 0$.

Example: Expand $(t-x)^{-1}$ in a series of Legendre polynomials.

If $|x| < t$, then

$$(t-x)^{-1} = t^{-1}(1-xt^{-1})^{-1} = \sum_{n=0}^{\infty} \frac{x^n}{t^{n+1}}.$$

By Theorem 66,

$$(t-x)^{-1} = \sum_{n=0}^{\infty} (2n+1)Q_n(t)P_n(x),$$

in which

$$(2n+1)Q_n(t) = \sum_{k=0}^{\infty} \frac{(n+2k)!(2n+1)}{t^{n+2k+1}2^{n+2k}k!(\tfrac{3}{2})_{n+k}},$$

or

$$Q_n(t) = \frac{n!}{2^n t^{n+1}(\tfrac{3}{2})_n} \sum_{k=0}^{\infty} \frac{(1+n)_{2k}t^{-2k}}{2^{2k}k!(\tfrac{3}{2}+n)_k},$$

$$Q_n(t) = \frac{n!}{2^n t^{n+1}(\tfrac{3}{2})_n}\, {}_2F_1\left[\begin{matrix} \tfrac{1}{2}(1+n),\ \tfrac{1}{2}(2+n); \\ \tfrac{3}{2}+n; \end{matrix}\ \frac{1}{t^2}\right]. \tag{4}$$

It will be found that the $Q_n(t)$ of (4) is a second solution of the differential equation for $P_n(t)$. See Ex. 3 below.

EXERCISES

1. Start with the defining relation for $P_n(x)$ at the beginning of this chapter. Use the fact that

$$(1-2xt+t^2)^{-\frac{1}{2}} = [1-(x+\sqrt{x^2-1})t]^{-\frac{1}{2}}[1-(x-\sqrt{x^2-1})t]^{-\frac{1}{2}}$$

and thus derive the result

$$P_n(x) = \sum_{k=0}^{n} \frac{(\frac{1}{2})_k(\frac{1}{2})_{n-k}(x + \sqrt{x^2 - 1})^{n-k}(x - \sqrt{x^2 - 1})^k}{k!(n-k)!},$$

in which it is to be noted that $x - \sqrt{x^2 - 1} = (x + \sqrt{x^2 - 1})^{-1}$.

2. Use the result in Ex. 1 to show that

$$P_n(x) = \frac{(\frac{1}{2})_n(x + \sqrt{x^2 - 1})^n}{n!} {}_2F_1\left[\begin{matrix} -n, \frac{1}{2}; \\ \frac{1}{2} - n; \end{matrix} \quad (x - \sqrt{x^2 - 1})^2\right].$$

3. In Section 93, equation (4), page 166, is

$$P_n(x) = \frac{(\frac{1}{2})_n(2x)^n}{n!} {}_2F_1\left[\begin{matrix} -\frac{1}{2}n, -\frac{1}{2}n + \frac{1}{2}; \\ \frac{1}{2} - n; \end{matrix} \quad \frac{1}{x^2}\right].$$

We know from Section 34 that the ${}_2F_1$ equation has two linearly independent solutions:

$${}_2F_1(a, b; c; z)$$

and

$$z^{1-c}F(a + 1 - c, b + 1 - c; 2 - c; z).$$

Combine these facts to conclude that the differential equation

$$(1 - t^2)y'' - 2ty' + n(n + 1)y = 0$$

has the two linearly independent solutions $y_1 = P_n(t)$ and $y_2 = Q_n(t)$, where $Q_n(t)$ is as given in equation (4) of Section 102.

4. Show that

$$\sum_{n=0}^{\infty} [xP_n'(x) - nP_n(x)]t^n = t^2(1 - 2xt + t^2)^{-\frac{3}{2}}$$

and

$$\sum_{n=0}^{\infty} \sum_{k=0}^{[n/2]} (2n - 4k + 1)P_{n-2k}(x)t^n = (1 - 2xt + t^2)^{-\frac{3}{2}}.$$

Thus conclude that

$$xP_n'(x) - nP_n(x) = \sum_{k=0}^{[(n-2)/2]} (2n - 4k - 3)P_{n-2-2k}(x).$$

5. Use Bateman's generating function (3), page 163, with $x = 0$, $t = 2y$ to conclude that

$${}_0F_1(-; 1; y)\, {}_0F_1(-; 1; -y) = {}_0F_3(-; 1, 1, \tfrac{1}{2}; -\tfrac{1}{4}y^2).$$

6. Use Brafman's generating function, page 168, to conclude that

$${}_2F_1\left[\begin{matrix} c, 1 - c; \\ 1; \end{matrix} \quad \frac{1 - t - \sqrt{1 + t^2}}{2}\right] {}_2F_1\left[\begin{matrix} c, 1 - c; \\ 1; \end{matrix} \quad \frac{1 + t - \sqrt{1 + t^2}}{2}\right]$$

$$= {}_4F_3\left[\begin{matrix} \frac{1}{2}c, \frac{1}{2}c + \frac{1}{2}, \frac{1}{2} - \frac{1}{2}c, 1 - \frac{1}{2}c; \\ 1, 1, \frac{1}{2}; \end{matrix} \quad -t^2\right].$$

7. Use equation (5), page 168, to obtain the results

$$\sin^n \beta \, P_n(\sin \beta) = \sum_{k=0}^{n} (-1)^k C_{n,k} \cos^k \beta \, P_k(\cos \beta),$$

$$P_n(x) = \sum_{k=0}^{n} (-1)^k C_{n,k} (2x)^{n-k} P_k(x),$$

$$P_n(1 - 2x^2) = \sum_{k=0}^{n} (-2x)^k C_{n,k} P_k(x).$$

8. Use the technique of Section 96 to derive other (see also, Weisner [1]) generating function relations for $P_n(x)$. For instance, obtain the results

$$\sum_{n=0}^{\infty} {}_1F_2(-n; 1, 1; y) P_n(x) t^n$$

$$= \rho^{-1} \, {}_0F_1\left[\begin{matrix} -; \\ 1; \end{matrix} \quad \frac{-yt(x - t - \rho)}{2\rho^2}\right] {}_0F_1\left[\begin{matrix} -; \\ 1; \end{matrix} \quad \frac{-yt(x - t + \rho)}{2\rho^2}\right]$$

in which $\rho = (1 - 2xt + t^2)^{\frac{1}{2}}$, and

$$\sum_{n=0}^{\infty} {}_2F_1(-n, c; 1; y) P_n(x) t^n$$

$$= \rho^{2c-1}(\rho^2 + xyt - yt^2)^{-c} \, {}_2F_1\left[\begin{matrix} \tfrac{1}{2}c, \tfrac{1}{2}c + \tfrac{1}{2}; \\ 1; \end{matrix} \quad \frac{y^2t^2(x^2 - 1)}{(\rho^2 + xyt - yt^2)^2}\right].$$

Also sum the series

$$\sum_{n=0}^{\infty} {}_3F_2(-n, c, 1 - c; 1, 1; y) P_n(x) t^n.$$

9. With $\rho = (1 - 2xt + t^2)^{\frac{1}{2}}$, show that

$$\rho^n P_n\left(\frac{1 - xt}{\rho}\right) = \sum_{k=0}^{n} (-1)^k C_{n,k} t^k P_k(x).$$

10. With the aid of the result in Ex. 7, page 31, show that

$$\int_{-1}^{1} (1 + x)^{\alpha-1}(1 - x)^{\beta-1} P_n(x)\, dx = 2^{\alpha+\beta-1} B(\alpha,\beta) \, {}_3F_2\left[\begin{matrix} -n, n + 1, \beta; \\ 1, \alpha + \beta; \end{matrix} \quad 1\right].$$

Investigate the three special cases $\alpha = 1$, $\beta = 1$, $\alpha + \beta = n + 1$.

11. Obtain from equation (5), page 168, the result

$$(1 + x)^{\frac{1}{2}n} P_n\left(\sqrt{\frac{1 + x}{2}}\right) = 2^{-\frac{1}{2}n} \sum_{k=0}^{n} C_{n,k} P_k(x)$$

and use it to evaluate (Bhonsle [1]) the integral

$$\int_{-1}^{1} (1 + x)^{\frac{1}{2}n} P_n\left(\sqrt{\frac{1 + x}{2}}\right) P_m(x)\, dx.$$

12. Evaluate

$$\int_0^1 x^n P_{n-2k}(x)\,dx = \tfrac{1}{2}\int_{-1}^1 x^n P_{n-2k}(x)\,dx,$$

and check your result by means of Theorem 65, page 181. Thus show that

$$\int_0^1 x^n P_{n-2k}(x)\,dx = \frac{n!}{2^n k!(\frac{3}{2})_{n-k}}$$

and, equivalently, that

$$\int_0^1 x^{n+2k} P_n(x)\,dx = \frac{(n+2k)!}{2^{n+2k} k!(\frac{3}{2})_{n+k}}.$$

13. Use formula (5), page 104, to obtain the result

$$\int_0^t \frac{x^n(t-x)^n\,dx}{(1-x^2)^{n+1}} = \left(\frac{t}{2}\right)^n Q_n\left(\frac{1}{t}\right),$$

where $Q_n(t)$ is the function given in (4), page 182.

14. Show that

$$P_n(x) = \frac{2^n(\frac{1}{2})_n(x-1)^n}{n!} F\left[\begin{matrix} -n, -n; \\ -2n; \end{matrix} \quad \frac{2}{1-x}\right].$$

15. Show that

$$P_n(x) = \frac{2^n(\frac{1}{2})_n(x+1)^n}{n!} F\left[\begin{matrix} -n, -n; \\ -2n; \end{matrix} \quad \frac{2}{1+x}\right].$$

16. Show that for $|t|$ sufficiently small

$$\sum_{n=0}^{\infty} (2n+1)P_n(x)t^n = (1-t^2)(1-2xt+t^2)^{-\frac{3}{2}}.$$

17. Use Theorem 48, page 137, with $c = 1$, x replaced by $\frac{1}{2}(1-x)$ and $\gamma_n = (\frac{1}{2})_n/n!$ to arrive at

$$(1-x)^n = 2^n(n!)^2 \sum_{k=0}^{n} \frac{(-1)^k(2k+1)P_k(x)}{(n-k)!(n+k+1)!}.$$

18. Use Theorem 48, page 137, to show that

$$\begin{aligned}(1-x)P_n'(x) + nP_n(x) &= nP_{n-1}(x) - (1-x)P_{n-1}'(x)\\ &= \sum_{k=0}^{n-1} P_k(x) - 2(1-x)\sum_{k=0}^{n-1} P_k'(x)\\ &= \sum_{k=0}^{n-1} (-1)^{n-k+1}(1+2k)P_k(x).\end{aligned}$$

19. Use Rodrigues' formula, page 162, and successive integrations by parts to derive the orthogonality property for $P_n(x)$ and to show that

$$\int_{-1}^{1} P_n^2(x)\, dx = \frac{2}{2n+1}.$$

20. Show that the polynomial $y_n(x) = (n!)^{-1}(1 - x^2)^{\frac{1}{2}n} P_n((1 - x^2)^{-\frac{1}{2}})$ has the generating relation

$$e^t\, {}_0F_1(-; 1; \tfrac{1}{4}x^2t^2) = \sum_{n=0}^{\infty} y_n(x) t^n$$

and that Theorem 45, page 133, is applicable to this $y_n(x)$. Translate the result into a property of $P_n(x)$, obtaining equation (7), page 159.

21. Let the polynomials $w_n(x)$ be defined by

$$e^{xt}\psi[t^2(x^2 - 1)] = \sum_{n=0}^{\infty} w_n(x) t^n,$$

with

$$\psi(u) = \sum_{n=0}^{\infty} \gamma_n u^n.$$

Show that

$$\sum_{n=0}^{\infty} (c)_n w_n(x) t^n = (1 - xt)^{-c} \sum_{k=0}^{\infty} (c)_{2k} \gamma_k \left[\frac{t^2(x^2 - 1)}{(1 - xt)^2}\right]^k$$

and thus obtain a result parallel to that in Theorem 46, page 134. Apply your new theorem to Legendre polynomials to derive equation (2), page 164.

CHAPTER 11

Hermite Polynomials

103. Definition of $H_n(x)$. We define the Hermite polynomials $H_n(x)$ by means of the relation

$$\exp(2xt - t^2) = \sum_{n=0}^{\infty} \frac{H_n(x)t^n}{n!}, \tag{1}$$

valid for all finite x and t. Since

$$\begin{aligned}\exp(2xt - t^2) &= \exp(2xt)\exp(-t^2)\\ &= \left(\sum_{n=0}^{\infty} \frac{(2x)^n t^n}{n!}\right)\left(\sum_{n=0}^{\infty} \frac{(-1)^n t^{2n}}{n!}\right)\\ &= \sum_{n=0}^{\infty}\sum_{k=0}^{[n/2]} \frac{(-1)^k (2x)^{n-2k} t^n}{k!(n-2k)!},\end{aligned}$$

it follows from (1) that

$$H_n(x) = \sum_{k=0}^{[n/2]} \frac{(-1)^k n!(2x)^{n-2k}}{k!(n-2k)!}. \tag{2}$$

Examination of equation (2) shows that $H_n(x)$ is a polynomial of degree precisely n in x and that

$$H_n(x) = 2^n x^n + \pi_{n-2}(x), \tag{3}$$

in which $\pi_{n-2}(x)$ is a polynomial of degree $(n-2)$ in x. From either (1) or (2) it follows that $H_n(x)$ is an even function of x for even n, an odd function of x for odd n:

(4) $$H_n(-x) = (-1)^n H_n(x).$$

From (2) it follows readily that

$$H_{2n}(0) = (-1)^n 2^{2n}(\tfrac{1}{2})_n; \qquad H_{2n+1}(0) = 0;$$
$$H'_{2n+1}(0) = (-1)^n 2^{2n+1}(\tfrac{3}{2})_n; \qquad H'_{2n}(0) = 0.$$

104. Recurrence relations. Since the generating function in Section 103 is of the form $G(2xt - t^2)$, the $H_n(x)$ must satisfy

(1) $$xH_n'(x) = nH'_{n-1}(x) + nH_n(x),$$

as we saw in Section 72.

Also the relation

(2) $$\exp(2xt - t^2) = \sum_{n=0}^{\infty} \frac{H_n(x)t^n}{n!}$$

yields at once

(3) $$2t \exp(2xt - t^2) = \sum_{n=0}^{\infty} \frac{H_n'(x)t^n}{n!}.$$

From (2) and (3) we get

$$\sum_{n=0}^{\infty} \frac{2H_n(x)t^{n+1}}{n!} = \sum_{n=0}^{\infty} \frac{H_n'(x)t^n}{n!}$$

which, with a shift of index on the left, yields $H_0'(x) = 0$, and for $n \geqq 1$,

(4) $$H_n'(x) = 2nH_{n-1}(x).$$

Iteration of (4) gives us

(5) $$D^s H_n(x) = \frac{2^s n! H_{n-s}(x)}{(n-s)!}; \qquad D \equiv \frac{d}{dx}.$$

Combination of (1) and (4) yields

(6) $$H_n(x) = 2xH_{n-1}(x) - H'_{n-1}(x).$$

We use (4) and (6) as our pair of independent differential recurrence relations. From this pair of equations we at once obtain both the pure recurrence relation

(7) $$H_n(x) = 2xH_{n-1}(x) - 2(n-1)H_{n-2}(x)$$

and Hermite's differential equation

(8) $$H_n''(x) - 2xH_n'(x) + 2nH_n(x) = 0.$$

Later equation (8) will be used to obtain an orthogonality property for $H_n(x)$. The differential equation is a natural starting point for the study of Hermite functions with unrestricted n.

The first few Hermite polynomials* are listed here for reference purposes:

$$H_0(x) = 1, \qquad H_1(x) = 2x, \qquad H_2(x) = 4x^2 - 2,$$
$$H_3(x) = 8x^3 - 12x, \qquad H_4(x) = 16x^4 - 48x^2 + 12,$$
$$H_5(x) = 32x^5 - 160x^3 + 120x, \qquad H_6(x) = 64x^6 - 480x^4 + 720x^2 - 120.$$

105. The Rodrigues formula. Examination of the defining relation

$$(1) \qquad \exp(2xt - t^2) = \sum_{n=0}^{\infty} \frac{H_n(x)t^n}{n!}$$

in the light of Maclaurin's theorem gives us at once

$$H_n(x) = \left[\frac{d^n}{dt^n} \exp(2xt - t^2)\right]_{t=0}.$$

The function $\exp(-x^2)$ is independent of t, so we may write

$$\exp(-x^2)H_n(x) = \left[\frac{d^n}{dt^n} \exp\{-(x - t)^2\}\right]_{t=0}.$$

Now put $x - t = w$. Then

$$\exp(-x^2)H_n(x) = (-1)^n\left[\frac{d^n}{dw^n} \exp(-w^2)\right]_{w=x}.$$

But it is ridiculous to differentiate with respect to w a function of w alone and afterward to put $w = x$. The w is superfluous. Therefore we write

$$\exp(-x^2)H_n(x) = (-1)^n D^n \exp(-x^2), \qquad D \equiv \frac{d}{dx};$$

or

$$(2) \qquad H_n(x) = (-1)^n \exp(x^2) D^n \exp(-x^2),$$

a formula of the same nature as Rodrigues' formula for Legendre polynomials.

---

*As is true for many special functions, the literature contains more than one notation. For Hermite polynomials the two most common are the one we use here and one which is employed widely in statistics. The latter notation is often distinguished from ours by the use of a script $\mathcal{H}$. The polynomials $\mathcal{H}_n(x)$ may be defined by

$$\exp(xt - \tfrac{1}{2}t^2) = \sum_{n=0}^{\infty} \frac{\mathcal{H}_n(x)t^n}{n!}.$$

This is the notation used by Jackson [1].

106. Other generating functions. Consider the sum

$$\sum_{n=0}^{\infty}\frac{(c)_nH_n(x)t^n}{n!} = \sum_{n=0}^{\infty}\sum_{k=0}^{[n/2]}\frac{(-1)^k(c)_n(2x)^{n-2k}t^n}{k!(n-2k)!}$$

$$= \sum_{n=0}^{\infty}\sum_{k=0}^{\infty}\frac{(-1)^k(c)_{n+2k}(2x)^nt^{n+2k}}{k!n!}$$

$$= \sum_{k=0}^{\infty}\sum_{n=0}^{\infty}\frac{(c+2k)_n(2xt)^n}{n!}\,\frac{(-1)^k(c)_{2k}t^{2k}}{k!}$$

$$= \sum_{k=0}^{\infty}\frac{(-1)^k(c)_{2k}t^{2k}}{k!(1-2xt)^{c+2k}}.$$

Using the fact that $(c)_{2k} = 2^{2k}(\frac{1}{2}c)_k(\frac{1}{2}c+\frac{1}{2})_k$, we thus arrive at the (divergent) generating function

$$(1)\quad (1-2xt)^{-c}\,{}_2F_0\left[\begin{matrix}\frac{1}{2}c,\ \frac{1}{2}c+\frac{1}{2}; \\ -;\end{matrix}\ \frac{-4t^2}{(1-2xt)^2}\right] \cong \sum_{n=0}^{\infty}\frac{(c)_nH_n(x)t^n}{n!},$$

published by Brafman [1]. The special case $c = 1$ was given by Truesdell [1] with the left member in a different form.

Brafman also summed the series

$$(2)\qquad \sum_{n=0}^{\infty}\frac{(c)_{[n/2]}H_n(x)t^n}{[n/2]!(\frac{1}{2})_{[n/2]}},$$

in which [] is our customary greatest integer symbol, thus obtaining a class (one for each c) of peculiar generating functions for the Hermite polynomials. In the special instance $c = \frac{1}{2}$, the series (2) had already been summed by Doetsch.

107. Integrals. Directly from the initial generating function

$$(1)\qquad \exp(2xt - t^2) = \sum_{n=0}^{\infty}\frac{H_n(x)t^n}{n!},$$

it follows by the usual theorem on the coefficient in a Taylor series that

$$(2)\qquad H_n(x) = \frac{n!}{2\pi i}\int^{(0+)} u^{-n-1}\exp(2xu - u^2)\,du,$$

where the contour of integration encircles the origin of the u-plane in the positive direction. From (2) we get, by using the contour $u = \exp(i\theta)$, the real integral representation

(3) $H_n(x) =$

$$\frac{n!}{\pi}\int_0^\pi \exp(2x\cos\theta - \cos 2\theta)\cos(2x\sin\theta - \sin 2\theta - n\theta)\,d\theta.$$

There are numerous interesting relations connecting the Hermite polynomials and the Legendre polynomials. We quote two fairly simple integral relations.

Curzon [1] obtained many relations between $H_n(x)$ and $P_n(x)$ with n usually not restricted to be integral. One of the simplest of his relations, one in which n is to be an integer, is (in our notation)

$$(4) \qquad P_n(x) = \frac{2}{n!\sqrt{\pi}}\int_0^\infty \exp(-t^2)t^n H_n(xt)\,dt.$$

Curzon also expresses the Hermite polynomial as a contour integral involving the Legendre polynominal.

A real integral relation giving $H_n(x)$ in terms of $P_n(x)$ was given by Rainville [5]. It is

$$(5) \qquad H_n(x) = 2^{n+1}\exp(x^2)\int_x^\infty \exp(-t^2)t^{n+1}P_n(x/t)\,dt.$$

Verification of (4) and (5) is left for the exercises.

108. The Hermite polynomial as a $_2F_0$. The formula

$$H_n(x) = \sum_{k=0}^{[n/2]} \frac{(-1)^k n!(2x)^{n-2k}}{k!(n-2k)!}$$

yields at once

$$H_n(x) = (2x)^n \sum_{k=0}^{[n/2]} \frac{(-n)_{2k}(-1)^k x^{-2k}}{2^{2k}k!},$$

or

$$H_n(x) = (2x)^n\, {}_2F_0\left(-\tfrac{1}{2}n, -\tfrac{1}{2}n + \tfrac{1}{2}; -; -\frac{1}{x^2}\right).$$

109. Orthogonality. Equation (8), page 188, is

$$(1) \qquad H_n''(x) - 2xH_n'(x) + 2nH_n(x) = 0,$$

which may be written

$$(2) \qquad [\exp(-x^2)H_n'(x)]' + 2n\exp(-x^2)H_n(x) = 0.$$

Along with (2) write

$$(3) \qquad [\exp(-x^2)H_m'(x)]' + 2m\exp(-x^2)H_m(x) = 0.$$

Then (2) and (3) combine to yield

$$\begin{aligned}2(n - m)\exp(-x^2)H_n(x)H_m(x) &= H_n(x)[\exp(-x^2)H_m'(x)]' - H_m(x)[\exp(-x^2)H_n'(x)]'.\\ &= [\exp(-x^2)\{H_n(x)H_m'(x) - H_n'(x)H_m(x)\}]'.\end{aligned}$$

It follows that

$$(4) \qquad 2(n - m)\int_a^b \exp(-x^2)H_n(x)H_m(x)\,dx = \Big[\exp(-x^2)\{H_n(x)H_m'(x) - H_n'(x)H_m(x)\}\Big]_a^b.$$

Since the product of any polynomial in x by $\exp(-x^2) \to 0$ as $x \to \infty$ or as $x \to -\infty$, we may conclude that

$$(5) \qquad \int_{-\infty}^{\infty} \exp(-x^2)H_n(x)H_m(x)\,dx = 0, \qquad m \neq n.$$

That is, the Hermite polynomials form an orthogonal set over the interval $(-\infty, \infty)$ with weight function $\exp(-x^2)$. Here the infinite limits cause no trouble because of the factor $\exp(-x^2)$.

From the definition

$$\exp(2xt - t^2) = \sum_{m=0}^{\infty} \frac{H_m(x)t^m}{m!}$$

we obtain

$$\exp(4xt - 2t^2) = \sum_{m=0}^{\infty}\sum_{k=0}^{m} \frac{H_k(x)H_{m-k}(x)t^m}{k!(m-k)!}$$

so that

$$\int_{-\infty}^{\infty} \exp(-x^2 + 4xt - 2t^2)\,dx = \sum_{m=0}^{\infty}\sum_{k=0}^{m}\int_{-\infty}^{\infty} \exp(-x^2)H_k(x)H_{m-k}(x)\,dx\,\frac{t^m}{k!(m-k)!}.$$

Because of (5) each term on the right vanishes except terms for which $k = m - k$. Then m must be even, $m = 2n$, and $k = n$. Therefore we have

$$\exp(2t^2)\int_{-\infty}^{\infty} \exp(-x^2 + 4xt - 4t^2)\,dx = \sum_{n=0}^{\infty}\int_{-\infty}^{\infty} \exp(-x^2)H_n^2(x)\,dx\,\frac{t^{2n}}{(n!)^2}.$$

Now

$$\exp(2t^2)\int_{-\infty}^{\infty} \exp(-x^2 + 4xt - 4t^2)\, dx$$

$$= \exp(2t^2)\int_{-\infty}^{\infty} \exp[-(x - 2t)^2]\, dx$$

$$= \exp(2t^2)\int_{-\infty}^{\infty} \exp(-y^2)\, dy = \sqrt{\pi}\exp(2t^2).$$

Therefore

$$\sum_{n=0}^{\infty}\int_{-\infty}^{\infty} \exp(-x^2)H_n^2(x)\, dx\, \frac{t^{2n}}{(n!)^2} = \sqrt{\pi}\exp(2t^2) = \sqrt{\pi}\sum_{n=0}^{\infty}\frac{2^n t^{2n}}{n!}$$

which yields

$$\text{(6)} \qquad \int_{-\infty}^{\infty} \exp(-x^2)H_n^2(x)\, dx = 2^n n!\sqrt{\pi}.$$

See also Ex. 7 at the end of this chapter for a simpler method of obtaining equation (6).

We now know that the $H_n(x)$ form an orthogonal set over $(-\infty, \infty)$ with the weight function $\exp(-x^2)$, and in the notation of Chapter 9, $g_n = 2^n n!\sqrt{\pi}$, and the leading coefficient in $H_n(x)$ is $h_n = 2^n$. The theory developed in Chapter 9 yields the following results.

THEOREM 67. *For the Hermite polynomials $H_n(x)$,*

$$\text{(7)} \quad \int_{-\infty}^{\infty} \exp(-x^2)x^k H_n(x)\, dx = 0, \qquad k = 0, 1, 2, \cdots, (n - 1);$$

(8) *The zeros of $H_n(x)$ are real and distinct;*

$$\text{(9)} \qquad \sum_{k=0}^{n}\frac{H_k(x)H_k(y)}{2^k k!} = \frac{H_{n+1}(y)H_n(x) - H_{n+1}(x)H_n(y)}{2^{n+1}n!(y - x)};$$

(10) *If $\int_{-\infty}^{\infty} \exp(-x^2)f^2(x)\, dx$ exists,*

$$\lim_{n\to\infty} (2^n n!)^{-\frac{1}{2}}\int_{-\infty}^{\infty} \exp(-x^2)f(x)H_n(x)\, dx = 0.$$

The three-term recurrence relation for $H_n(x)$ has already been obtained on page 188.

110. Expansion of polynomials. Any polynomial can be expanded in a series of Hermite polynomials, and the coefficients can be determined as in the general theory: if

(1) $$P(x) = \sum_{k=0}^{n} c_k H_k(x),$$

(2) $$2^k k! \sqrt{\pi}\, c_k = \int_{-\infty}^{\infty} \exp(-x^2) P(x) H_k(x)\, dx.$$

As with the Legendre polynomials, we find it desirable to bypass (2) by obtaining the expansion of x^n directly from a generating function. Since

(3) $$\exp(2xt - t^2) = \sum_{n=0}^{\infty} \frac{H_n(x)t^n}{n!}$$

it follows that

$$\exp(2xt) = \exp(t^2) \sum_{n=0}^{\infty} \frac{H_n(x)t^n}{n!},$$

or

$$\sum_{n=0}^{\infty} \frac{(2x)^n t^n}{n!} = \left(\sum_{n=0}^{\infty} \frac{t^{2n}}{n!}\right)\left(\sum_{n=0}^{\infty} \frac{H_n(x)t^n}{n!}\right)$$
$$= \sum_{n=0}^{\infty} \sum_{k=0}^{[n/2]} \frac{H_{n-2k}(x)t^n}{k!(n-2k)!}.$$

Hence

(4) $$x^n = \sum_{k=0}^{[n/2]} \frac{n! H_{n-2k}(x)}{2^n k!(n-2k)!}.$$

Let us employ (4) to expand the Legendre polynomial in a series of Hermite polynomials. Consider the series

$$\sum_{n=0}^{\infty} P_n(x)t^n = \sum_{n=0}^{\infty} \sum_{k=0}^{[n/2]} \frac{(-1)^k (\tfrac{1}{2})_{n-k} (2x)^{n-2k} t^n}{k!(n-2k)!}$$
$$= \sum_{n,k=0}^{\infty} \frac{(-1)^k (\tfrac{1}{2})_{n+k} (2x)^n t^{n+2k}}{k!n!}.$$

From (4) we have

(5) $$\frac{(2x)^n}{n!} = \sum_{s=0}^{[n/2]} \frac{H_{n-2s}(x)}{s!(n-2s)!}.$$

Hence we may write

$$\sum_{n=0}^{\infty} P_n(x)t^n = \sum_{n,k=0}^{\infty} \sum_{s=0}^{[n/2]} \frac{(-1)^k (\tfrac{1}{2})_{n+k} H_{n-2s}(x) t^{n+2k}}{k!s!(n-2s)!}$$
$$= \sum_{n,k,s=0}^{\infty} \frac{(-1)^k (\tfrac{1}{2})_{n+k+2s} H_n(x) t^{n+2k+2s}}{k!s!n!},$$

in which we have used Lemma 11, page 57. We need to collect powers of t in the last summation above. By Lemma 10, page 56, we may write

$$\sum_{n=0}^{\infty} P_n(x)t^n = \sum_{n,k=0}^{\infty} \sum_{s=0}^{k} \frac{(-1)^{k-s}(\frac{1}{2})_{n+k+s} H_n(x) t^{n+2k}}{s!(k-s)!n!}$$

$$= \sum_{n,k=0}^{\infty} \sum_{s=0}^{k} \frac{(-1)^s k!(\frac{1}{2}+n+k)_s(-1)^k(\frac{1}{2})_{n+k} H_n(x) t^{n+2k}}{s!(k-s)!k!n!}$$

$$= \sum_{n,k=0}^{\infty} \frac{{}_2F_0(-k, \frac{1}{2}+n+k; -; 1)(-1)^k(\frac{1}{2})_{n+k} H_n(x) t^{n+2k}}{k!n!}.$$

We use Lemma 11 again to obtain

$$\sum_{n=0}^{\infty} P_n(x)t^n = \sum_{n=0}^{\infty} \sum_{k=0}^{[n/2]} \frac{{}_2F_0(-k, \frac{1}{2}+n-k; -; 1)(-1)^k(\frac{1}{2})_{n-k} H_{n-2k}(x) t^n}{k!(n-2k)!}.$$

The final result is

$$(6) \quad P_n(x) = \sum_{k=0}^{[n/2]} \frac{{}_2F_0(-k, \frac{1}{2}+n-k; -; 1)(-1)^k(\frac{1}{2})_{n-k} H_{n-2k}(x)}{k!(n-2k)!}.$$

Next let us expand the Hermite polynomial in a series of Legendre polynomials. By Theorem 65, page 181,

$$(7) \qquad \frac{(2x)^n}{n!} = \sum_{s=0}^{[n/2]} \frac{(2n-4s+1)P_{n-2s}(x)}{s!(\frac{3}{2})_{n-s}}.$$

Now

$$\sum_{n=0}^{\infty} \frac{H_n(x)t^n}{n!} = \sum_{n=0}^{\infty} \sum_{k=0}^{[n/2]} \frac{(-1)^k(2x)^{n-2k}t^n}{k!(n-2k)!}$$

$$= \sum_{n,k=0}^{\infty} \frac{(-1)^k(2x)^n t^{n+2k}}{k!n!}$$

$$= \sum_{n,k=0}^{\infty} \sum_{s=0}^{[n/2]} \frac{(-1)^k(2n-4s+1)P_{n-2s}(x)t^{n+2k}}{k!s!(\frac{3}{2})_{n-s}}$$

$$= \sum_{n,k,s=0}^{\infty} \frac{(-1)^k(2n+1)P_n(x)t^{n+2k+2s}}{s!k!(\frac{3}{2})_{n+s}}.$$

Again we collect powers of t:

$$\sum_{n=0}^{\infty} \frac{H_n(x)t^n}{n!}$$

$$= \sum_{n,k=0}^{\infty} \sum_{s=0}^{k} \frac{(-1)^{k-s}(2n+1)P_n(x)t^{n+2k}}{s!(k-s)!(\frac{3}{2})_{n+s}}$$

$$= \sum_{n,k=0}^{\infty} \sum_{s=0}^{k} \frac{(-1)^s k!}{s!(k-s)!(\frac{3}{2}+n)_s} \cdot \frac{(-1)^k(2n+1)P_n(x)t^{n+2k}}{k!(\frac{3}{2})_n}$$

$$= \sum_{n,k=0}^{\infty} \frac{{}_1F_1(-k; \frac{3}{2}+n; 1)(-1)^k(2n+1)P_n(x)t^{n+2k}}{k!(\frac{3}{2})_n}$$

$$= \sum_{n=0}^{\infty} \sum_{k=0}^{[n/2]} \frac{{}_1F_1(-k; \frac{3}{2}+n-2k; 1)(-1)^k(2n-4k+1)P_{n-2k}(x)t^n}{k!(\frac{3}{2})_{n-2k}}.$$

Therefore,

(8) $H_n(x) =$

$$\sum_{k=0}^{[n/2]} \frac{{}_1F_1(-k; \frac{3}{2}+n-2k; 1)(-1)^k n!(2n-4k+1)P_{n-2k}(x)}{k!(\frac{3}{2})_{n-2k}}.$$

The expansion of functions other than polynomials into series of Hermite polynomials is omitted here. Theorems exist similar to the ones relative to expansions in series of Legendre polynomials.

111. More generating functions. We wish to obtain for $H_n(x)$ a property similar to that for $P_n(x)$ expressed in equation (7), page 169. Consider the series

$$\sum_{k=0}^{\infty} \sum_{n=0}^{\infty} \frac{H_{n+k}(x)t^n v^k}{k!n!} = \sum_{n=0}^{\infty} \sum_{k=0}^{n} \frac{H_n(x)t^{n-k}v^k}{k!(n-k)!}$$

$$= \sum_{n=0}^{\infty} \frac{H_n(x)(t+v)^n}{n!}$$

$$= \exp[2x(t+v) - (t+v)^2]$$

$$= \exp(2xt - t^2) \exp[2(x-t)v - v^2]$$

$$= \exp(2xt - t^2) \sum_{k=0}^{\infty} \frac{H_k(x-t)v^k}{k!}.$$

By equating coefficients of $v^k/k!$, we obtain

$$(1) \qquad \sum_{n=0}^{\infty} \frac{H_{n+k}(x)t^n}{n!} = \exp(2xt - t^2)H_k(x - t).$$

As a first example in the use of equation (1) let us derive what is sometimes called a bilinear generating function. Consider the series

$$\sum_{n=0}^{\infty} \frac{H_n(x)H_n(y)t^n}{n!} = \sum_{n=0}^{\infty} \sum_{k=0}^{[n/2]} \frac{(-1)^k(2x)^{n-2k}H_n(y)t^n}{k!(n-2k)!}$$

$$= \sum_{n,k=0}^{\infty} \frac{(-1)^k(2x)^n H_{n+2k}(y)t^{n+2k}}{k!n!}$$

$$= \sum_{k=0}^{\infty} \sum_{n=0}^{\infty} \frac{H_{n+2k}(y)(2xt)^n}{n!} \cdot \frac{(-1)^k t^{2k}}{k!}$$

$$= \sum_{k=0}^{\infty} \frac{\exp(4xyt - 4x^2t^2)H_{2k}(y - 2xt)(-1)^k t^{2k}}{k!},$$

by (1). Since

$$H_{2k}(v) = \sum_{s=0}^{k} \frac{(-1)^s(2k)!(2v)^{2k-2s}}{s!(2k-2s)!},$$

and $(2k)! = 2^{2k}k!(\frac{1}{2})_k$, it follows that

$$\sum_{n=0}^{\infty} \frac{H_n(x)H_n(y)t^n}{n!}$$

$$= \exp(4xyt - 4x^2t^2) \sum_{k=0}^{\infty} \sum_{s=0}^{k} \frac{(-1)^{k+s}2^{2k}(\frac{1}{2})_k(2y - 4xt)^{2k-2s}t^{2k}}{s!(2k-2s)!}$$

$$= \exp(4xyt - 4x^2t^2) \sum_{k,s=0}^{\infty} \frac{(-1)^k 2^{2k+2s}(\frac{1}{2})_{k+s}(2y - 4xt)^{2k}t^{2k+2s}}{s!(2k)!}$$

$$= \exp(4xyt - 4x^2t^2) \sum_{k,s=0}^{\infty} \frac{(-1)^k 2^{2s}(\frac{1}{2})_{k+s}(2y - 4xt)^{2k}t^{2k+2s}}{s!k!(\frac{1}{2})_k}$$

$$= \exp(4xyt - 4x^2t^2) \sum_{k=0}^{\infty} \sum_{s=0}^{\infty} \frac{(\frac{1}{2} + k)_s 2^{2s}t^{2s}}{s!} \cdot \frac{(-1)^k t^{2k}(2y - 4xt)^{2k}}{k!}$$

$$= \exp(4xyt - 4x^2t^2) \sum_{k=0}^{\infty} \frac{(-1)^k t^{2k}(2y - 4xt)^{2k}}{k!(1 - 4t^2)^{\frac{1}{2}+k}}$$

$$= (1 - 4t^2)^{-\frac{1}{2}} \exp(4xyt - 4x^2t^2) \exp\left[\frac{-4t^2(y - 2xt)^2}{1 - 4t^2}\right].$$

The exponential factors may be combined and the preceding identity written as

$$(2)\qquad \sum_{n=0}^{\infty} \frac{H_n(x)H_n(y)t^n}{n!} = (1-4t^2)^{-\frac{1}{2}} \exp\left[y^2 - \frac{(y-2xt)^2}{1-4t^2}\right],$$

a generating relation known for about a century.

We can apply equation (1) to any known generating relation and sometimes obtain a new result. On page 190 we obtained the relation

$$(3)\qquad (1-2xt)^{-c}\,{}_2F_0\left[\begin{matrix}\tfrac{1}{2}c,\ \tfrac{1}{2}c+\tfrac{1}{2}; \\ -;\end{matrix}\ \frac{-4t^2}{(1-2xt)^2}\right] \cong \sum_{n=0}^{\infty} \frac{(c)_n H_n(x)t^n}{n!}.$$

To (3) we apply (1) in the following manner. Consider the series

$$\sum_{k=0}^{\infty} \frac{(c)_k H_k(x-t)(-ty)^k}{k!}$$

$$\cong \sum_{k=0}^{\infty}\sum_{n=0}^{\infty} \frac{(c)_k \exp(-2xt+t^2)H_{n+k}(x)t^n(-ty)^k}{k!n!}$$

$$\cong \exp(-2xt+t^2)\sum_{n=0}^{\infty}\sum_{k=0}^{\infty} \frac{(-1)^k(c)_k y^k H_{n+k}(x)t^{n+k}}{k!n!}$$

$$\cong \exp(-2xt+t^2)\sum_{n=0}^{\infty}\sum_{k=0}^{n} \frac{(-1)^k(c)_k y^k H_n(x)t^n}{k!(n-k)!}.$$

$$\cong \exp(-2xt+t^2)\sum_{n=0}^{\infty} \frac{{}_2F_0(-n, c; -; y)H_n(x)t^n}{n!}.$$

Because of (3) it now follows that

$$\sum_{n=0}^{\infty} \frac{{}_2F_0(-n, c; -; y)H_n(x)t^n}{n!} \cong \exp(2xt-t^2)\sum_{k=0}^{\infty} \frac{(c)_k H_k(x-t)(-yt)^k}{k!}$$

$$\cong \exp(2xt-t^2)[1+2yt(x-t)]^{-c}\cdot$$

$${}_2F_0\left[\begin{matrix}\tfrac{1}{2}c,\ \tfrac{1}{2}c+\tfrac{1}{2}; \\ -;\end{matrix}\ \frac{-4y^2t^2}{(1+2xyt-2yt^2)^2}\right],$$

a relation obtained by Brafman [2] with contour integration as the main tool.

EXERCISES

1. Use the fact that

$$\exp(2xt - t^2) = \exp(2xt - x^2t^2)\exp[t^2(x^2 + 1)]$$

to obtain the expansion

$$H_n(x) = \sum_{k=0}^{[n/2]} \frac{n!H_{n-2k}(1)x^{n-2k}(x^2 + 1)^k}{k!(n - 2k)!}.$$

2. Use the expansion of x^n in a series of Hermite polynomials to show that

$$\int_{-\infty}^{\infty} \exp(-x^2)x^nH_{n-2k}(x)\,dx = 2^{-2k}n!\frac{\sqrt{\pi}}{k!}.$$

Note in particular the special case $k = 0$.

3. Use the integral evaluation in equation (4), page 192 to obtain the result

$$\int_0^{\infty} \exp(-x^2)H_{2k}(x)H_{2s+1}(x)\,dx = (-1)^{k+s}2^{2k+2s}(\tfrac{1}{2})_k(\tfrac{3}{2})_s/(2s + 1 - 2k).$$

4. By evaluating the integral on the right, using equation (2), page 187, and term-by-term integration, show that

(A) $$P_n(x) = \frac{2}{n!\sqrt{\pi}}\int_0^{\infty} \exp(-t^2)t^nH_n(xt)\,dt,$$

which is Curzon's integral for $P_n(x)$, equation (4), page 191.

5. Let $v_n(x)$ denote the right member of equation (A) of Ex. 4. Prove (A) by showing that

$$\sum_{n=0}^{\infty} v_n(x)y^n = (1 - 2xy + y^2)^{-\frac{1}{2}}.$$

6. Evaluate the integral on the right in

(B) $$H_n(x) = 2^{n+1}\exp(x^2)\int_x^{\infty} \exp(-t^2)t^{n+1}P_n\left(\frac{x}{t}\right)dt$$

by using

$$(2t)^nP_n\left(\frac{x}{t}\right) = \sum_{k=0}^{[n/2]} \frac{n!(x^2 - t^2)^k(2x)^{n-2k}}{(k!)^2(n - 2k)!}$$

derived from equation (1), page 164, and term-by-term integration to prove the validity of (B), which is equation (5), page 191.

7. Use the Rodrigues formula

$$\exp(-x^2)H_n(x) = (-1)^nD^n\exp(-x^2); \qquad D = \frac{d}{dx}$$

and iterated integration by parts to show that

$$\begin{aligned}\int_{-\infty}^{\infty} \exp(-x^2)H_n(x)H_m(x)\,dx &= 0, \qquad m \neq n\\ &= 2^nn!\sqrt{\pi}, \qquad m = n.\end{aligned}$$

CHAPTER 12

Laguerre Polynomials

112. The polynomial $L_n^{(\alpha)}(x)$. Let us consider a naturally terminating ${}_1F_1$. We define, for n a non-negative integer,

$$(1) \qquad L_n^{(\alpha)}(x) = \frac{(1+\alpha)_n}{n!}\, {}_1F_1(-n; 1+\alpha; x).$$

The factor $(1+\alpha)_n/n!$ is inserted for convenience only. The polynomials (1) are called Laguerre, generalized Laguerre, or Sonine polynomials. The special case $\alpha = 0$ receives much individual attention and is known either as the Laguerre or the simple Laguerre polynomial. When $\alpha = 0$, α is usually omitted from the symbol:

$$(2) \qquad L_n(x) = L_n^{(0)}(x) = {}_1F_1(-n; 1; x).$$

We shall work with $L_n^{(\alpha)}(x)$, but for reference purposes a list of properties of $L_n(x)$ is included at the end of the chapter.

The notation in (1) is quite standard with the one exception that some authors permit α to depend upon n; others do not. We shall insist that α be independent of n because for the polynomials (1) so many properties which are valid for α independent of n fail (Shively [1]) to be valid for α dependent upon n.

It should be apparent to the reader by the time he has finished reading this book, if not before, that for a polynomial $\varphi_n(x)$ of hypergeometric character, the way in which the index n enters the parameters of the ${}_pF_q$ involved has a vital effect upon the properties of the polynomial. For a mathematician to use the same name for the two polynomials

$$\frac{(1+\alpha)_n}{n!}\,{}_1F_1(-n;1+\alpha;x)$$

and

$$\frac{(1+c+n)_n}{n!}\,{}_1F_1(-n;1+c+n;x),$$

in which α and c are independent of n, is roughly the equivalent of a layman's using the same name for an eagle and a kitten.

From (1) it follows at once that

$$(3)\qquad L_n^{(\alpha)}(x) = \sum_{k=0}^{n}\frac{(-1)^k(1+\alpha)_n x^k}{k!(n-k)!(1+\alpha)_k},$$

from which we see that the $L_n^{(\alpha)}(x)$ form a simple set of polynomials, the coefficient of x^n being $(-1)^n/n!$.

From (3) we obtain

$$L_0^{(\alpha)}(x)=1,\qquad L_1^{(\alpha)}(x)=1+\alpha-x,$$

$$L_2^{(\alpha)}(x)=\tfrac{1}{2}(1+\alpha)(2+\alpha)-(2+\alpha)x+\tfrac{1}{2}x^2,$$

$$L_3^{(\alpha)}(x)=\tfrac{1}{6}(1+\alpha)(2+\alpha)(3+\alpha)-\tfrac{1}{2}(2+\alpha)(3+\alpha)x+\tfrac{1}{2}(3+\alpha)x^2-\tfrac{1}{6}x^3.$$

113. Generating functions. Directly from (3) of the preceding section we obtain

$$\sum_{n=0}^{\infty}\frac{L_n^{(\alpha)}(x)t^n}{(1+\alpha)_n} = \sum_{n=0}^{\infty}\sum_{k=0}^{n}\frac{(-1)^k x^k t^n}{k!(n-k)!(1+\alpha)_k}$$

$$= \left(\sum_{n=0}^{\infty}\frac{t^n}{n!}\right)\left(\sum_{n=0}^{\infty}\frac{(-1)^n x^n t^n}{n!(1+\alpha)_n}\right).$$

Hence the Laguerre polynomials have the generating function indicated in

$$(1)\qquad e^t\,{}_0F_1(-;1+\alpha;-xt) = \sum_{n=0}^{\infty}\frac{L_n^{(\alpha)}(x)t^n}{(1+\alpha)_n}.$$

Since any ${}_0F_1$ is a Bessel function, we are led also to write the left member of (1) in the less pretty form

$$(2)\qquad \Gamma(1+\alpha)(xt)^{-\alpha/2}e^tJ_\alpha(2\sqrt{xt}) = \sum_{n=0}^{\infty}\frac{L_n^{(\alpha)}(x)t^n}{(1+\alpha)_n}.$$

A set of other generating functions for these polynomials is easily found. Let c be arbitrary and proceed as follows:

$$\sum_{n=0}^{\infty} \frac{(c)_n L_n^{(\alpha)}(x) t^n}{(1+\alpha)_n} = \sum_{n=0}^{\infty} \sum_{k=0}^{n} \frac{(c)_n (-x)^k t^n}{k!(n-k)!(1+\alpha)_k}$$

$$= \sum_{n=0}^{\infty} \sum_{k=0}^{\infty} \frac{(c)_{n+k} (-x)^k t^{n+k}}{k! n! (1+\alpha)_k}$$

$$= \sum_{k=0}^{\infty} \sum_{n=0}^{\infty} \frac{(c+k)_n t^n}{n!} \cdot \frac{(c)_k (-xt)^k}{k!(1+\alpha)_k}$$

$$= \sum_{k=0}^{\infty} {}_1F_0(c+k; -; t) \frac{(c)_k (-xt)^k}{k!(1+\alpha)_k}$$

$$= \sum_{k=0}^{\infty} \frac{(c)_k (-xt)^k}{k!(1+\alpha)_k (1-t)^{c+k}}.$$

We thus arrive at the generating relation (see also pages 134–135)

$$(3) \qquad \frac{1}{(1-t)^c} {}_1F_1\left[\begin{matrix} c; \\ 1+\alpha; \end{matrix} \; \frac{-xt}{1-t}\right] = \sum_{n=0}^{\infty} \frac{(c)_n L_n^{(\alpha)}(x) t^n}{(1+\alpha)_n}.$$

Equation (3) is a special case of a result due to Chaundy [1]. Note the commonly quoted special case with $c = 1 + \alpha$:

$$(4) \qquad \frac{1}{(1-t)^{1+\alpha}} \exp\left(\frac{-xt}{1-t}\right) = \sum_{n=0}^{\infty} L_n^{(\alpha)}(x) t^n.$$

114. Recurrence relations. We have already seen in Chapter 8 that the very form of the generating functions (1) and (4) of the preceding section leads at once to the relations (with $D = d/dx$)

$$(1) \qquad x\,D\,L_n^{(\alpha)}(x) = nL_n^{(\alpha)}(x) - (\alpha + n)L_{n-1}^{(\alpha)}(x),$$

$$(2) \qquad D\,L_n^{(\alpha)}(x) = D\,L_{n-1}^{(\alpha)}(x) - L_{n-1}^{(\alpha)}(x),$$

$$(3) \qquad D\,L_n^{(\alpha)}(x) = -\sum_{k=0}^{n-1} L_k^{(\alpha)}(x).$$

Elimination of the derivatives from (1) and (2) yields the pure recurrence relation

$$(4) \quad nL_n^{(\alpha)}(x) = (2n - 1 + \alpha - x)L_{n-1}^{(\alpha)}(x) - (n - 1 + \alpha)L_{n-2}^{(\alpha)}(x).$$

We already know three $(2p + q)$ contiguous function relations for the ${}_1F_1$. From equations (15), (18), and (20) of Section 48, using $p = 1$, $q = 1$, $\alpha_1 = -n$, $\beta_1 = 1 + \alpha$, we obtain

$$(5)\quad (-n-\alpha)\,{}_1F_1(-n;1+\alpha;x) = -n\,{}_1F_1(-n+1;1+\alpha;x) - \alpha\,{}_1F_1(-n;\alpha;x),$$

$$(6)\quad (-n+x)\,{}_1F_1(-n;1+\alpha;x) = -n\,{}_1F_1(-n+1;1+\alpha;x) + \frac{(n+\alpha+1)x}{1+\alpha}\,{}_1F_1(-n;2+\alpha;x),$$

$$(7)\quad {}_1F_1(-n;1+\alpha;x) = {}_1F_1(-n-1;1+\alpha;x) + \frac{x}{1+\alpha}\,{}_1F_1(-n;2+\alpha;x).$$

Since

$${}_1F_1(-n;1+\alpha;x) = \frac{n!L_n^{(\alpha)}(x)}{(1+\alpha)_n},$$

equations (5), (6), (7) may be converted into the mixed recurrence relations

$$(8)\qquad L_n^{(\alpha)}(x) = L_{n-1}^{(\alpha)}(x) + L_n^{(\alpha-1)}(x),$$

$$(9)\qquad (n-x)L_n^{(\alpha)}(x) = (\alpha+n)L_{n-1}^{(\alpha)}(x) - xL_n^{(\alpha+1)}(x),$$

$$(10)\qquad (1+\alpha+n)L_n^{(\alpha)}(x) = (n+1)L_{n+1}^{(\alpha)}(x) + xL_n^{(\alpha+1)}(x).$$

Next a shift of index in (10) yields

$$xL_{n-1}^{(\alpha+1)}(x) = (\alpha+n)L_{n-1}^{(\alpha)}(x) - nL_n^{(\alpha)}(x) = -x\,D\,L_n^{(\alpha)}(x),$$

by (1) above. Hence we have

$$(11)\qquad D\,L_n^{(\alpha)}(x) = -L_{n-1}^{(\alpha+1)}(x).$$

Comparison of (3) and (11) shows that

$$(12)\qquad L_n^{(\alpha+1)}(x) = \sum_{k=0}^{n} L_k^{(\alpha)}(x).$$

115. The Rodrigues formula. Let us return to the expanded form

$$(1)\qquad L_n^{(\alpha)}(x) = \sum_{k=0}^{n} \frac{(1+\alpha)_n(-x)^k}{k!(n-k)!(1+\alpha)_k}.$$

Since

$$D^{n-k}x^{n+\alpha} = \frac{(1+\alpha)_n x^{k+\alpha}}{(1+\alpha)_k},$$

we may write

$$L_n^{(\alpha)}(x) = \frac{x^{-\alpha}}{n!}\sum_{k=0}^{n}\frac{n!(-1)^k D^{n-k}x^{n+\alpha}}{k!(n-k)!}$$

$$= \frac{x^{-\alpha}}{n!}\sum_{k=0}^{n}(-1)^k C_{n,k} D^{n-k}x^{n+\alpha},$$

involving the binomial coefficient $C_{n,k}$. Now $D^k e^{-x} = (-1)^k e^{-x}$; so, we may conclude that

(2) $$L_n^{(\alpha)}(x) = \frac{e^x x^{-\alpha}}{n!}\sum_{k=0}^{n} C_{n,k}[D^{n-k}x^{n+\alpha}][D^k e^{-x}].$$

In view of Leibnitz' rule for the nth derivative of a product, equation (2) yields

(3) $$L_n^{(\alpha)}(x) = \frac{x^{-\alpha}e^x}{n!}D^n[e^{-x}x^{n+\alpha}],$$

the desired formula of Rodrigues type.

116. The differential equation. Since the Laguerre polynomial is a constant multiple of a ${}_1F_1$, we may obtain the differential equation

(1) $$x\,D^2\,L_n^{(\alpha)}(x) + (1 + \alpha - x)\,D\,L_n^{(\alpha)}(x) + nL_n^{(\alpha)}(x) = 0$$

from the general theory. Equation (1) is also easy to derive by eliminating $L_{n-1}^{(\alpha)}(x)$ from the two differential recurrence relations (1) and (2) of Section 114.

The three-term pure recurrence relation (4), page 202, suggests that we look for an orthogonality property of the Laguerre polynomials. Either the differential equation or the Rodrigues formula leads us quickly to the desired result.

117. Orthogonality. The preceding differential equation for $L_n^{(\alpha)}(x)$ may be put in the form

(1) $$D[x^{\alpha+1}e^{-x}DL_n^{(\alpha)}(x)] + nx^\alpha e^{-x}L_n^{(\alpha)}(x) = 0; \qquad D \equiv \frac{d}{dx},$$

as is easily verified. Equation (1) together with

(2) $$D[x^{\alpha+1}e^{-x}DL_m^{(\alpha)}(x)] + mx^\alpha e^{-x}L_m^{(\alpha)}(x) = 0$$

leads at once to

$$(m - n)x^\alpha e^{-x}L_n^{(\alpha)}(x)L_m^{(\alpha)}(x)$$
$$= L_m^{(\alpha)}(x)\,D\,[x^{\alpha+1}e^{-x}\,D\,L_n^{(\alpha)}(x)] - L_n^{(\alpha)}(x)\,D\,[x^{\alpha+1}e^{-x}\,D\,L_m^{(\alpha)}(x)]$$
$$= D\,[x^{\alpha+1}e^{-x}\{L_m^{(\alpha)}(x)\,D\,L_n^{(\alpha)}(x) - L_n^{(\alpha)}(x)\,D\,L_m^{(\alpha)}(x)\}].$$

Therefore we have

$$(3) \qquad (m-n)\int_a^b x^\alpha e^{-x} L_n^{(\alpha)}(x) L_m^{(\alpha)}(x)\,dx$$

$$= \Big[x^{\alpha+1}e^{-x}\{L_m^{(\alpha)}(x) D\, L_n^{(\alpha)}(x) - L_n^{(\alpha)}(x) D\, L_m^{(\alpha)}(x)\}\Big]_a^b.$$

The product of e^{-x} and any polynomial in $x \to 0$ as $x \to \infty$. Furthermore, $x^{\alpha+1} \to 0$ as $x \to 0$ if Re $(\alpha) > -1$, so equation (3) yields the orthogonality property

$$(4) \qquad \int_0^\infty x^\alpha e^{-x} L_n^{(\alpha)}(x) L_m^{(\alpha)}(x)\,dx = 0, \qquad m \neq n,\ \mathrm{Re}(\alpha) > -1.$$

Equation (4) shows that if $\mathrm{Re}(\alpha) > -1$, the polynomials $L_n^{(\alpha)}(x)$ form an orthogonal set over the interval $(0, \infty)$ with weight function $x^\alpha e^{-x}$. We now need to evaluate the integral on the left in (4) for $m = n$. For the sake of variety we use the Rodrigues formula

$$(5) \qquad L_n^{(\alpha)}(x) = \frac{x^{-\alpha}e^x}{n!} D^n[e^{-x}x^{n+\alpha}], \qquad D = \frac{d}{dx},$$

to evaluate the integral on the left in (4) both for $m = n$ and $m \neq n$.

Because of (5) we may write

$$\int_0^\infty x^\alpha e^{-x} L_n^{(\alpha)}(x) L_m^{(\alpha)}(x)\,dx = \frac{1}{n!}\int_0^\infty D^n(e^{-x}x^{n+\alpha}) L_m^{(\alpha)}(x)\,dx,$$

and then integrate by parts n times to obtain

$$(6) \qquad \int_0^\infty x^\alpha e^{-x} L_n^{(\alpha)}(x) L_m^{(\alpha)}(x)\,dx = \frac{(-1)^n}{n!}\int_0^\infty e^{-x}x^{n+\alpha}[D^n L_m^{(\alpha)}(x)]\,dx,$$

for $\mathrm{Re}(\alpha) > -1$. At each integration by parts the integrated portion,

$$[D^{n-k}(e^{-x}x^{n+\alpha})][D^{k-1}L_m^{(\alpha)}(x)], \qquad 0 < k \leqq n,$$

vanishes both at $x = 0$ and as $x \to \infty$.

Since $L_m^{(\alpha)}(x)$ is of degree m, $D^n L_m^{(\alpha)}(x) = 0$ for $n > m$. Therefore the integral on the left in (6) vanishes for $n > m$. Since that integral is symmetric in n and m, it also vanishes for $n < m$, which completes our second proof of (4).

We know that

$$D^n L_n^{(\alpha)}(x) = D^n\left[\frac{(-1)^n x^n}{n!} + \pi_{n-1}\right] = (-1)^n.$$

Hence, for $m = n$, equation (6) yields

$$\int_0^\infty x^\alpha e^{-x}[L_n^{(\alpha)}(x)]^2 dx = \frac{1}{n!}\int_0^\infty e^{-x}x^{n+\alpha}\, dx,$$

or

$$(7)\qquad \int_0^\infty x^\alpha e^{-x}[L_n^{(\alpha)}(x)]^2\, dx = \frac{\Gamma(1+\alpha+n)}{n!}, \qquad \mathrm{Re}(\alpha) > -1.$$

In the notation of the theory of orthogonal polynomials, Chapter 9, we have

$$g_n = \frac{\Gamma(1+\alpha+n)}{n!}, \qquad h_n = \frac{(-1)^n}{n!}.$$

THEOREM 68. *If* $\alpha > -1$, *the Laguerre polynomials have the following properties:*

$$(8)\qquad \int_0^\infty x^\alpha e^{-x}L_n^{(\alpha)}(x)x^k\, dx = 0, \qquad k = 0, 1, 2, \cdots, (n-1);$$

(9) *The zeros of* $L_n^{(\alpha)}(x)$ *are positive and distinct;*

$$(10)\ \sum_{k=0}^{n} \frac{k!L_k^{(\alpha)}(x)L_k^{(\alpha)}(y)}{(1+\alpha)_k} = \frac{(n+1)!}{(1+\alpha)_n}\,\frac{L_{n+1}^{(\alpha)}(y)L_n^{(\alpha)}(x) - L_{n+1}^{(\alpha)}(x)L_n^{(\alpha)}(y)}{x-y};$$

(11) *If* $\int_0^\infty x^\alpha e^{-x}f^2(x)\, dx$ *exists,*

$$\lim_{n\to\infty}\left[\frac{(1+\alpha)_n}{n!}\right]^{-\frac{1}{2}}\int_0^\infty x^\alpha e^{-x}f(x)L_n^{(\alpha)}(x)\, dx = 0.$$

The three-term recurrence relation for $L_n^{(\alpha)}(x)$ has already been obtained; it is equation (4), page 202.

118. Expansion of polynomials. Since the $L_n^{(\alpha)}(x)$ form an orthogonal set, the classical technique for expanding a polynomial by the method indicated in Theorem 56, page 151, is available. As usual we prefer to treat the problem by obtaining first the expansion of x^n and then using generating function techniques whenever we can.

Equation (1), page 201, yields

$$(1)\qquad {}_0F_1(-;1+\alpha;-xt) = e^{-t}\sum_{n=0}^{\infty}\frac{L_n^{(\alpha)}(x)t^n}{(1+\alpha)_n}.$$

Therefore

$$\sum_{n=0}^{\infty} \frac{(-x)^n t^n}{(1+\alpha)_n n!} = \sum_{n=0}^{\infty} \sum_{k=0}^{n} \frac{(-1)^{n-k} L_k^{(\alpha)}(x) t^n}{(n-k)!(1+\alpha)_k}$$

from which it follows that

$$x^n = \sum_{k=0}^{n} \frac{(-1)^k n!(1+\alpha)_n L_k^{(\alpha)}(x)}{(n-k)!(1+\alpha)_k}. \tag{2}$$

Let us employ (2) in expanding the Hermite polynomial in a series of Laguerre polynomials. Consider the series

$$\sum_{n=0}^{\infty} \frac{H_n(x)t^n}{n!} = \exp(2xt - t^2) = \sum_{n,s=0}^{\infty} \frac{(-1)^s (2x)^n t^{n+2s}}{s!n!}$$

$$= \sum_{n,s=0}^{\infty} \sum_{k=0}^{n} \frac{(-1)^{k+s} 2^n (1+\alpha)_n L_k^{(\alpha)}(x) t^{n+2s}}{s!(n-k)!(1+\alpha)_k}$$

$$= \sum_{n,k,s=0}^{\infty} \frac{(-1)^{k+s} 2^{n+k} (1+\alpha)_{n+k} L_k^{(\alpha)}(x) t^{n+k+2s}}{s!n!(1+\alpha)_k}$$

$$= \sum_{n,k=0}^{\infty} \sum_{s=0}^{[n/2]} \frac{(-1)^{k+s} 2^{n+k-2s} (1+\alpha)_{n+k-2s} L_k^{(\alpha)}(x) t^{n+k}}{s!(n-2s)!(1+\alpha)_k}$$

$$= \sum_{n,k=0}^{\infty} {}_2F_2\left[\begin{matrix} -\tfrac{1}{2}n, -\tfrac{1}{2}(n-1); \\ -\tfrac{1}{2}(\alpha+n+k), -\tfrac{1}{2}(\alpha+n+k-1); \end{matrix} -\tfrac{1}{4}\right] \times \frac{(-1)^k 2^{n+k} (1+\alpha)_{n+k} L_k^{(\alpha)}(x) t^{n+k}}{n!(1+\alpha)_k}$$

$$= \sum_{n=0}^{\infty} \sum_{k=0}^{n} {}_2F_2\left[\begin{matrix} -\tfrac{1}{2}(n-k), -\tfrac{1}{2}(n-k-1); \\ -\tfrac{1}{2}(\alpha+n), -\tfrac{1}{2}(\alpha+n-1); \end{matrix} -\tfrac{1}{4}\right] \times \frac{(-1)^k 2^n (1+\alpha)_n L_k^{(\alpha)}(x) t^n}{(n-k)!(1+\alpha)_k}.$$

From the above we may conclude that

$$H_n(x) = 2^n (1+\alpha)_n \sum_{k=0}^{n} {}_2F_2\left[\begin{matrix} -\tfrac{1}{2}(n-k), -\tfrac{1}{2}(n-k-1); \\ -\tfrac{1}{2}(\alpha+n), -\tfrac{1}{2}(\alpha+n-1); \end{matrix} -\tfrac{1}{4}\right] \times \frac{(-n)_k L_k^{(\alpha)}(x)}{(1+\alpha)_k}. \tag{3}$$

Next let us expand the Legendre polynomial in a series of Laguerre polynomials. Consider the series

$$\sum_{n=0}^{\infty} P_n(x)t^n$$

$$= \sum_{n,s=0}^{\infty} \frac{(-1)^s(\tfrac{1}{2})_{n+s}(2x)^n t^{n+2s}}{s!n!}$$

$$= \sum_{n,s=0}^{\infty} \sum_{k=0}^{n} \frac{(-1)^{k+s}2^n(\tfrac{1}{2})_{n+s}(1+\alpha)_n L_k^{(\alpha)}(x)t^{n+2s}}{s!(n-k)!(1+\alpha)_k}$$

$$= \sum_{n,k,s=0}^{\infty} \frac{(-1)^{k+s}2^{n+k}(\tfrac{1}{2})_{n+k+s}(1+\alpha)_{n+k}L_k^{(\alpha)}(x)t^{n+k+2s}}{s!n!(1+\alpha)_k}$$

$$= \sum_{n,k=0}^{\infty} \sum_{s=0}^{[n/2]} \frac{(-1)^{k+s}2^{n+k-2s}(\tfrac{1}{2})_{n+k-s}(1+\alpha)_{n+k-2s}L_k^{(\alpha)}(x)t^{n+k}}{s!(n-2s)!(1+\alpha)_k}$$

$$= \sum_{n,k=0}^{\infty} {}_2F_3\left[\begin{matrix} -\tfrac{1}{2}n, -\tfrac{1}{2}(n-1); \\ \tfrac{1}{2}-n-k, -\tfrac{1}{2}(\alpha+n+k), -\tfrac{1}{2}(\alpha+n+k-1); \end{matrix} \tfrac{1}{4}\right] \times \frac{(-1)^k 2^{n+k}(\tfrac{1}{2})_{n+k}(1+\alpha)_{n+k}L_k^{(\alpha)}(x)t^{n+k}}{n!(1+\alpha)_k}$$

$$= \sum_{n=0}^{\infty} \sum_{k=0}^{n} {}_2F_3\left[\begin{matrix} -\tfrac{1}{2}(n-k), -\tfrac{1}{2}(n-k-1); \\ \tfrac{1}{2}-n, -\tfrac{1}{2}(\alpha+n), -\tfrac{1}{2}(\alpha+n-1); \end{matrix} \tfrac{1}{4}\right] \times \frac{(-1)^k 2^n(\tfrac{1}{2})_n(1+\alpha)_n L_k^{(\alpha)}(x)t^n}{(n-k)!(1+\alpha)_k}.$$

We may therefore write

(4) $P_n(x)$

$$= \frac{2^n(\tfrac{1}{2})_n(1+\alpha)_n}{n!} \sum_{k=0}^{n} {}_2F_3\left[\begin{matrix} -\tfrac{1}{2}(n-k), -\tfrac{1}{2}(n-k-1); \\ \tfrac{1}{2}-n, -\tfrac{1}{2}(\alpha+n), -\tfrac{1}{2}(\alpha+n-1); \end{matrix} \tfrac{1}{4}\right] \times \frac{(-n)_k L_k^{(\alpha)}(x)}{(1+\alpha)_k}.$$

The Laguerre polynomial can be expanded in series of either Legendre or Hermite polynomials by employing precisely the technique used above with the aid of the pertinent expansions of x^n

from Chapters 10 and 11. This is left for the reader to do; the results may be found in Exs. 2 and 3 at the end of this chapter.

119. Special properties. The generating functions of Section 113 lead to certain simple finite sum properties of the Laguerre polynomials. For instance, from

$$(1) \qquad (1-t)^{-1-\alpha}\exp\left(\frac{-xt}{1-t}\right) = \sum_{n=0}^{\infty} L_n^{(\alpha)}(x)t^n,$$

and

$$(1-t)^{-1-\alpha}\exp\left(\frac{-xt}{1-t}\right) = (1-t)^{-(\alpha-\beta)}(1-t)^{-1-\beta}\exp\left(\frac{-xt}{1-t}\right),$$

it follows at once that

$$(2) \qquad L_n^{(\alpha)}(x) = \sum_{k=0}^{n} \frac{(\alpha-\beta)_k L_{n-k}^{(\beta)}(x)}{k!}$$

for arbitrary α and β.

From equation (1) and the fact that

$$(1-t)^{-1-\alpha}\exp\left(\frac{-xt}{1-t}\right)(1-t)^{-1-\beta}\exp\left(\frac{-yt}{1-t}\right)$$
$$= (1-t)^{-1-(\alpha+\beta+1)}\exp\left(\frac{-(x+y)t}{1-t}\right)$$

it follows that

$$(3) \qquad L_n^{(\alpha+\beta+1)}(x+y) = \sum_{k=0}^{n} L_k^{(\alpha)}(x)L_{n-k}^{(\beta)}(y).$$

The generating relation

$$(4) \qquad e^t\,{}_0F_1(-;1+\alpha;-xt) = \sum_{n=0}^{\infty} \frac{L_n^{(\alpha)}(x)t^n}{(1+\alpha)_n}$$

together with the fact that

$$e^t{}_0F_1(-;1+\alpha;-xyt) = e^{(1-y)t}e^{yt}\,{}_0F_1(-;1+\alpha;-x(yt))$$

yields

$$\sum_{n=0}^{\infty} \frac{L_n^{(\alpha)}(xy)t^n}{(1+\alpha)_n} = \left(\sum_{n=0}^{\infty} \frac{(1-y)^n t^n}{n!}\right)\left(\sum_{n=0}^{\infty} \frac{L_n^{(\alpha)}(x)y^n t^n}{(1+\alpha)_n}\right)$$

from which we get

$$(5) \qquad L_n^{(\alpha)}(xy) = \sum_{k=0}^{n} \frac{(1+\alpha)_n(1-y)^{n-k}y^k L_k^{(\alpha)}(x)}{(n-k)!(1+\alpha)_k}.$$

For (5) see also Ex. 1, page 145.

We know that for arbitrary c,

$$(6)\qquad (1-t)^{-c}\,{}_1F_1\left[\begin{matrix} c; \\ 1+\alpha; \end{matrix}\ \frac{-xt}{1-t}\right] = \sum_{n=0}^{\infty} \frac{(c)_n L_n^{(\alpha)}(x) t^n}{(1+\alpha)_n}.$$

By Kummer's first formula, page 125, we have

$$(7)\qquad {}_1F_1\left[\begin{matrix} c; \\ 1+\alpha; \end{matrix}\ \frac{-xt}{1-t}\right] = \exp\left(\frac{-xt}{1-t}\right) {}_1F_1\left[\begin{matrix} 1+\alpha-c; \\ 1+\alpha; \end{matrix}\ \frac{xt}{1-t}\right].$$

Using (6) and (7) we write

$$\sum_{n=0}^{\infty} \frac{(c)_n L_n^{(\alpha)}(x) t^n}{(1+\alpha)_n}$$

$$= (1-t)^{-c} \exp\left(\frac{-xt}{1-t}\right) {}_1F_1\left[\begin{matrix} 1+\alpha-c; \\ 1+\alpha; \end{matrix}\ \frac{xt}{1-t}\right]$$

$$= (1-t)^{-1-(2c-\alpha-2)} \exp\left(\frac{-xt}{1-t}\right) (1-t)^{-(1+\alpha-c)}\,{}_1F_1\left[\begin{matrix} 1+\alpha-c; \\ 1+\alpha; \end{matrix}\ \frac{xt}{1-t}\right]$$

$$= \left[\sum_{n=0}^{\infty} L_n^{(2c-\alpha-2)}(x) t^n\right]\left[\sum_{n=0}^{\infty} \frac{(1+\alpha-c)_n L_n^{(\alpha)}(-x) t^n}{(1+\alpha)_n}\right],$$

with the aid of (1) and (6). We conclude that for arbitrary c (not zero or a negative integer)

$$(8)\qquad L_n^{(\alpha)}(x) = \frac{(1+\alpha)_n}{(c)_n} \sum_{k=0}^{n} \frac{(1+\alpha-c)_k L_k^{(\alpha)}(-x) L_{n-k}^{(2c-\alpha-2)}(x)}{(1+\alpha)_k}.$$

In equation (8) the two special choices $c = 1 + \frac{1}{2}\alpha + \frac{1}{2}m$ and $c = 1 + \alpha + m$, for non-negative integral m, are particularly recommended. See Exs. 6 and 7 at the end of this chapter.

We next seek for Laguerre polynomials a relation analogous to equation (7) of Section 95 on Legendre polynomials and equation (1) of Section 111 on Hermite polynomials. Consider the series

$$\sum_{k=0}^{\infty}\sum_{n=0}^{\infty} \frac{(n+k)!\,L_{n+k}^{(\alpha)}(x) t^n v^k}{k!\,n!} = \sum_{n=0}^{\infty}\sum_{k=0}^{n} \frac{n!\,t^{n-k} v^k L_n^{(\alpha)}(x)}{k!(n-k)!}$$

$$= \sum_{n=0}^{\infty} L_n^{(\alpha)}(x)(t+v)^n = (1-t-v)^{-1-\alpha} \exp\left(\frac{-x(v+t)}{1-t-v}\right).$$

We wish to expand the right member above in powers of v in another way. Now

$$(1 - t - v)^{-1-\alpha} = (1 - t)^{-1-\alpha}\left[1 - \frac{v}{1 - t}\right]^{-1-\alpha}$$

and

$$\exp\left[\frac{-x(v + t)}{1 - t - v}\right] = \exp\left(\frac{-xt}{1 - t}\right) \exp\left[\frac{-xv}{(1 - t)(1 - t - v)}\right]$$

$$= \exp\left(\frac{-xt}{1 - t}\right) \exp\left[\frac{\dfrac{-x}{1 - t} \cdot \dfrac{v}{1 - t}}{1 - \dfrac{v}{1 - t}}\right].$$

Hence we may write

$$\sum_{k=0}^{\infty} \sum_{n=0}^{\infty} \frac{(n + k)! L_{n+k}^{(\alpha)}(x) t^n v^k}{k! n!}$$
$$= (1 - t)^{-1-\alpha} \exp\left(\frac{-xt}{1 - t}\right) \sum_{k=0}^{\infty} L_k^{(\alpha)}\left(\frac{x}{1 - t}\right)\left(\frac{v}{1 - t}\right)^k.$$

We find, by comparing coefficients of v^k, that

$$(9) \quad \sum_{n=0}^{\infty} \frac{(n + k)! L_{n+k}^{(\alpha)}(x) t^n}{k! n!} = (1 - t)^{-1-\alpha-k} \exp\left(\frac{-xt}{1 - t}\right) L_k^{(\alpha)}\left(\frac{x}{1 - t}\right),$$

a relation which is useful in discovering generating functions.

120. Other generating functions. Consider the series

$$\sum_{n=0}^{\infty} \frac{n! L_n^{(\alpha)}(x) L_n^{(\alpha)}(y) t^n}{(1 + \alpha)_n} = \sum_{n=0}^{\infty} \sum_{k=0}^{n} \frac{(-1)^k n! y^k L_n^{(\alpha)}(x) t^n}{k!(n - k)!(1 + \alpha)_k}$$

$$= \sum_{n,k=0}^{\infty} \frac{(-1)^k (n + k)! y^k L_{n+k}^{(\alpha)}(x) t^{n+k}}{k! n! (1 + \alpha)_k}$$

$$= \sum_{k=0}^{\infty} \sum_{n=0}^{\infty} \frac{(n + k)! L_{n+k}^{(\alpha)}(x) t^n}{k! n!} \cdot \frac{(-1)^k y^k t^k}{(1 + \alpha)_k}.$$

For the moment let

$$\psi = \sum_{n=0}^{\infty} \frac{n! L_n^{(\alpha)}(x) L_n^{(\alpha)}(y) t^n}{(1 + \alpha)_n}.$$

We may now use equation (9) above to conclude that

$$\psi = (1-t)^{-1-\alpha}\exp\left(\frac{-xt}{1-t}\right)\sum_{k=0}^{\infty}\frac{(1-t)^{-k}L_k^{(\alpha)}\left(\frac{x}{1-t}\right)(-yt)^k}{(1+\alpha)_k}$$

$$= (1-t)^{-1-\alpha}\exp\left(\frac{-xt}{1-t}\right)\exp\left(\frac{-yt}{1-t}\right){}_0F_1\left[\begin{matrix} -; \\ 1+\alpha; \end{matrix}\quad \frac{xyt}{(1-t)^2}\right].$$

THEOREM 69. *If $|t| < 1$ and α is not a negative integer,*

$$(1-t)^{-1-\alpha}\exp\left[\frac{-(x+y)t}{1-t}\right]{}_0F_1\left[\begin{matrix} -; \\ 1+\alpha; \end{matrix}\quad \frac{xyt}{(1-t)^2}\right]$$

$$= \sum_{n=0}^{\infty}\frac{n!L_n^{(\alpha)}(x)L_n^{(\alpha)}(y)t^n}{(1+\alpha)_n}.$$

Theorem 69 has been known for a long time.

Applying formula (9) of the preceding section to the relation

$$(1)\qquad (1-t)^{-c}\,{}_1F_1\left[\begin{matrix} c; \\ 1+\alpha; \end{matrix}\quad \frac{-xt}{1-t}\right] = \sum_{k=0}^{\infty}\frac{(c)_kL_k^{(\alpha)}(x)t^k}{(1+\alpha)_k},$$

which is (3) of Section 113, yields a result obtained by contour integral methods by Brafman [2]. We shall perform this transformation to exhibit the technique. In (1) replace x by $x(1-t)^{-1}$, t by $(-yt)(1-t)^{-1}$, and multiply both sides by $(1-t)^{-1-\alpha}\exp[-xt(1-t)^{-1}]$. The result is

$$(1-t)^{-1-\alpha}\left[1+\frac{yt}{1-t}\right]^{-c}\exp\left(\frac{-xt}{1-t}\right){}_1F_1\left[\begin{matrix} c; \\ 1+\alpha; \end{matrix}\quad \frac{xyt(1-t)^{-2}}{1+yt(1-t)^{-1}}\right]$$

$$= \sum_{k=0}^{\infty}\frac{(c)_k(1-t)^{-1-\alpha-k}\exp\left(\frac{-xt}{1-t}\right)L_k^{(\alpha)}\left(\frac{x}{1-t}\right)(-1)^ky^kt^k}{(1+\alpha)_k}$$

$$= \sum_{n,k=0}^{\infty}\frac{(c)_k(n+k)!L_{n+k}^{(\alpha)}(x)t^n(-1)^ky^kt^k}{(1+\alpha)_kk!n!}$$

$$= \sum_{n=0}^{\infty}\sum_{k=0}^{n}\frac{(c)_kn!L_n^{(\alpha)}(x)(-1)^ky^kt^n}{k!(n-k)!(1+\alpha)_k}$$

$$= \sum_{n=0}^{\infty}\sum_{k=0}^{n}\frac{(-n)_k(c)_ky^kL_n^{(\alpha)}(x)t^n}{k!(1+\alpha)_k}$$

$$= \sum_{n=0}^{\infty} {}_2F_1\left[\begin{matrix} -n, c; \\ 1+\alpha; \end{matrix} \quad y\right] L_n^{(\alpha)}(x)t^n.$$

We may rearrange the resulting relation into the form

$$(1-t)^{-1+c-\alpha}(1-t+yt)^{-c}\exp\left(\frac{-xt}{1-t}\right) {}_1F_1\left[\begin{matrix} c; \\ 1+\alpha; \end{matrix} \quad \frac{xyt}{(1-t)(1-t+yt)}\right]$$

$$= \sum_{n=0}^{\infty} {}_2F_1\left[\begin{matrix} -n, c; \\ 1+\alpha; \end{matrix} \quad y\right] L_n^{(\alpha)}(x)t^n,$$

a bilateral generating function involving the Laguerre polynomial and a certain terminating ${}_2F_1$.

121. The simple Laguerre polynomials. When $\alpha = 0$ the resultant polynomial is denoted by $L_n(x)$. It is called the simple Laguerre polynomial or just the Laguerre polynomial when no confusion with $L_n^{(\alpha)}(x)$ is anticipated. Because $L_n(x)$ arises frequently, we now list properties of that polynomial for convenience of the reader. Each of the following results, except for (27), may be obtained by putting $\alpha = 0$ in a result already known for $L_n^{(\alpha)}(x)$.

(1) $$L_n(x) = {}_1F_1(-n; 1; x),$$

(2) $$L_n(x) = \sum_{k=0}^{n} \frac{(-1)^k n! x^k}{(k!)^2(n-k)!},$$

(3) $$e^t \, {}_0F_1(-; 1; -xt) = \sum_{n=0}^{\infty} \frac{L_n(x)t^n}{n!},$$

(4) $$(1-t)^{-c} \, {}_1F_1\left[\begin{matrix} c; \\ 1; \end{matrix} \quad \frac{-xt}{1-t}\right] = \sum_{n=0}^{\infty} \frac{(c)_n L_n(x)t^n}{n!},$$

(4a) $$(1-t)^{-1}\exp\left(\frac{-xt}{1-t}\right) = \sum_{n=0}^{\infty} L_n(x)t^n,$$

(5) $$xL_n'(x) = nL_n(x) - nL_{n-1}(x),$$

(6) $$L_n'(x) = L_{n-1}'(x) - L_{n-1}(x),$$

(7) $$L_n'(x) = -\sum_{k=0}^{n-1} L_k(x),$$

(8) $$nL_n(x) = (2n - 1 - x)L_{n-1}(x) - (n - 1)L_{n-2}(x),$$

(9) $$L_n(x) = \frac{e^x}{n!}\frac{d^n}{dx^n}(x^n e^{-x}),$$

(10) $$xL_n''(x) + (1 - x)L_n'(x) + nL_n(x) = 0,$$

(11) $$(m - n)\int_a^b e^{-x}L_n(x)L_m(x)\,dx = \Big[xe^{-x}\{L_m(x)L_n'(x) - L_n(x)L_m'(x)\}\Big]_a^b,$$

(12) $$\int_0^\infty e^{-x}L_n(x)L_m(x)\,dx = 0, \qquad m \neq n,$$

(13) $$\int_0^\infty e^{-x}L_n^2(x)\,dx = 1,$$

(14) $$\int_0^\infty e^{-x}x^kL_n(x)\,dx = 0; \qquad k = 0, 1, 2, \cdots, (n - 1),$$

(15) $$\int_0^\infty e^{-x}x^nL_n(x)\,dx = (-1)^n n!,$$

(16) The zeros of $L_n(x)$ are positive and distinct,

(17) $$\sum_{k=0}^{n} L_k(x)L_k(y) = (n+1)(x-y)^{-1}[L_{n+1}(y)L_n(x) - L_{n+1}(x)L_n(y)],$$

(18) *If* $\int_0^\infty e^{-x}f^2(x)\,dx$ *exists,*

$$\lim_{n\to\infty}\int_0^\infty e^{-x}f(x)L_n(x)\,dx = 0,$$

(19) $$x^n = \sum_{k=0}^{n}\frac{(-1)^k(n!)^2L_k(x)}{k!(n-k)!},$$

(20) $$H_n(x) = 2^n n!\sum_{k=0}^{n} {}_2F_2\left[\begin{matrix}-\tfrac{1}{2}(n-k),\ -\tfrac{1}{2}(n-k-1); \\ -\tfrac{1}{2}n,\ -\tfrac{1}{2}(n-1);\end{matrix}\ -\tfrac{1}{4}\right]\cdot \frac{(-n)_kL_k(x)}{k!}; \qquad n > 1,$$

(21) $$P_n(x) = 2^n(\tfrac{1}{2})_n \sum_{k=0}^{n} {}_2F_3\left[\begin{matrix} -\tfrac{1}{2}(n-k),\ -\tfrac{1}{2}(n-k-1); \\ \tfrac{1}{2}-n,\ -\tfrac{1}{2}n,\ -\tfrac{1}{2}(n-1); \end{matrix}\ \tfrac{1}{4}\right] \frac{(-n)_k L_k(x)}{k!}; \qquad n > 1,$$

(22) $$L_n(x) = \sum_{k=0}^{n} {}_2F_2\left[\begin{matrix} -\tfrac{1}{2}(n-k),\ -\tfrac{1}{2}(n-k-1); \\ \tfrac{1}{2}(1+k),\ \tfrac{1}{2}(2+k); \end{matrix}\ \tfrac{1}{4}\right] \frac{(-n)_k H_k(x)}{2^k(k!)^2},$$

(23) $$L_n(x) = \sum_{k=0}^{n} {}_2F_3\left[\begin{matrix} -\tfrac{1}{2}(n-k),\ -\tfrac{1}{2}(n-k-1); \\ \tfrac{3}{2}+k,\ \tfrac{1}{2}(1+k),\ \tfrac{1}{2}(2+k); \end{matrix}\ \tfrac{1}{4}\right] \frac{(-n)_k 2^k P_k(x)}{(2k)!},$$

(24) $$L_n(xy) = \sum_{k=0}^{n} C_{n,k}(1-y)^{n-k} y^k L_k(x); \qquad C_{n,k} = \frac{n!}{k!(n-k)!},$$

(25) $$\sum_{n=0}^{\infty} \frac{(n+k)! L_{n+k}(x) t^n}{k!\,n!} = (1-t)^{-1-k} \exp\left(\frac{-xt}{1-t}\right) L_k\left(\frac{x}{1-t}\right),$$

(26) $$(1-t)^{-1} \exp\left[\frac{-(x+y)t}{1-t}\right] {}_0F_1\left[\begin{matrix} -; \\ 1; \end{matrix}\ \frac{xyt}{(1-t)^2}\right] = \sum_{n=0}^{\infty} L_n(x) L_n(y) t^n,$$

(27) $$(1-2xt+t^2)^{-\frac{1}{2}} \exp\left[\frac{yt(t-x)}{1-2xt+t^2}\right] {}_0F_1\left[\begin{matrix} -; \\ 1; \end{matrix}\ \frac{y^2t^2(x^2-1)}{4(1-2xt+t^2)^2}\right] = \sum_{n=0}^{\infty} L_n(y) P_n(x) t^n,$$

(28) $$(1-t)^{c-1}(1-t+yt)^{-c} \exp\left(\frac{-xt}{1-t}\right) \cdot {}_1F_1\left[\begin{matrix} c; \\ 1; \end{matrix}\ \frac{xyt}{(1-t)(1-t+yt)}\right] = \sum_{n=0}^{\infty} {}_2F_1\left[\begin{matrix} -n,\ c; \\ 1; \end{matrix}\ y\right] L_n(x) t^n.$$

EXERCISES

1. Show that

$$H_{2n}(x) = (-1)^n 2^{2n} n! L_n^{(-\frac{1}{2})}(x^2),$$
$$H_{2n+1}(x) = (-1)^n 2^{2n+1} n! x L_n^{(\frac{1}{2})}(x^2).$$

2. Use Theorem 65, page 181, and the method of Section 118 above to derive the result

$$L_n^{(\alpha)}(x) = \sum_{k=0}^{n} {}_2F_3\left[\begin{matrix} -\frac{1}{2}(n-k),\ -\frac{1}{2}(n-k-1); \\ \frac{3}{2}+k,\ \frac{1}{2}(1+\alpha+k),\ \frac{1}{2}(2+\alpha+k); \end{matrix}\ \frac{1}{4}\right] \frac{(-1)^k(1+\alpha)_n(2k+1)P_k(x)}{2^k(n-k)!(\frac{3}{2})_k(1+\alpha)_k}.$$

3. Use formula (4), page 194, and the method of Section 118 to derive the result

$$L_n^{(\alpha)}(x) = \sum_{k=0}^{n} {}_2F_2\left[\begin{matrix} -\frac{1}{2}(n-k),\ -\frac{1}{2}(n-k-1); \\ \frac{1}{2}(1+\alpha+k),\ \frac{1}{2}(2+\alpha+k); \end{matrix}\ \frac{1}{4}\right] \frac{(-1)^k(1+\alpha)_n H_k(x)}{k!(n-k)!2^k(1+\alpha)_k}.$$

4. Use the results in Section 56, page 102, to show that

$$\int_0^t L_n[x(t-x)]\,dx = \frac{(-1)^n H_{2n+1}(\frac{1}{2}t)}{2^{2n}(\frac{3}{2})_n}.$$

5. Use the results in Section 56, page 102, to show that

$$\int_0^t \frac{H_{2n}(\sqrt{x(t-x)})\,dx}{\sqrt{x(t-x)}} = (-1)^n \pi 2^{2n} (\tfrac{1}{2})_n L_n(\tfrac{1}{4}t^2).$$

6. Show that if m is a non-negative integer and α is not a negative integer,

$$L_n^{(\alpha)}(x) = \frac{(1+\alpha)_n}{(1+\frac{1}{2}\alpha+\frac{1}{2}m)_n} \sum_{k=0}^{n} \frac{(\frac{1}{2}\alpha-\frac{1}{2}m)_k L_k^{(\alpha)}(-x) L_{n-k}^{(m)}(x)}{(1+\alpha)_k}.$$

7. Show that if m is a non-negative integer and α is not a negative integer,

$$L_n^{(\alpha)}(x) = \frac{(1+\alpha)_n(1+\alpha)_m}{(1+\alpha)_{m+n}} \sum_{k=0}^{n} \frac{(-m)_k L_k^{(\alpha)}(-x) L_{n-k}^{(\alpha+2m)}(x)}{(1+\alpha)_k}.$$

8. Use integration by parts and equation (2), page 202, to show that

$$\int_x^\infty e^{-y} L_n^{(\alpha)}(y)\,dy = e^{-x}[L_n^{(\alpha)}(x) - L_{n-1}^{(\alpha)}(x)].$$

9. Show that

$$\int_0^t x^\alpha (t-x)^{\beta-1} L_n^{(\alpha)}(x)\,dx = \frac{\Gamma(1+\alpha)\Gamma(\beta)}{\Gamma(1+\alpha+\beta)} \cdot \frac{(1+\alpha)_n t^{\alpha+\beta}}{(1+\alpha+\beta)_n} L_n^{(\alpha+\beta)}(t).$$

10. Show that the Laplace transform of $L_n(t)$ is

$$\int_0^\infty e^{-st} L_n(t)\,dt = \frac{1}{s}\left(1 - \frac{1}{s}\right)^n.$$

11. Show by the convolution theorem for Laplace transforms, or otherwise, that

$$\int_0^t L_n(t-x)L_m(x)\,dx = \int_0^t L_{m+n}(x)\,dx = L_{m+n}(t) - L_{m+n+1}(t).$$

12. Evaluate the integral

$$\int_0^\infty x^\alpha e^{-x}[L_n^{(\alpha)}(x)]^2\,dx$$

of (7), page 206, by the following method. From (4), Section 113, show that

$$\sum_{n=0}^\infty \int_0^\infty x^\alpha e^{-x}[L_n^{(\alpha)}(x)]^2 dx\, t^{2n} = (1-t)^{-2-2\alpha}\int_0^\infty x^\alpha \exp\left[\frac{-x(1+t)}{1-t}\right] dx$$

$$= (1-t^2)^{-1-\alpha}\Gamma(1+\alpha) = \sum_{n=0}^\infty \frac{\Gamma(1+\alpha+n)t^{2n}}{n!}.$$

CHAPTER 13

The Sheffer Classification and Related Topics

122. Differential operators and polynomial sets. Let $\varphi_n(x)$; $n = 0, 1, 2, \cdots$, be any simple set of polynomials and let $D \equiv d/dx$. Let us define the set (not necessarily a simple set) of polynomials $T_n(x)$, $n \geqq 0$, by

$$T_0(x)D\varphi_1(x) = \varphi_0(x), \tag{1}$$

$$T_n(x)D^{n+1}\varphi_{n+1}(x) = \varphi_n(x) - \sum_{k=0}^{n-1} T_k(x)D^{k+1}\varphi_{n+1}(x), \; n \geqq 1. \tag{2}$$

Because $\varphi_n(x)$ is of degree precisely n for each n, it follows that $T_n(x)$ is uniquely defined and is of degree $\leqq n$. Note that $D\varphi_1$ is constant, as is φ_0, so $T_0(x)$ is constant. For $n \geqq 1$, each $T_n(x)$ is defined by (2) in terms of previous elements of the set, $T_k(x)$ for $0 \leqq k \leqq (n-1)$. Because $D^{n+1}\varphi_{n+1}$ is constant and the degree of $T_k(x)\, D^{k+1}\varphi_{n+1}(x)$ exceeds the degree of $T_k(x)$ by exactly $(n-k)$, each member of (2) has degree at most n.

THEOREM 70. *For the simple set of polynomials $\varphi_n(x)$ there exists a unique differential operator of the form*

$$J(x, D) = \sum_{k=0}^{\infty} T_k(x)\, D^{k+1} \tag{3}$$

in which $T_k(x)$ is a polynominal of degree $\leqq k$, for which

$$J(x, D)\varphi_n(x) = \varphi_{n-1}(x), \qquad n \geqq 1. \tag{4}$$

It is important that J be independent of n.

Proof: The requirement (4) demands that, for $n \geqq 1$,

$$\sum_{k=0}^{n-1} T_k(x)\, D^{k+1}\varphi_n(x) = \varphi_{n-1}(x),$$

which is merely a restatement of (1) and (2). Equations (1) and (2), as we saw, determine $T_k(x)$ uniquely.

We say that the polynomial set $\varphi_n(x)$ *belongs to the operator* J and that J is the operator associated with the set $\varphi_n(x)$. There is only one such operator associated with a given $\varphi_n(x)$,but there are infinitely many sets of polynomials belonging to the same operator.

THEOREM 71. *A necessary and sufficient condition that two simple sets of polynomials $\varphi_n(x)$ and $\psi_n(x)$ belong to the same operator J is that there exists a sequence of numbers b_k, independent of n, such that*

$$\psi_n(x) = \sum_{k=0}^{n} b_k\varphi_{n-k}(x). \tag{5}$$

Proof: Assume (5) to hold. There exists an operator J to which $\varphi_n(x)$ belongs. That $\psi_n(x)$ belongs to the same operator follows from

$$J\psi_n(x) = \sum_{k=0}^{n} b_k J\varphi_{n-k}(x) = \sum_{k=0}^{n-1} b_k\varphi_{n-1-k} = \psi_{n-1}(x).$$

Next assume that $\varphi_n(x)$ and $\psi_n(x)$ belong to the same operator J. We need to show that the b_k of (5) exist. We know, because $\varphi_n(x)$ and $\psi_n(x)$ are simple sets of polynomials, that there exists the relation

$$\psi_n(x) = \sum_{k=0}^{n} A(k, n)\varphi_{n-k}(x), \tag{6}$$

but, in general, the coefficients $A(k, n)$ depend upon n as well as on k. Since $\varphi_n(x)$ and $\psi_n(x)$ belong to J, we may apply J to each member of (6) and obtain

$$\psi_{n-1}(x) = \sum_{k=0}^{n-1} A(k, n)\varphi_{n-1-k}(x), \qquad n \geqq 1. \tag{7}$$

Recall that $J\varphi_0(x) = 0$, which is the reason that the term $A(n, n)\varphi_0(x)$ dropped out when the operator J was applied to (6). We may shift index from n to $(n + 1)$ in (7) to get

$$\psi_n(x) = \sum_{k=0}^{n} A(k, n + 1)\varphi_{n-k}(x); \qquad n \geqq 0. \tag{8}$$

Comparing (6) and (8), we see that

$$A(k, n) = A(k, n + 1)$$

for all k, n. Then $A(k, n) = b_k$, independent of n.

Not every operator of the form (3) is associated with some polynomial set in the sense we have defined. For the operator J of the form (3) to be associated with some simple set, it is necessary and sufficient that J transform every polynomial of degree precisely n into a polynomial of degree precisely $(n - 1)$.

EXAMPLE: Determine the operator associated with the set $\varphi_n(x) = H_n(x)/(n!)^2$, in which $H_n(x)$ is the Hermite polynomial.

Here $\varphi_0(x) = 1$, $\varphi_1(x) = 2x$, $\varphi_2(x) = x^2 - \frac{1}{2}$, $\varphi_3(x) = \frac{2}{9}x^3 - \frac{1}{3}x$, etc. We seek an operator J of the form

$$J = \sum_{k=0}^{\infty} T_k(x)\, D^{k+1}$$

such that $J\varphi_n = \varphi_{n-1}$ for $n \geqq 1$. Then

$$T_0(x)\, D\varphi_1 = \varphi_0, \qquad \text{or} \qquad T_0(x) \cdot 2 = 1,$$

so that $T_0(x) = \frac{1}{2}$. Next we have

$$[T_0(x)\, D + T_1(x)\, D^2]\varphi_2(x) = \varphi_1(x),$$

or

$$[\tfrac{1}{2}D + T_1(x)\, D^2](x^2 - \tfrac{1}{2}) = 2x.$$

Then

$$\tfrac{1}{2}(2x) + T_1(x)(2) = 2x,$$

so that $T_1(x) = \frac{1}{2}x$. In turn

$$[\tfrac{1}{2}D + \tfrac{1}{2}x\, D^2 + T_2(x)\, D^3](\tfrac{2}{9}x^3 - \tfrac{1}{3}x) = x^2 - \tfrac{1}{2},$$

or

$$\tfrac{1}{2}(\tfrac{2}{3}x^2 - \tfrac{1}{3}) + \tfrac{1}{2}x(\tfrac{4}{3}x) + T_2(x)(\tfrac{4}{3}) = x^2 - \tfrac{1}{2},$$

from which $T_2(x) = -\frac{1}{4}$.

If we continue the above procedure, we find that $T_3(x) = 0$, $T_4(x) = 0$, and we begin to suspect that J may terminate. Let us therefore define

$$J_1 = \tfrac{1}{2}D + \tfrac{1}{2}x\, D^2 - \tfrac{1}{4}D^3, \tag{9}$$

operate on $\varphi_n(x)$ with J_1 and see whether the result is $\varphi_{n-1}(x)$.

Now

$$J_1\varphi_n(x) = J_1\frac{H_n(x)}{(n!)^2} = \frac{1}{(n!)^2}[\tfrac{1}{2}H_n' + \tfrac{1}{2}xH_n'' - \tfrac{1}{4}H_n''']$$

$$= \frac{-1}{4(n!)^2}[H_n''' - 2xH_n'' - 2H_n'].$$

From Hermite's differential equation

$$H_n'' - 2xH_n' + 2nH_n = 0$$

we obtain

$$H_n''' - 2xH_n'' - 2H_n' + 2nH_n' = 0$$

so that we have

$$J_1\varphi_n(x) = \frac{-[-2nH_n']}{4(n!)^2} = \frac{H_n'}{2n!(n-1)!}.$$

But we also know that $H_n'(x) = 2nH_{n-1}(x)$. Hence

$$J_1\varphi_n(x) = \frac{2nH_{n-1}(x)}{2n!(n-1)!} = \frac{H_{n-1}(x)}{[(n-1)!]^2} = \varphi_{n-1}(x),$$

as desired. Therefore $\varphi_n(x) = H_n(x)/(n!)^2$ belongs to the operator J_1 of (9).

The example above yields a specific result of interest to us later, but its simplicity may be misleading. It is wise to keep in mind that the operator associated with a given simple set of polynomials need not terminate and, indeed, that there is no reason to think that the $T_k(x)$ in the operator of equation (3) can be determined other than successively by iteration of equation (2).

123. Sheffer's A-type classification. Sheffer [1] used the operators discussed in Section 122 to classify polynomial sets. The first of his classifications will now be discussed.

DEFINITION: Let $\varphi_n(x)$ be a simple set of polynomials and let $\varphi_n(x)$ belong to the operator

$$(1) \qquad J(x, D) = \sum_{k=0}^{\infty} T_k(x)\, D^{k+1},$$

with $T_k(x)$ of degree $\leq k$, in the sense of Section 122. If the maximum degree of the coefficients $T_k(x)$ is m, we say that the set $\varphi_n(x)$ is of Sheffer A-type m. If the degree of $T_k(x)$ is unbounded as $k \to \infty$, we say that $\varphi_n(x)$ is of Sheffer A-type ∞.

EXAMPLE: The set $\varphi_n(x) = H_n(x)/(n!)^2$ is of Sheffer A-type one, since $\varphi_n(x)$ belongs to the operator J_1 of (9), Section 122.

Explicit polynomial sets of type m for each m are obtained, with corresponding generating functions, by Huff and Rainville [1]. They show, among other things, that if $y_n(x)$ is defined by

$$(2) \qquad \varphi(t)\ {}_0F_m(-;\ \beta_1, \beta_2, \cdots, \beta_m;\ \sigma x t) = \sum_{n=0}^{\infty} y_n(x) t^n,$$

with σ constant and $\varphi(t)$ analytic and not zero at $t = 0$, then $y_n(x)$ is of Sheffer A-type m. The Sheffer [1] paper contains a study of polynomials of any A-type, but the most satisfying results are those bearing on polynomials of type zero.

124. Polynomials of Sheffer A-type zero. Let $\varphi_n(x)$ be of Sheffer A-type zero. Then $\varphi_n(x)$ belongs to an operator

$$(1) \qquad J(D) = \sum_{k=0}^{\infty} c_k D^{k+1}$$

in which the c_k are constants. Note that $c_0 \neq 0$, since $J\varphi_n = \varphi_{n-1}$. Furthermore, since c_k is independent of x for every k, a function $J(t)$ exists with the formal power-series expansion

$$(2) \qquad J(t) = \sum_{k=0}^{\infty} c_k t^{k+1}, \qquad c_0 \neq 0.$$

Let $H(t)$ be the formal inverse of $J(t)$; that is,

$$(3) \qquad J\big(H(t)\big) = H\big(J(t)\big) = t.$$

THEOREM 72. *A necessary and sufficient condition that $\varphi_n(x)$ be of Sheffer A-type zero is that $\varphi_n(x)$ possess the generating function indicated in*

$$(4) \qquad A(t) \exp[xH(t)] = \sum_{n=0}^{\infty} \varphi_n(x) t^n,$$

in which $H(t)$ and $A(t)$ have (at least the formal) expansions

$$(5) \qquad H(t) = \sum_{n=0}^{\infty} h_n t^{n+1}, \qquad h_0 \neq 0,$$

$$(6) \qquad A(t) = \sum_{n=0}^{\infty} a_n t^n, \qquad a_0 \neq 0.$$

Proof: Assume (4), (5), and (6). Then the $\varphi_n(x)$ form a simple set of polynomials. The function $H(t)$ has a formal inverse. Call it $J(t)$, defined by (3),

$$J(t) = \sum_{k=0}^{\infty} c_k t^{k+1}, \qquad c_0 \neq 0. \tag{7}$$

Let, as usual, $D = d/dx$. Then, by (4),

$$\begin{aligned} J(D) \sum_{n=0}^{\infty} \varphi_n(x)t^n &= J(D)\,\{A(t)\exp[xH(t)]\} \\ &= A(t)J\big(H(t)\big)\exp[xH(t)] \\ &= tA(t)\exp[xH(t)]. \end{aligned}$$

Hence

$$\begin{aligned} \sum_{n=0}^{\infty} J(D)\,\varphi_n(x)t^n &= \sum_{n=0}^{\infty} \varphi_n(x)t^{n+1} \\ &= \sum_{n=1}^{\infty} \varphi_{n-1}(x)t^n, \end{aligned}$$

from which $J(D)\varphi_0(x) = 0$ and $J(D)\varphi_n(x) = \varphi_{n-1}(x)$, $n \geqq 1$. Since $J(D)$ is independent of x, $\varphi_n(x)$ is of Sheffer A-type zero.

Next assume that $\varphi_n(x)$ is of Sheffer A-type zero. Let $\varphi_n(x)$ belong to the operator J. Then

$$J = J(D) = \sum_{k=0}^{\infty} c_k D^{k+1}, \qquad c_0 \neq 0, \tag{8}$$

and the c_k are independent of x because $\varphi_n(x)$ is of A-type zero. The function $J(t)$ has a formal inverse $H(t)$ defined by (3) above. Consider the polynomials $\psi_n(x)$, a simple set, defined by

$$\exp[xH(t)] = \sum_{n=0}^{\infty} \psi_n(x)t^n. \tag{9}$$

We know already that $\psi_n(x)$ is of Sheffer A-type zero and that it belongs to the operator $J(D)$. By Theorem 71, page 219, there exists a sequence a_k, independent of n, such that

$$\varphi_n(x) = \sum_{k=0}^{n} a_k \psi_{n-k}(x). \tag{10}$$

Define $A(t)$ by

$$A(t) = \sum_{n=0}^{\infty} a_n t^n.$$

Then

$$A(t)\exp[xH(t)] = \left(\sum_{n=0}^{\infty} a_n t^n\right)\left(\sum_{n=0}^{\infty} \psi_n(x) t^n\right)$$

$$= \sum_{n=0}^{\infty}\sum_{k=0}^{n} a_k \psi_{n-k}(x) t^n$$

$$= \sum_{n=0}^{\infty} \varphi_n(x) t^n,$$

which is (4).

The generating function in Theorem 72 is one of Boas and Buck kind (Section 76) with $\gamma_n = 1/n!$. It follows that Theorem 50, page 141, may be applied to any $\varphi_n(x)$ of Sheffer A-type zero.

Sheffer [1] obtained many properties of polynomial sets of A-type zero. Here we state (with some modifications in notation) only a few of his results.

THEOREM 73. *Let $\varphi_n(x)$ be of Sheffer A-type zero; let $\varphi_n(x)$ belong to the operator $J(D)$ and have the generating function in (4) of Theorem 72. There exist sequences α_k and ϵ_k, independent of x and n, such that for $n \geqq 1$,*

$$\sum_{k=0}^{n-1} (\alpha_k + x\epsilon_k) J^{k+1}(D)\, \varphi_n(x) = n\varphi_n(x), \tag{11}$$

$$\sum_{k=0}^{\infty} \alpha_k t^k = \frac{A'(t)}{A(t)}, \tag{12}$$

$$\sum_{k=0}^{\infty} \epsilon_k t^k = H'(t). \tag{13}$$

Of course $\epsilon_k = (k+1)h_k$, in terms of the h_k of equation (5), page 222.

THEOREM 74. *With the assumptions of Theorem 73 there exist sequences μ_k and ν_k, independent of x and n, such that*

$$\sum_{k=0}^{n-1} (\mu_k + x\nu_k) D^{k+1}\varphi_n(x) = n\varphi_n(x), \tag{14}$$

$$\sum_{k=0}^{\infty} \mu_k t^{k+1} = \frac{uA'(u)}{A(u)}, \qquad u = J(t), \tag{15}$$

$$\sum_{k=0}^{\infty} \nu_k t^{k+1} = uH'(u), \qquad u = J(t). \tag{16}$$

THEOREM 75. *A necessary and sufficient condition that $\varphi_n(x)$ be of Sheffer A-type zero is that there exist sequences α_k and ϵ_k, independent of x and n, such that*

$$\sum_{k=0}^{n-1} (\alpha_k + x\epsilon_k)\varphi_{n-1-k}(x) = n\varphi_n(x). \tag{17}$$

The α_k and ϵ_k of Theorem 75 are those of Theorem 73.

THEOREM 76. *A necessary and sufficient condition that $\varphi_n(x)$ be of Sheffer A-type zero is that there exists a sequence h_k, independent of x and n, such that*

$$\sum_{k=0}^{n-1} h_k\varphi_{n-1-k}(x) = \varphi_n'(x), \qquad n \geqq 1. \tag{18}$$

Proof of Theorems 73–76: These theorems all follow from Theorem 72. For instance, consider the series

$$\begin{aligned}\sum_{n=0}^{\infty} n\varphi_n(x)t^n &= t\frac{\partial}{\partial t}\{A(t)\exp[xH(t)]\}\\ &= t[A'(t) + xH'(t)A(t)]\exp[xH(t)]\\ &= t\left[\frac{A'(t)}{A(t)} + xH'(t)\right]A(t)\exp[xH(t)].\end{aligned}$$

The functions $A'(t)/A(t)$ and $H'(t)$ have series expansions; let them be (12) and (13), thus defining α_k and ϵ_k. Then

$$\begin{aligned}\sum_{n=1}^{\infty} n\varphi_n(x)t^n &= t\left[\sum_{n=0}^{\infty}(\alpha_n + x\epsilon_n)t^n\right]\left[\sum_{n=0}^{\infty}\varphi_n(x)t^n\right]\\ &= \sum_{n=0}^{\infty}\sum_{k=0}^{n}(\alpha_k + x\epsilon_k)\varphi_{n-k}(x)t^{n+1}\\ &= \sum_{n=1}^{\infty}\sum_{k=0}^{n-1}(\alpha_k + x\epsilon_k)\varphi_{n-1-k}(x)t^n.\end{aligned}$$

Since

$$J^{k+1}(D)\,\varphi_n(x) = J^k(D)\,\varphi_{n-1}(x) = \cdots = \varphi_{n-1-k}(x),$$

the above argument yields both Theorems 73 and 75.

Theorem 74 is a result of rearrangement of terms in the result of Theorem 73.

Theorem 76 follows from the fact that

$$\begin{aligned}\sum_{n=1}^{\infty} \varphi_n'(x)t^n &= \frac{\partial}{\partial x}\{A(t) \exp[xH(t)]\} \\ &= H(t)A(t) \exp[xH(t)] \\ &= \left(\sum_{n=0}^{\infty} h_n t^{n+1}\right)\left(\sum_{n=0}^{\infty} \varphi_n(x)t^n\right) \\ &= \sum_{n=1}^{\infty} \sum_{k=0}^{n-1} h_k \varphi_{n-1-k}(x)t^n,\end{aligned}$$

and the fact that from the equality of the first and last series above, it follows that

$$y = \sum_{n=0}^{\infty} \varphi_n(x)t^n$$

must satisfy the equation $Dy = H(t)y$.

We shall meet many instances of polynomial sets of Sheffer A-type zero. In examining a new set of polynomials, it seems desirable always to obtain its Sheffer A-type. For Sheffer's classifications of B-types and C-types and for more results on the A-type, see Sheffer [1] and also the exercises at the end of this chapter.

125. An extension of Sheffer's classification. It can be seen from the preceding section that much of the value of the A-type classification lies in the existence of the generating relation

$$A(t) \exp[xH(t)] = \sum_{n=0}^{\infty} \varphi_n(x)t^n$$

for sets $\varphi_n(x)$ of A-type zero. An essential characteristic of the function $y = \exp[xH(t)]$ is that, with $D = d/dx$,

$$Dy = H(t)y.$$

It is natural to expect that much can be retained if D is replaced by some other differential operator σ and $\exp(z)$ by some other function $F(z)$ such that

$$\sigma F(z) = F(z).$$

Let $D = d/dx$, $\theta = xD$, and define the differential operator σ by

$$\sigma = D \prod_{i=1}^{q} (\theta + b_i - 1), \tag{1}$$

in which the b_i are constants and no b_i is either zero or a negative integer. Note that application of σ reduces the degree of any polynomial by exactly one.

THEOREM 77. *Let $\varphi_n(x)$ be a simple set of polynomials. There exists a unique set of polynomials $T_k(x)$, $T_k(x)$ of degree $\leqq k$, such that the differential operator*

$$J(x, \sigma) = \sum_{k=0}^{\infty} T_k(x)\sigma^{k+1} \tag{2}$$

has the property that

$$J(x, \sigma)\varphi_n(x) = \varphi_{n-1}(x), \qquad n \geqq 1. \tag{3}$$

It is to be noted that $J(x, \sigma)\varphi_0(x) = 0$ and that $J(x, \sigma)$ is independent of n.

Proof: Given $\varphi_n(x)$, define polynomials $T_n(x)$ successively by

$$T_0(x)\sigma\varphi_1(x) = \varphi_0(x), \tag{4}$$

$$T_n(x)\sigma^{n+1}\varphi_{n+1}(x) = \varphi_n(x) - \sum_{k=0}^{n-1} T_k(x)\sigma^{k+1}\varphi_{n+1}(x), \qquad n \geqq 1. \tag{5}$$

The definition of $T_n(x)$ is unique and the degree of $T_n(x)$ is $\leqq n$, by the same arguments as were used in the Sheffer classification. Furthermore,

$$J(x, \sigma)\varphi_n(x) = \sum_{k=0}^{n-1} T_k(x)\sigma^{k+1}\varphi_n(x) = \varphi_{n-1}(x)$$

by (5) with n replaced by $(n - 1)$.

Again we say that the polynomial set $\varphi_n(x)$ belongs to the operator $J(x, \sigma)$ and that $J(x, \sigma)$ is the operator associated with $\varphi_n(x)$. There is only one operator associated with $\varphi_n(x)$, but infinitely many sets of polynomials belong to a specified permissible (it must transform every polynomial of degree n into one of degree $n - 1$) operator $J(x, \sigma)$.

THEOREM 78. *A necessary and sufficient condition that two simple sets of polynomials $\varphi_n(x)$ and $\psi_n(x)$ belong to the same operator $J(x, \sigma)$ is that there exists a sequence of numbers b_k, independent of n, such that*

$$\psi_n(x) = \sum_{k=0}^{n} b_k\varphi_{n-k}(x). \tag{6}$$

Proof: Parallel the proof of Theorem 71 in every respect.

DEFINITION: Let $\varphi_n(x)$ be a simple set of polynomials, and let $\varphi_n(x)$ belong to the operator

$$J(x, \sigma) = \sum_{k=0}^{\infty} T_k(x)\sigma^{k+1}, \tag{2}$$

with $T_k(x)$ of degree $\leqq k$ and $J(x, \sigma)\varphi_n(x) = \varphi_{n-1}(x)$. If the maximum degree of the coefficients $T_k(x)$ is m, we say that the set $\varphi_n(x)$ is of σ-type m; if the degree of $T_k(x)$ is unbounded, as $k \to \infty$, we say the set $\varphi_n(x)$ is of σ-type ∞.

125. Polynomials of σ-type zero. Let $\varphi_n(x)$ be of σ-type zero. Then $\varphi_n(x)$ belongs to an operator

$$J(\sigma) = \sum_{k=0}^{\infty} c_k \sigma^{k+1} \tag{1}$$

in which the c_k are constants and $c_0 \neq 0$, since $J(\sigma)\varphi_n(x) = \varphi_{n-1}(x)$. Here again we are using $D = d/dx$, $\theta = xD$,

$$\sigma = D \prod_{i=1}^{q} (\theta + b_i - 1), \tag{2}$$

as in the preceding section. Since c_k is independent of x, a function $J(t)$ exists with an inverse $H(t)$:

$$J\big(H(t)\big) = H\big(J(t)\big) = t,$$

$$J(t) = \sum_{n=0}^{\infty} c_n t^{n+1}, \qquad c_0 \neq 0, \tag{3}$$

$$H(t) = \sum_{n=0}^{\infty} h_n t^{n+1}, \qquad h_0 \neq 0. \tag{4}$$

THEOREM 79. *A necessary and sufficient condition that $\varphi_n(x)$ be of σ-type zero, with*

$$\sigma = D \prod_{i=1}^{q} (\theta + b_i - 1), \tag{5}$$

is that $\varphi_n(x)$ possess the generating function in

$$A(t)\, {}_0F_q\big(-; b_1, b_2, \cdots, b_q; xH(t)\big) = \sum_{n=0}^{\infty} \varphi_n(x) t^n, \tag{6}$$

in which $H(t)$ is given by (4) *and $A(t)$ has the formal expansion*

$$A(t) = \sum_{n=0}^{\infty} a_n t^n, \qquad a_0 \neq 0. \tag{7}$$

Proof: Assume (6) together with (4), (5), and (7). Then the $\varphi_n(x)$ form a simple set of polynomials. The function $H(t)$ has a formal inverse; call it $J(t)$ defined by (3). As we saw in Chapter 5, page 75, the function

$$y = {}_0F_q(-; b_1, \cdots, b_q; z)$$

is a solution of the differential equation

$$\left[\theta_z \prod_{i=1}^{q} (\theta_z + b_i - 1) - z\right] y = 0; \qquad \theta_z = z\frac{d}{az}. \tag{8}$$

Put $z = xH(t)$ and hold t constant. Then $\theta = x(d/dx) = \theta_z$ and (8) becomes

$$\theta \prod_{i=1}^{q} (\theta + b_i - 1)y = xH(t)y. \tag{9}$$

But $\theta = xD$, so that (5) and (9) combine to yield

$$\sigma\ {}_0F_q(-; b_1, \cdots, b_q; xH(t)) = H(t)\ {}_0F_q(-; b_1, \cdots, b_q; xH(t)). \tag{10}$$

We now operate with $J(\sigma)$ on both members of (6):

$$\begin{aligned} J(\sigma) \sum_{n=0}^{\infty} \varphi_n(x)t^n &= J(\sigma)A(t)\ {}_0F_q(-; b_1, \cdots, b_q; xH(t)) \\ &= A(t)J(H(t))\ {}_0F_q(-; b_1, \cdots, b_q; xH(t)) \\ &= t\sum_{n=0}^{\infty} \varphi_n(x)t^n = \sum_{n=1}^{\infty} \varphi_{n-1}(x)t^n. \end{aligned}$$

Therefore $J(\sigma)\varphi_0(x) = 0$ and $J(\sigma)\varphi_n(x) = \varphi_{n-1}(x)$ for $n \geqq 1$. Since $J(\sigma)$ is independent of x, $\varphi_n(x)$ is of σ-type zero.

Next assume that $\varphi_n(x)$ is of σ-type zero. Let $\varphi_n(x)$ belong to the operator $J(\sigma)$. The function $J(t)$ has a formal inverse $H(t)$. Consider the polynomials $\psi_n(x)$, a simple set, defined by

$${}_0F_q(-; b_1, \cdots, b_q; xH(t)) = \sum_{n=0}^{\infty} \psi_n(x)t^n. \tag{11}$$

We know that $\psi_n(x)$ is of σ-type zero and that it belongs to the operator $J(\sigma)$. By Theorem 78, page 227, there exists a sequence a_k, independent of n, such that

$$\varphi_n(x) = \sum_{k=0}^{n} a_k\psi_{n-k}(x). \tag{12}$$

Define $A(t)$ by (7). Then, using (12), we find that

$$\begin{aligned} \sum_{n=0}^{\infty} \varphi_n(x)t^n &= \left(\sum_{n=0}^{\infty} a_nt^n\right)\left(\sum_{n=0}^{\infty} \psi_n(x)t^n\right) \\ &= A(t)\ {}_0F_q(-; b_1, \cdots, b_q; xH(t)), \end{aligned}$$

as desired.

It is of interest to note that polynomials of any Sheffer A-type may be polynomials of σ-type zero, if σ is properly chosen. It is also easy to show that a polynomial of Sheffer A-type unity or larger may not be of σ-type zero for any choice of σ.

As an example, consider polynomials $y_n(x)$ defined by

$$e^t\,{}_0F_q(-;1,1,\cdots,1;-xt) = \sum_{n=0}^{\infty} y_n(x)t^n. \tag{13}$$

Here $H(t) = -t$ and $J(t) = -t$. Let $\sigma = D\theta^q$. Then the $y_n(x)$ are of σ-type zero for that σ, but are of Sheffer A-type q, because σ is a polynominal in x and D and is of degree precisely q in x.

Because the generating function in Theorem 79 is of the Boas and Buck kind (Section 76), Theorem 50, page 141, applies to all polynomials of σ-type zero.

THEOREM 80. *With σ specified as in* (5) *of Theorem 79, page 228 a necessary and sufficient condition that $\varphi_n(x)$ be of σ-type zero is that there exists a sequence h_k, independent of x and n, such that*

$$\sum_{k=0}^{n-1} h_k\varphi_{n-1-k}(x) = \sigma\varphi_n(x). \tag{14}$$

Proof: If $\varphi_n(x)$ is of σ-type zero, it follows from Theorem 79 that

$$\begin{aligned}\sum_{n=0}^{\infty} \sigma\varphi_n(x)t^n &= \sigma A(t)\,{}_0F_q(-;b_1,\cdots,b_q;xH(t))\\ &= H(t)A(t)\,{}_0F_q(-;b_1,\cdots,b_q;xH(t))\\ &= \left(\sum_{n=0}^{\infty} h_n t^{n+1}\right)\left(\sum_{n=0}^{\infty}\varphi_n(x)t^n\right)\\ &= \sum_{n=1}^{\infty}\sum_{k=0}^{n-1} h_k\varphi_{n-1-k}(x)t^n,\end{aligned}$$

so that (14) is satisfied. If (14) is satisfied, let

$$y(x,t) = \sum_{n=0}^{\infty}\varphi_n(x)t^n.$$

Then, by (14),

$$\sigma y = \sum_{n=1}^{\infty}\sum_{k=0}^{n-1} h_k\varphi_{n-1-k}(x)t^n = H(t)y(x,t).$$

Hence $y(x,t)$ is one of the solutions of the ${}_0F_q$ equation (Chapter 5, page 75), with $z = xH(t)$. But

(15) $$_0F_q(-;b_1,\cdots,b_q;xH(t))$$

is the only one of those solutions which is nonzero and analytic at $x = 0$, $(z = 0)$. Hence y is a constant (function of t not x) multiple of (15), which concludes the proof.

Our main interest in both the Sheffer and the σ-type classifications centers on the generating functions in Theorems 72 and 79. Upon meeting a new simple set of polynomials, we try to determine its Sheffer A-type and to see whether the set is of σ-type zero for some σ. These tools will be used at times in later chapters.

EXERCISES

1. Prove Theorem A (Sheffer): If $\varphi_n(x)$ is of Sheffer A-type zero, $g_n(m, x) = D^m\varphi_{n+m}(x)$ is also of Sheffer A-type zero and belongs to the same operator as does $\varphi_n(x)$.

2. Prove Theorem B: If $\varphi_n(x)$ is of Sheffer A-type zero,

$$\psi_n(x) = \varphi_n(x)\left[\prod_{i=1}^{m}(1+\rho_i)_n\right]^{-1}$$

is of Sheffer A-type m.

3. Show that

$$\frac{H_n(x)}{n!}$$

is of Sheffer A-type zero, obtain the associated functions $J(t)$, $H(t)$, $A(t)$, and draw what conclusions you can from Theorems 73–76.

4. Show that $L_n^{(\alpha)}(x)$ is of Sheffer A-type zero, and proceed as in Ex. 3.

5. Show that the Newtonian polynomials

$$N_n(x) = \frac{(-1)^n(-x)_n}{n!}$$

are of Sheffer A-type zero, and proceed as in Ex. 3.

6. Show that

$$\varphi_n(x) = \frac{L_n^{(\alpha)}(x)}{(1+\alpha)_n}$$

is of Sheffer A-type unity but that with σ chosen to be

$$\sigma = D(\theta + \alpha)$$

the polynomials $\varphi_n(x)$ are of σ-type zero.

7. Determine the Sheffer operator associated with the set

$$\varphi_n(x) = \frac{L_n(x)}{(n!)^2},$$

and thus show that $\varphi_n(x)$ is of Sheffer A-type 2.

8. Prove that if we know a generating function

$$y(x, t) = \sum_{n=0}^{\infty} \varphi_n(x)t^n$$

for the simple set of polynomials $\varphi_n(x)$ belonging to a Sheffer operator $J(x, D)$, no matter what the A-type of $\varphi_n(x)$, we can obtain a generating function (sum the series)

$$\sum_{n=0}^{\infty} \psi_n(x)t^n$$

for any other polynomial $\psi_n(x)$ belonging to the same operator $J(x, D)$.

9. Obtain a theorem analogous to that of Ex. 8 but with Sheffer A-type replaced by σ-type.

10. Show that if $P_n(x)$ is the Legendre polynomial,

$$\varphi_n(x) = \frac{(1 + x^2)^{\frac{1}{2}n}}{n!} P_n\left(\frac{x}{\sqrt{1 + x^2}}\right)$$

is a simple set of polynomials of Sheffer A-type zero.

11. Let $P_n(x)$ be the Legendre polynomial. Choose $\sigma = D\theta$ and show that the polynomials

$$\varphi_n(x) = \frac{(x - 1)^n}{(n!)^2} P_n\left(\frac{x + 1}{x - 1}\right)$$

are of σ-type zero for that σ.

12. Prove that if the operator $J(x, D)$ is such that a simple set of polynomials $\psi_n(x)$ belongs to it in the Sheffer sense, then $J(x, D)$ transforms every polynomial of degree precisely n into a polynomial of degree precisely $(n - 1)$.

CHAPTER 14

Pure Recurrence Relations

127. Sister Celine's technique. Years ago it seemed customary upon entering the study of a new set of polynomials to seek recurrence relations, pure or mixed (with or without derivatives involved) by essentially a hit-and-miss process. Manipulative skill was used and, if there was enough of it, some relations emerged; others might easily have been lurking around a corner without being discovered. For polynomials of hypergeometric (${}_pF_q$) character, some systemization resulted from the existence of the contiguous function relations published in 1945. The interesting problem of the pure recurrence relation for hypergeometric polynomials received probably its first systematic attack at the hands of Sister Mary Celine Fasenmyer in a Michigan thesis in 1945. She introduced the tool in her study of a certain class of hypergeometric polynomials, for which see Section 149, page 290, or Fasenmyer [1].

In Fasenmyer [2], Sister Celine illustrated her technique by obtaining pure recurrence relations for Bateman's $Z_n(t)$, touched upon in this book in Section 146, page 285, and for one of her own sets of polynomials (see Section 149, page 290).

As a first example we shall find the pure recurrence relation for the simplest of Sister Celine's polynomials,

$$f_n(x) = {}_2F_2(-n, n + 1; 1, \tfrac{1}{2}; x), \tag{1}$$

$$f_n(x) = \sum_{k=0}^{n} \frac{(-1)^k (n+k)! x^k}{(k!)^2 (\frac{1}{2})_k (n-k)!}. \tag{2}$$

For convenience we use ∞ hereafter as the upper limit of summation for series such as that on the right in (2). We shall deal with hypergeometric polynomials which always terminate naturally. We also use a common convention that $f_n(x)$ is defined as zero whenever the subscript is negative. This convention permits us to avoid discussing separately the smallest values of n, $n = 0, 1, 2$ for the polynomials now to be treated.

First rewrite (2) as

$$f_n(x) = \sum_{k=0}^{\infty} \epsilon(k, n). \tag{3}$$

Sister Celine's technique is to express $f_{n-1}(x)$, $f_{n-2}(x)$, $xf_{n-1}(x)$, etc., as series involving $\epsilon(k, n)$, and then to find a combination of coefficients which vanishes identically. Note that, by (2),

$$f_{n-1}(x) = \sum_{k=0}^{\infty} \frac{(-1)^k (n-1+k)! x^k}{(k!)^2 (\frac{1}{2})_k (n-1-k)!},$$

so that

$$f_{n-1}(x) = \sum_{k=0}^{\infty} \frac{n-k}{n+k} \epsilon(k, n). \tag{4}$$

In the same manner it follows that

$$f_{n-2}(x) = \sum_{k=0}^{\infty} \frac{(n-k)(n-k-1)}{(n+k)(n+k-1)} \epsilon(k, n), \tag{5}$$

$$f_{n-3}(x) = \sum_{k=0}^{\infty} \frac{(n-k)(n-k-1)(n-k-2)}{(n+k)(n+k-1)(n+k-2)} \epsilon(k, n). \tag{6}$$

Let us turn to $xf_{n-1}(x)$, $xf_{n-2}(x)$, etc. We find that

$$xf_{n-1}(x) = \sum_{k=0}^{\infty} \frac{(-1)^k (n-1+k)! x^{k+1}}{(k!)^2 (\frac{1}{2})_k (n-1-k)!} = \sum_{k=1}^{\infty} \frac{(-1)^{k-1} (n-2+k)! x^k}{[(k-1)!]^2 (\frac{1}{2})_{k-1} (n-k)!},$$

or

$$xf_{n-1}(x) = \sum_{k=0}^{\infty} \frac{-k^2(\frac{1}{2}+k-1)}{(n+k)(n+k-1)} \epsilon(k, n). \tag{7}$$

In like manner,

$$xf_{n-2}(x) = \sum_{k=0}^{\infty} \frac{-k^2(\frac{1}{2}+k-1)(n-k)}{(n+k)(n+k-1)(n+k-2)} \epsilon(k, n). \tag{8}$$

In equations (3)–(8) the coefficients of $\epsilon(k, n)$ have a lowest common denominator $(n + k)(n + k - 1)(n + k - 2)$. When that denominator is used in each coefficient, the maximum degree with respect to k of the numerators is four. Then there exist constants (functions of n but not of k or x) A, B, C, D, E such that

$$(9) \quad f_n(x) + (A + Bx)f_{n-1}(x) + (C + Dx)f_{n-2}(x) + Ef_{n-3}(x) = 0$$

is an identity. By using equations (3)–(8) and reducing all coefficients of $\epsilon(k, n)$ to the least common denominator, we see that (9) is equivalent to the following identity in k:

$$(10) \quad \begin{aligned} &(n+k)(n+k-1)(n+k-2)+A(n-k)(n+k-1)(n+k-2) \\ &- Bk^2(k - \tfrac{1}{2})(n + k - 2) + C(n - k)(n - k - 1)(n + k - 2) \\ &- Dk^2(k - \tfrac{1}{2})(n - k) + E(n - k)(n - k - 1)(n - k - 2) = 0. \end{aligned}$$

The identity (10) easily yields five equations for the determination of A, B, C, D, E. It is wisest to get those equations by a judicious mixture of employing specific k values and equating coefficients of powers of k to zero. In (10), for instance, the choices $k = n$, $k = n - 1$, coefficients of k^4, $k = 1 - n$, $k = 2 - n$, yield simple equations. By that or some other elementary method, it is found that

$$A = -\frac{3n - 2}{n}, \quad B = \frac{4}{n}, \quad C = \frac{3n - 4}{n}, \quad D = \frac{4}{n}, \quad E = -\frac{n - 2}{n}.$$

With the aid of (9) we may conclude that the polynomials $f_n(x)$ of (1) satisfy the relation

$$(11) \quad \begin{aligned} nf_n(x) &- (3n - 2 - 4x)f_{n-1}(x) \\ &+ (3n - 4 + 4x)f_{n-2}(x) - (n - 2)f_{n-3}(x) = 0. \end{aligned}$$

As a second example, consider the polynomials

$$(12) \quad \varphi_n(x) = {}_2F_2(-n, 1 + \beta; 1, 1 + \alpha; x).$$

The $\varphi_n(x)$ interest us because they permit independent verification in two special cases. If $\beta = \alpha$, $\varphi_n(x)$ becomes the simple Laguerre polynomial $L_n(x)$. If $\beta = 0$, $\varphi_n(x)$ becomes $n!L_n^{(\alpha)}(x)/(1 + \alpha)_n$. For each of those polynomials we already know the pure recurrence relation.

Now

$$\varphi_n(x) = \sum_{k=0}^{n} \frac{(-1)^k n!(1 + \beta)_k x^k}{(k!)^2(1 + \alpha)_k(n - k)!}.$$

Put

$$(13)\qquad \gamma_n(x) = \frac{\varphi_n(x)}{n!}.$$

Then

$$(14)\qquad \gamma_n(x) = \sum_{k=0}^{\infty} \frac{(-1)^k(1+\beta)_k x^k}{(k!)^2(1+\alpha)_k(n-k)!} = \sum_{k=0}^{\infty} \epsilon(k, n).$$

Proceeding as in the earlier example, we find that

$$(15)\qquad \gamma_{n-1}(x) = \sum_{k=0}^{\infty} (n-k)\epsilon(k, n),$$

$$(16)\qquad \gamma_{n-2}(x) = \sum_{k=0}^{\infty} (n-k)(n-k-1)\epsilon(k,n),$$

$$(17)\qquad \gamma_{n-3}(x) = \sum_{k=0}^{\infty} (n-k)(n-k-1)(n-k-2)\epsilon(k, n),$$

$$(18)\qquad x\gamma_{n-1}(x) = \sum_{k=0}^{\infty} \frac{-k^2(\alpha+k)}{\beta+k}\epsilon(k, n),$$

$$(19)\qquad x\gamma_{n-2}(x) = \sum_{k=0}^{\infty} \frac{-k^2(\alpha+k)(n-k)}{\beta+k}\epsilon(k, n).$$

It follows that there exists a relation

$$(20)\qquad \gamma_n(x)+(A+Bx)\gamma_{n-1}(x)+(C+Dx)\gamma_{n-2}(x)+E\gamma_{n-3}(x)=0$$

and that the A, B, C, D, E are determined by the identity in k:

$$(21)\qquad \beta+k+A(\beta+k)(n-k)-Bk^2(\alpha+k)+C(\beta+k)(n-k)(n-k-1)$$
$$-Dk^2(\alpha+k)(n-k)+E(\beta+k)(n-k)(n-k-1)(n-k-2)=0.$$

From (21) it is an elementary matter to show that

$$A = \frac{-3n^2+3n-1-\alpha(2n-1)}{n^2(\alpha+n)}, \qquad B = \frac{\beta+n}{n^2(\alpha+n)},$$

$$C = \frac{\alpha+3n-3}{n^2(\alpha+n)}, \qquad D = \frac{-1}{n^2(\alpha+n)}, \qquad E = \frac{-1}{n^2(\alpha+n)}.$$

Thus the polynomials $\gamma_n(x)$ satisfy the relation

$$(22)\qquad n^2(\alpha+n)\gamma_n(x)-[3n^2-3n+1+\alpha(2n-1)-(\beta+n)x]\gamma_{n-1}(x)$$
$$+(\alpha+3n-3-x)\gamma_{n-2}(x)-\gamma_{n-3}(x)=0.$$

But, by (13), $\gamma_n(x) = \varphi_n(x)/n!$. Hence the polynomials

$$\varphi_n(x) = {}_2F_2(-n, 1 + \beta; 1, 1 + \alpha; x) \tag{12}$$

satisfy the recurrence relation

$$\begin{aligned} n(\alpha+n)\varphi_n(x) - [3n^2-3n+1+\alpha(2n-1)-(\beta+n)x]\varphi_{n-1}(x) \\ +(n-1)(\alpha+3n-3-x)\varphi_{n-2}(x)-(n-1)(n-2)\varphi_{n-3}(x)=0. \end{aligned} \tag{23}$$

It is not difficult to rewrite (23) as

$$\begin{aligned} (\alpha+n)[n\varphi_n(x)-(2n-1-x)\varphi_{n-1}(x)+(n-1)\varphi_{n-2}(x)] \\ -(n-1)[(n-1)\varphi_{n-1}(x)-(2n-3-x)\varphi_{n-2}(x)+(n-2)\varphi_{n-3}(x)] \\ +(\beta-\alpha)x\varphi_{n-1}(x)=0, \end{aligned} \tag{24}$$

in which form it becomes evident that if $\beta = \alpha$, (24) is an iteration of the known relation

$$nL_n(x) - (2n - 1 - x)L_{n-1}(x) + (n - 1)L_{n-2}(x) = 0,$$

as it should be, since, when $\beta = \alpha$, $\varphi_n(x)$ reduces to $L_n(x)$.

If the relation (23) is rewritten in the form

$$\begin{aligned} n[(\alpha+n)\varphi_n(x)-(2n-1+\alpha-x)\varphi_{n-1}(x)+(n-1)\varphi_{n-2}(x)] \\ -(n-1)[(\alpha+n-1)\varphi_{n-1}(x)-(2n-3+\alpha-x)\varphi_{n-2}(x)+(n-2)\varphi_{n-3}(x)] \\ +\beta x\varphi_{n-1}(x)=0, \end{aligned} \tag{25}$$

it is a simple matter to obtain a check for $\beta = 0$, for which value $\varphi_n(x)$ degenerates into $n!L_n^{(\alpha)}(x)/(1 + \alpha)_n$.

In both the preceding examples, the polynomials were expressed in ascending powers of x. That form has nothing to do with the success of the method. As a third example, consider the polynomials

$$g_n(x) = (2x)^n\, {}_3F_1\left[\begin{matrix} -\tfrac{1}{2}n, -\tfrac{1}{2}n + \tfrac{1}{2}, 1 + \alpha; \\ 1 + \beta; \end{matrix} -\frac{1}{x^2}\right], \tag{26}$$

or

$$g_n(x) = \sum_{k=0}^{[n/2]} \frac{(-1)^k n!(1 + \alpha)_k(2x)^{n-2k}}{k!(1 + \beta)_k(n - 2k)!}. \tag{27}$$

Put

$$\lambda_n(x) = \frac{g_n(x)}{n!}. \tag{28}$$

Then

$$\lambda_n(x) = \sum_{k=0}^{\infty} \frac{(-1)^k(1 + \alpha)_k(2x)^{n-2k}}{k!(1 + \beta)_k(n - 2k)!} = \sum_{k=0}^{\infty} \epsilon(k, n). \tag{29}$$

Form the series

(30) $$(2x)\lambda_{n-1}(x) = \sum_{k=0}^{\infty} \frac{(-1)^k(1+\alpha)_k(2x)^{n-2k}}{k!(1+\beta)_k(n-1-2k)!} = \sum_{k=0}^{\infty} (n-2k)\epsilon(k,n),$$

$$\lambda_{n-2}(x) = \sum_{k=0}^{\infty} \frac{(-1)^k(1+\alpha)_k(2x)^{n-2-2k}}{k!(1+\beta)_k(n-2-2k)!}$$

$$= \sum_{k=1}^{\infty} \frac{(-1)^{k-1}(1+\alpha)_{k-1}(2x)^{n-2k}}{(k-1)!(1+\beta)_{k-1}(n-2k)!},$$

or

(31) $$\lambda_{n-2}(x) = \sum_{k=0}^{\infty} \frac{-k(\beta+k)}{(\alpha+k)}\epsilon(k,n),$$

and, in the same manner,

(32) $$(2x)^2\lambda_{n-2}(x) = \sum_{k=0}^{\infty} (n-2k)(n-2k-1)\epsilon(k,n),$$

(33) $$(2x)\lambda_{n-3}(x) = \sum_{k=0}^{\infty} \frac{-k(n-2k)(\beta+k)}{\alpha+k}\epsilon(k,n).$$

It follows from the series (29)–(33) that there exists a relation of the form

(34) $$\lambda_n(x)+A(2x)\lambda_{n-1}(x)+[B+C(2x)^2]\lambda_{n-2}(x)+D(2x)\lambda_{n-3}(x)=0,$$

in which the constants A, B, C, D are determined by the identity in k:

(35) $$\alpha+k+A(n-2k)(\alpha+k)-Bk(\beta+k)$$
$$+C(n-2k)(n-2k-1)(\alpha+k)-Dk(n-2k)(\beta+k)=0.$$

The identity (35) yields

$$A = -\frac{2\beta+2n-1}{n(2\beta+n)}, \qquad B = \frac{2(2\alpha+n)}{n(2\beta+n)},$$

$$C = \frac{1}{n(2\beta+n)}, \qquad D = \frac{-2}{n(2\beta+n)}.$$

Hence the $\lambda_n(x)$ satisfy

(36) $$n(2\beta+n)\lambda_n(x)-(2\beta+2n-1)(2x)\lambda_{n-1}(x)$$
$$+[2(2\alpha+n)+(2x)^2]\lambda_{n-2}(x)-2(2x)\lambda_{n-3}(x)=0.$$

Since, by (28), $\lambda_n(x) = g_n(x)/n!$, we find that the polynomials

$$(26)\qquad g_n(x) = (2x)^n\ {}_3F_1\left[\begin{matrix} -\tfrac{1}{2}n,\ -\tfrac{1}{2}n+\tfrac{1}{2},\ 1+\alpha; \\ 1+\beta; \end{matrix}\ -\frac{1}{x^2}\right]$$

satisfy the pure recurrence relation

$$(37)\qquad (2\beta+n)g_n(x)-(2\beta+2n-1)(2x)g_{n-1}(x) \\ +(n-1)[2(2\alpha+n)+(2x)^2]g_{n-2}(x)-2(n-1)(n-2)(2x)g_{n-3}(x)=0.$$

When $\beta = \alpha$ the $g_n(x)$ degenerates into the Hermite polynomial $H_n(x)$, for which we already know the relation

$$(38)\qquad H_n(x) - 2xH_{n-1}(x) + 2(n-1)H_{n-2}(x) = 0.$$

The relation (37) may be put in the form

$$(39)\qquad (2\beta+n)[g_n(x) - 2xg_{n-1}(x) + 2(n-1)g_{n-2}(x)] \\ - (n-1)(2x)[g_{n-1}(x) - 2xg_{n-2}(x) + 2(n-2)g_{n-3}(x)] \\ + 4(\alpha-\beta)(n-1)g_{n-2}(x) = 0.$$

It is now evident that if $\beta = \alpha$, (39) is an iteration of (38).

While we are on the topic of the $g_n(x)$ of (26), it may be of interest to express $g_n(x)$ in terms of the Hermite polynomials of which $g_n(x)$ is a generalization.

Using (27) we find that

$$\sum_{n=0}^{\infty} \frac{g_n(x)t^n}{n!} = \sum_{n=0}^{\infty}\sum_{k=0}^{[n/2]} \frac{(-1)^k(1+\alpha)_k(2x)^{n-2k}t^n}{k!(1+\beta)_k(n-2k)!}$$

$$= \left(\sum_{n=0}^{\infty} \frac{(2x)^n t^n}{n!}\right)\left(\sum_{n=0}^{\infty} \frac{(-1)^n(1+\alpha)_n t^{2n}}{n!(1+\beta)_n}\right),$$

so that the $g_n(x)$ possess the generating function in

$$(40)\qquad \exp(2xt)\ {}_1F_1(1+\alpha; 1+\beta; -t^2) = \sum_{n=0}^{\infty} \frac{g_n(x)t^n}{n!}.$$

By Kummer's first formula, page 125.

$${}_1F_1(1+\alpha; 1+\beta; -t^2) = \exp(-t^2){}_1F_1(\beta-\alpha; 1+\beta; t^2).$$

Therefore (40) yields

$$(41)\qquad \exp(2xt - t^2)\ {}_1F_1(\beta-\alpha; 1+\beta; t^2) = \sum_{n=0}^{\infty} \frac{g_n(x)t^n}{n!}.$$

Then

$$\sum_{n=0}^{\infty} \frac{g_n(x)t^n}{n!} = \left(\sum_{n=0}^{\infty} \frac{H_n(x)t^n}{n!}\right)\left(\sum_{n=0}^{\infty} \frac{(\beta-\alpha)_n t^{2n}}{n!(1+\beta)_n}\right),$$

which yields

$$g_n(x) = \sum_{k=0}^{[n/2]} \frac{n!(\beta - \alpha)_k H_{n-2k}(x)}{k!(n - 2k)!(1 + \beta)_k}. \tag{42}$$

Shively [1] used Sister Celine's technique to obtain differential recurrence relations, and Bedient [1] employed her method to get mixed recurrence relations. See also Exs. 8-11 at the end of this chapter.

We already know from Chapter 9 that any orthogonal set of polynomials necessarily satisfies a particular type of pure recurrence relation, for which see Section 83. At once we see that the technique discussed here is a potent method for showing that a specific polynomial set is not orthogonal with respect to any weight function over any interval. See, for instance, Ex. 12 at the end of this chapter.

128. A mild extension. Consider the polynomials $\sigma_n(x)$ defined by

$$\sigma_n(x) = \sum_{k=0}^{n} \frac{(-1)^k n! H_k(x)}{(k!)^2(n - k)!}, \tag{1}$$

in which $H_k(x)$ is the Hermite polynomial of Chapter 11. We shall encounter the $\sigma_n(x)$ in our study of symbolic relations in the next chapter. It is not difficult to show that in ascending powers of x the $\sigma_n(x)$ has the following appearance:

$$\sigma_n(x) = \sum_{k=0}^{n} {}_2F_2\left[\begin{matrix} -\tfrac{1}{2}(n-k), \; -\tfrac{1}{2}(n-k-1); \\ \tfrac{1}{2}(1+k), \; \tfrac{1}{2}(2+k); \end{matrix} \; -1\right] \frac{(-n)_k(2x)^k}{(k!)^2}. \tag{2}$$

If the reader wishes to attack the problem of finding a pure recurrence relation for $\sigma_n(x)$ on the basis of (2) rather than of (1), he is welcome to the task. Here we prefer to modify Sister Celine's technique to fit the situation in which a polynomial is expressed not in a series of powers of x but in a series of other polynomials.

In equation (1) put

$$\frac{\sigma_n(x)}{n!} = s_n(x), \qquad \frac{(-1)^k H_k(x)}{(k!)^2} = v_k(x), \tag{3}$$

so that

$$s_n(x) = \sum_{k=0}^{n} \frac{v_k(x)}{(n - k)!}. \tag{4}$$

From the known relation

$$(5)\qquad H_k(x) - 2xH_{k-1}(x) + 2(k-1)H_{k-2}(x) = 0,$$

we determine a recurrence relation for $v_k(x)$, use that to find a relation for $s_n(x)$, and finally use (3) to transform our result into a relation for $\sigma_n(x)$.

In (5) put $H_k(x) = (-1)^k(k!)^2v_k(x)$. The resulting relation for $v_k(x)$ is

$$(6)\qquad k^2(k-1)v_k(x) + 2x(k-1)v_{k-1}(x) + 2v_{k-2}(x) = 0.$$

First we set out to use (4) to obtain the series

$$(7)\qquad \sum_{k=0}^{\infty} \frac{k^2(k-1)v_k(x)}{(n-k)!}.$$

From (4) we get

$$(8)\qquad s_{n-1}(x) = \sum_{k=0}^{\infty} \frac{v_k(x)}{(n-1-k)!} = \sum_{k=0}^{\infty} \frac{(n-k)v_k(x)}{(n-k)!},$$

$$(9)\qquad s_{n-2}(x) = \sum_{k=0}^{\infty} \frac{(n-k)(n-k-1)v_k(x)}{(n-k)!},$$

$$(10)\qquad s_{n-3}(x) = \sum_{k=0}^{\infty} \frac{(n-k)(n-k-1)(n-k-2)v_k(x)}{(n-k)!}.$$

The four series (4), (8), (9), and (10) contain numerator coefficients of not more than third degree in k, the degree of the numerator in (7). Therefore there exist constants A, B, C, D, such that

$$(11)\quad As_n(x) + Bs_{n-1}(x) + Cs_{n-2}(x) + Ds_{n-3}(x) = \sum_{k=0}^{\infty} \frac{k^2(k-1)v_k(x)}{(n-k)!}.$$

Indeed (11) implies the identity in k:

$$(12)\qquad \begin{aligned} A + B(n-k) + C(n-k)(n-k-1) \\ + D(n-k)(n-k-1)(n-k-2) = k^2(k-1), \end{aligned}$$

from which it follows readily that

$$A = n^2(n-1), \qquad B = -(n-1)(3n-2), \qquad C = 3n-4, \qquad D = -1.$$

Hence we have

$$(13)\quad \begin{aligned} n^2(n-1)s_n(x) - (n-1)(3n-2)s_{n-1}(x) + (3n-4)s_{n-2}(x) \\ - s_{n-3}(x) = \sum_{k=0}^{\infty} \frac{k^2(k-1)v_k(x)}{(n-k)!}. \end{aligned}$$

Next we wish to construct, corresponding to the second term in (6), the series

$$\sum_{k=0}^{\infty} \frac{(k-1)(2x)v_{k-1}(x)}{(n-k)!}, \tag{14}$$

using the convention that $v_{-1}(x) \equiv 0$.

Now

$$s_{n-1}(x) = \sum_{k=0}^{\infty} \frac{v_k(x)}{(n-1-k)!} = \sum_{k=0}^{\infty} \frac{v_{k-1}(x)}{(n-k)!}. \tag{15}$$

Furthermore,

$$\begin{aligned} s_{n-2}(x) &= \sum_{k=0}^{\infty} \frac{v_k(x)}{(n-2-k)!} \\ &= \sum_{k=0}^{\infty} \frac{v_{k-1}(x)}{(n-1-k)!} = \sum_{k=0}^{\infty} \frac{(n-k)v_{k-1}(x)}{(n-k)!}. \end{aligned} \tag{16}$$

Hence there exist constants E and F such that

$$2xEs_{n-1}(x) + 2xFs_{n-2}(x) = \sum_{k=0}^{\infty} \frac{(k-1)(2x)v_{k-1}(x)}{(n-k)!}. \tag{17}$$

From the identity

$$E + F(n-k) = k - 1,$$

we obtain $E = n - 1$, $F = -1$, so that

$$(n-1)(2x)s_{n-1}(x) - (2x)s_{n-2}(x) = \sum_{k=0}^{\infty} \frac{(k-1)(2x)v_{k-1}(x)}{(n-k)!}. \tag{18}$$

Finally,

$$2s_{n-2}(x) = \sum_{k=0}^{\infty} \frac{2v_k(x)}{(n-2-k)!} = \sum_{k=0}^{\infty} \frac{2v_{k-2}(x)}{(n-k)!}. \tag{19}$$

We now add corresponding members of (13), (18), and (19) to get

$$\begin{aligned} n^2(n-1)s_n(x) &- (n-1)(3n-2)s_{n-1}(x) + (3n-4)s_{n-2}(x) - s_{n-3}(x) \\ &+ (n-1)(2x)s_{n-1}(x) - (2x)s_{n-2}(x) + 2s_{n-2}(x) \\ &= \sum_{k=0}^{\infty} \frac{k^2(k-1)v_k(x) + (k-1)(2x)v_{k-1}(x) + 2v_{k-2}(x)}{(n-k)!} = 0, \end{aligned}$$

because of equation (6). Therefore $s_n(x)$ satisfies the equation

$$\begin{aligned} n^2(n-1)s_n(x) &- (n-1)(3n-2-2x)s_{n-1}(x) \\ &+ (3n-2-2x)s_{n-2}(x) - s_{n-3}(x) = 0. \end{aligned} \tag{20}$$

Since, by (3), $s_n(x) = \dfrac{\sigma_n(x)}{n!}$, we find that the polynomials

$$(1) \qquad \sigma_n(x) = \sum_{k=0}^{n} \frac{(-1)^k n! H_k(x)}{(k!)^2 (n-k)!}$$

satisfy the pure recurrence relation

$$(21) \qquad n\sigma_n(x) - (3n - 2 - 2x)\sigma_{n-1}(x) + (3n - 2 - 2x)\sigma_{n-2}(x) - (n-2)\sigma_{n-3}(x) = 0.$$

EXERCISES

1. Show that Bateman's polynomial (see Section 146, page 285)

$$Z_n(t) = {}_2F_2(-n, n+1; 1, 1; t)$$

has the recurrence relation

$$n^2(2n-3)Z_n(t) - (2n-1)[3n^2 - 6n + 2 - 2(2n-3)t]Z_{n-1}(t) + (2n-3)[3n^2 - 6n + 2 + 2(2n-1)t]Z_{n-2}(t) - (2n-1)(n-2)^2 Z_{n-3}(t) = 0.$$

2. Show that Sister Celine's polynomial $f_n(a; -; x)$, or

$$f_n(x) = {}_3F_2(-n, n+1, a; 1, \tfrac{1}{2}; x)$$

has the recurrence relation

$$nf_n(x) - [3n - 2 - 4(n - 1 + a)x]f_{n-1}(x) + [3n - 4 - 4(n - 1 - a)x]f_{n-2}(x) - (n-2)f_{n-3}(x) = 0.$$

3. Show that Rice's polynomial (see Section 147, page 287)

$$H_n = H_n(\zeta, p, v) = {}_3F_2(-n, n+1, \zeta; 1, p; v)$$

satisfies the relation

$$\begin{aligned} &n(2n-3)(p+n-1)H_n \\ &\quad -(2n-1)[(n-2)(p-n+1)+2(n-1)(2n-3)-2(2n-3)(\zeta+n-1)v]H_{n-1} \\ &\quad +(2n-3)[2(n-1)^2-n(p-n+1)+2(2n-1)(\zeta-n+1)v]H_{n-2} \\ &\quad +(n-2)(2n-1)(p-n+1)H_{n-3} = 0. \end{aligned}$$

4. Show that the polynomial

$$f_n(x) = {}_1F_2(-n; 1+\alpha, 1+\beta; x),$$

which is intimately related to Bateman's $J_n^{u,v}$ of Section 146, page 287, satisfies the relation

$$(\alpha + n)(\beta + n)f_n(x) - [3n^2 - 3n + 1 + (2n-1)(\alpha+\beta) + \alpha\beta - x]f_{n-1}(x) + (n-1)(3n - 3 + \alpha + \beta)f_{n-2}(x) - (n-1)(n-2)f_{n-3}(x) = 0.$$

5. Define the polynomial $w_n(x)$ by

$$w_n(x) = \sum_{k=0}^{n} \frac{(-1)^k n! L_k(x)}{(k!)^2 (n-k)!}$$

in terms of the simple Laguerre polynomial $L_k(x)$. Show that $w_n(x)$ possesses the pure recurrence relation

$$n^2w_n(x) - [(n-1)(4n-3)+x]w_{n-1}(x) + (6n^2-19n+16+x)w_{n-2}(x) - (n-2)(4n-9)w_{n-3}(x) + (n-2)(n-3)w_{n-4}(x) = 0.$$

6. Show that the polynomial $w_n(x)$ of Ex. 5 may be written

$$w_n(x) = \sum_{k=0}^{n} {}_1F_1\left[\begin{matrix} -n+k; \\ \\ 1+k; \end{matrix}\; 1\right]\frac{(-n)_k(-x)^k}{(k!)^3}.$$

7. Define the polynomial $v_n(x)$ by (see Section 131, page 251)

$$v_n(x) = \sum_{k=0}^{n} \frac{(-1)^k n! P_k(x)}{(k!)^2(n-k)!}$$

in terms of the Legendre polynomial $P_k(x)$. Show that $v_n(x)$ satisfies the recurrence relation

$$n^2v_n(x) - [4n^2-5n+2-(2n-1)x]v_{n-1}(x) + [6n^2-15n+11-(4n-5)x]v_{n-2}(x) - (n-2)(4n-7-2x)v_{n-3}(x) + (n-2)(n-3)v_{n-4}(x) = 0.$$

8. Show that the $v_n(x)$ of Ex. 7 satisfies the relations

$$(1-x^2)v_n''(x) - 2xv_n'(x) + n(n+1)v_n(x) = 2n^2v_{n-1}(x) - n(n-1)v_{n-2}(x)$$

and

$$(1-x^2)v_n'(x) + nxv_n(x) = [(2n-1)x-1]v_{n-1}(x) - (n-1)xv_{n-2}(x) + (1-x^2)v'_{n-1}(x).$$

9. Let

$$\gamma(k,n) = \frac{(-1)^k(\tfrac{1}{2})_{n-k}(2x)^{n-2k}}{k!(n-2k)!},$$

so that the Legendre polynomial of Chapter 10 may be written

$$P_n(x) = \sum_{k=0}^{\infty} \gamma(k,n).$$

Show that

$$xP_{n-1}(x) = \sum_{k=0}^{\infty} \frac{(n-2k)\gamma(k,n)}{2n-2k-1}, \qquad P_{n-2}(x) = \sum_{k=0}^{\infty} \frac{-2k\gamma(k,n)}{2n-2k-1},$$

$$xP_n'(x) = \sum_{k=0}^{\infty} (n-2k)\gamma(k,n), \qquad P'_{n+1}(x) = \sum_{k=0}^{\infty} (1+2n-2k)\gamma(k,n),$$

$$P'_{n-1}(x) = \sum_{k=0}^{\infty} -2k\gamma(k,n), \text{ etc.}$$

Use Sister Celine's method to discover the various differential recurrence relations and the pure recurrence relation for $P_n(x)$.

10. Apply Sister Celine's method to discover relations satisfied by the Hermite polynomials of Chapter 11.

11. Find the various relations of Section 114 on Laguerre polynomials by using Sister Celine's technique.

12. Consider the pseudo-Laguerre polynomials (Boas and Buck [2; 16]) $f_n(x)$ defined for nonintegral λ by

$$f_n(x) = \frac{(-\lambda)_n}{n!}\, {}_1F_1(-n; 1 + \lambda - n; x) = \sum_{k=0}^{n} \frac{(-\lambda)_{n-k} x^k}{k!(n-k)!}.$$

Show that the polynomials $f_n(x)$ are not orthogonal with respect to any weight function over any interval because no relation of the form

$$f_n(x) = (A_n + B_n x) f_{n-1}(x) + C_n f_{n-2}(x)$$

is possible. Obtain the pure recurrence relation

$$n f_n(x) = (x + n - 1 - \lambda) f_{n-1}(x) - x f_{n-2}(x).$$

For the polynomials in each of Exs. 13–17, use Sister Celine's technique to discover the pure recurrence relation and whatever mixed relations exist.

13. The Bessel polynomials of Section 150.
14. Bedient's polynomials R_n of Section 151.
15. Bedient's polynomials G_n of Section 151.
16. Shively's polynomials R_n of Section 152.
17. Shively's polynomials σ_n of Section 152.

CHAPTER 15

Symbolic Relations

129. Notation. Many relations involving finite series of polynomials can be put into particularly neat form by use of an old symbolic notation. Whenever $\doteq$ is used to replace $=$, it is to be understood that exponents will be lowered to subscripts on any symbol which is undefined here except with subscripts. For instance, we already know, page 213, that the simple Laguerre polynomial $L_n(x)$ satisfies the relation

$$\frac{x^n}{n!} = \sum_{k=0}^{n} \frac{(-1)^k n! L_k(x)}{k!(n-k)!}. \tag{1}$$

In symbolic notation, since L without a subscript is not defined here, we may write equation (1) as

$$\frac{x^n}{n!} \doteq \{1 - L(x)\}^n. \tag{2}$$

In Section 126 we obtained a recurrence relation for a polynomial $\sigma_n(x)$ defined by

$$\sigma_n(x) = \sum_{k=0}^{n} \frac{(-1)^k n! H_k(x)}{(k!)^2(n-k)!}. \tag{3}$$

In symbolic notation (3) is replaced by

$$\sigma_n(x) \doteq L_n(H(x)). \tag{4}$$

A final example: if $H_n(x)$ is the Hermite polynomial and $P_n(x)$

the Legendre polynomial, then the meaning of $\{H(x) - P(x)\}^2$ is given by

$$\{H(x) - P(x)\}^2 \doteq H_2(x)P_0(x) - 2H_1(x)P_1(x) + H_0(x)P_2(x).$$

130. Symbolic relations among classical polynomials. In addition to (2) of Section 129, we have already encountered earlier in this book a few relations which fit nicely into symbolic form. For instance,

$$L_n(xy) \doteq \{1 - y + yL(x)\}^n, \tag{1}$$

$$P_n(x) \doteq \{2x - P(x)\}^n, \tag{2}$$

$$P_n(1 - 2x^2) \doteq \{1 - 2xP(x)\}^n, \tag{3}$$

$$\rho^n P_n\left(\frac{1 - xt}{\rho}\right) \doteq \{1 - tP(x)\}^n; \qquad \rho = (1 - 2xt + t^2)^{\frac{1}{2}}. \tag{4}$$

For the most part the relations with which we deal here are included because they are amusing or particularly pretty. It would be unwise, however, to pass up the subject as one of no other value. Szegö [2] made good use of equation (4). The symbolic notation also suggests the study of some interesting polynomials which may not otherwise be noticed.

We shall now prove

$$H_n(P(x)) \doteq \{H(2x) - 2P(x)\}^n, \tag{5}$$

$$2^n L_n(P(x)) \doteq \{L(x - 1) + L(x + 1)\}^n, \tag{6}$$

$$H_n(H(x)) \doteq 5^{\frac{1}{2}n} H_n\left(\frac{2x}{\sqrt{5}}\right), \tag{7}$$

as examples and leave other relations as exercises for the reader.

To prove (5), consider the series

$$\sum_{n=0}^{\infty} \frac{H_n(P(x))t^n}{n!} \doteq \sum_{n=0}^{\infty} \sum_{k=0}^{[n/2]} \frac{(-1)^k 2^{n-2k} P_{n-2k}(x) t^n}{k!(n - 2k)!}$$

$$= \left(\sum_{n=0}^{\infty} \frac{(-1)^n t^{2n}}{n!}\right)\left(\sum_{n=0}^{\infty} \frac{2^n P_n(x) t^n}{n!}\right)$$

$$= \exp(-t^2) \exp(2xt)\ {}_0F_1(-; 1; t^2(x^2 - 1))$$

$$= \exp(4xt - t^2) \exp(-2xt)\ {}_0F_1(-; 1; t^2(x^2 - 1)).$$

Then

$$\sum_{n=0}^{\infty} \frac{H_n(P(x))t^n}{n!} \doteqdot \left(\sum_{n=0}^{\infty} \frac{H_n(2x)t^n}{n!}\right)\left(\sum_{n=0}^{\infty} \frac{(-2)^n P_n(x)t^n}{n!}\right)$$

$$= \sum_{n=0}^{\infty} \sum_{k=0}^{n} \frac{(-2)^k H_{n-k}(2x) P_k(x) t^n}{k!(n-k)!}.$$

It follows that

$$H_n(P(x)) \doteqdot \sum_{k=0}^{n} \frac{n! H_{n-k}(2x)(-2)^k P_k(x)}{k!(n-k)!},$$

which is (5).

To prove (6), consider the series

$$\sum_{n=0}^{\infty} \frac{2^n L_n(P(x))t^n}{n!} \doteqdot \sum_{n=0}^{\infty} \sum_{k=0}^{n} \frac{(-1)^k P_k(x)(2t)^n}{(k!)^2(n-k)!}$$

$$= \left(\sum_{n=0}^{\infty} \frac{(2t)^n}{n!}\right)\left(\sum_{n=0}^{\infty} \frac{(-1)^n P_n(x)(2t)^n}{(n!)^2}\right)$$

$$= e^{2t}\, {}_0F_1(-;1;-t(x-1))\, {}_0F_1(-;1;-t(x+1))$$

$$= e^{t}{}_0F_1(-;1;-t(x-1))e^{t}\, {}_0F_1(-;1;-t(x+1))$$

$$= \left(\sum_{n=0}^{\infty} \frac{L_n(x-1)t^n}{n!}\right)\left(\sum_{n=0}^{\infty} \frac{L_n(x+1)t^n}{n!}\right)$$

$$= \sum_{n=0}^{\infty} \sum_{k=0}^{n} \frac{L_k(x-1)L_{n-k}(x+1)t^n}{k!(n-k)!},$$

from which (6) follows by equating coefficients of t^n.

Since

$$H_n(H(x)) \doteqdot \sum_{k=0}^{[n/2]} \frac{(-1)^k n! 2^{n-2k} H_{n-2k}(x)}{k!(n-2k)!},$$

we find that

$$\sum_{n=0}^{\infty} \frac{H_n(H(x))t^n}{n!} \doteqdot \left(\sum_{n=0}^{\infty} \frac{(-1)^n t^{2n}}{n!}\right)\left(\sum_{n=0}^{\infty} \frac{H_n(x)(2t)^n}{n!}\right)$$

$$= \exp(-t^2)\exp(4xt - 4t^2)$$

$$= \exp(4xt - 5t^2)$$

$$= \exp\left[\frac{4x}{\sqrt{5}}(t\sqrt{5}) - (t\sqrt{5})^2\right]$$

$$= \sum_{n=0}^{\infty} \frac{H_n(2x/\sqrt{5})5^{\frac{1}{2}n}t^n}{n!}$$

which yields equation (7).

131. Polynomials of symbolic form $L_n(y(x))$. Suppose $y_n(x)$ is a previously studied simple set of polynomials. We proceed to start the study of the polynomials $L_n(y(x))$.

THEOREM 81. *A necessary and sufficient condition that*

$$f_n(x) \doteqdot L_n(y(x))$$

be of Sheffer A-type zero is that $y_n(x)/n!$ be of Sheffer A-type zero.

Proof: Consider the series

$$\sum_{n=0}^{\infty} f_n(x)t^n = \sum_{n=0}^{\infty}\sum_{k=0}^{n} \frac{(-1)^k n! y_k(x)t^n}{(k!)^2(n-k)!}$$

$$= \sum_{n,k=0}^{\infty} \frac{(-1)^k(n+k)! y_k(x)t^{n+k}}{(k!)^2 n!}$$

$$= \sum_{k=0}^{\infty}\sum_{n=0}^{\infty} \frac{(1+k)_n t^n}{n!} \cdot \frac{(-1)^k y_k(x)t^k}{k!}$$

$$= \sum_{k=0}^{\infty} \frac{(-1)^k y_k(x)t^k}{k!(1-t)^{k+1}}.$$

By Theorem 72, page 222, if $y_n(x)/n!$ is of Sheffer A-type zero

$$\sum_{k=0}^{\infty} \frac{y_k(x)t^k}{k!} = A(t)\exp[xH(t)].$$

Then also

$$\sum_{n=0}^{\infty} f_n(x)t^n = (1-t)^{-1}A\left(\frac{-t}{1-t}\right)\exp\left[xH\left(\frac{-t}{1-t}\right)\right]$$

$$= B(t)\exp[xh(t)],$$

so that $f_n(x)$ is of Sheffer A-type zero also. Since

$$\sum_{n=0}^{\infty} f_n(x)\left(\frac{-t}{1-t}\right)^n = (1-t)\sum_{k=0}^{\infty} \frac{y_k(x)t^k}{k!},$$

the same method of proof works in the other direction.

Because of Theorem 79, page 228, it can be seen that the same proof works if Sheffer A-type zero is replaced by σ-type zero for any specified σ.

THEOREM 82. *Theorem 81 holds true if Sheffer A-type zero is everywhere replaced by σ-type zero.*

Theorems 73–76 and Theorem 80 are then applicable to polynomials of the form $L_n(y(x))$ for properly selected $y_n(x)$.

Let us consider the polynomials

$$\sigma_n(x) \doteqdot L_n(H(x)) \tag{1}$$

for which we obtained a pure recurrence relation in Section 128, page 243. Since $\dfrac{H_n(x)}{n!}$ is of Sheffer A-type zero, it follows that $\sigma_n(x)$ is of Sheffer A-type zero. Indeed, because

$$\sum_{n=0}^{\infty} \frac{H_n(x)t^n}{n!} = \exp(2xt - t^2),$$

we get

$$\sum_{n=0}^{\infty} \sigma_n(x)t^n = (1-t)^{-1} \exp\left[\frac{-2xt}{1-t} - \frac{t^2}{(1-t)^2}\right]. \tag{2}$$

In order to apply the theorems of Section 124, we note that, for the polynomials $\sigma_n(x)$,

$$A(t) = (1-t)^{-1} \exp\left[\frac{-t^2}{(1-t)^2}\right], \quad H(t) = \frac{-2t}{1-t}, \quad J(t) = \frac{-t}{2-t}. \tag{3}$$

It is easy to show that

$$A(t) = \sum_{k=0}^{\infty} {}_2F_2\left[\begin{matrix} -\frac{1}{2}k, \ -\frac{1}{2}k + \frac{1}{2}; \\ 1, \ \frac{1}{2}; \end{matrix} \ -1\right] t^k, \tag{4}$$

$$H(t) = \sum_{k=0}^{\infty} (-2)t^{k+1}, \tag{5}$$

$$J(t) = \sum_{k=0}^{\infty} \frac{-1}{2^{k+1}} t^{k+1}, \tag{6}$$

$$\frac{A'(t)}{A(t)} = \sum_{k=0}^{\infty} (1 - k - k^2)t^k, \tag{7}$$

$$(8) \qquad H'(t) = \sum_{k=0}^{\infty} [-2(k+1)]t^k,$$

$$(9) \qquad \frac{uA'(u)}{A(u)} = -\tfrac{1}{2}t - \tfrac{1}{2}t^2 + \tfrac{1}{4}t^3;\ u = J(t),$$

$$(10) \qquad uH'(u) = t - \tfrac{1}{2}t^2;\ u = J(t).$$

It follows that, in the notation of Theorems 73-76,

$$(11) \qquad \alpha_k = 1 - k - k^2,\ \epsilon_k = -2(k+1),$$

$$(12) \qquad \mu_0 = -\tfrac{1}{2},\ \mu_1 = -\tfrac{1}{2},\ \mu_2 = \tfrac{1}{4},\ \mu_k = 0 \text{ for } k \geqq 3,$$
$$\nu_0 = 1,\ \nu_1 = -\tfrac{1}{2},\ \nu_k = 0 \text{ for } k \geqq 2,$$

$$(13) \qquad h_k = -2.$$

We may apply Theorems 73-76 to obtain properties of $\sigma_n(x)$. For instance, Theorem 76 yields

$$(14) \qquad \sigma_n'(x) = -2\sum_{k=0}^{n-1} \sigma_{n-1-k}(x),\ n \geqq 1.$$

Theorem 74 yields the differential equation

$$\sigma_n'''(x) - 2(1+x)\sigma_n''(x) + 2(2x-1)\sigma_n'(x) - 4n\sigma_n(x) = 0.$$

Next let us consider the polynomials

$$(15) \qquad v_n(x) \doteqdot L_n(P(x))$$

for which a pure recurrence relation was obtained in Ex. 7, page 244. We already know one other property, relation (6), page 247, for these polynomials. Two generating functions for $v_n(x)$ are easily found. Since

$$\sum_{n=0}^{\infty} v_n(x)t^n = \sum_{n=0}^{\infty}\sum_{k=0}^{n} \frac{(-1)^k n! P_k(x) t^n}{(k!)^2 (n-k)!}$$

$$= \sum_{n,k=0}^{\infty} \frac{(-1)^k (n+k)! P_k(x) t^{n+k}}{(k!)^2 n!}$$

$$= \sum_{k=0}^{\infty}\sum_{n=0}^{\infty} \frac{(1+k)_n t^n}{n!} \cdot \frac{(-1)^k P_k(x) t^k}{k!}$$

$$= \sum_{k=0}^{\infty} \frac{(-1)^k P_k(x) t^k}{k!(1-t)^{1+k}},$$

we obtain

$$(16)\quad (1-t)^{-1}\exp\left(\frac{-xt}{1-t}\right) {}_0F_1\left[\begin{matrix} -; & \dfrac{t^2(x^2-1)}{4(1-t)^2} \\ 1; & \end{matrix}\right] = \sum_{n=0}^{\infty} v_n(x)t^n.$$

Next

$$\sum_{n=0}^{\infty} \frac{v_n(x)t^n}{n!} = \sum_{n,k=0}^{\infty} \frac{(-1)^k P_k(x) t^{n+k}}{(k!)^2 n!},$$

and we may conclude that

$$(17)\quad e^t\,{}_0F_1(-;1;-\tfrac{1}{2}t(x-1))\,{}_0F_1(-;1;-\tfrac{1}{2}t(x+1)) = \sum_{n=0}^{\infty} \frac{v_n(x)t^n}{n!}.$$

It is also not difficult to derive such results as

$$(18)\quad (1-x^2)v_n''(x) - 2xv_n'(x) + n(n+1)v_n(x) = 2n^2 v_{n-1}(x) - n(n-1)v_{n-2}(x),$$

$$(19)\quad (1-x^2)v_n'(x) + nxv_n(x) = [(2n-1)x-1]v_{n-1}(x) - (n-1)xv_{n-2}(x) + (1-x^2)v_{n-1}'(x),$$

$$(20)\quad v_n(x) = \frac{1}{\pi}\int_0^{\pi} L_n(x+\sqrt{x^2-1}\cos\beta)\,d\beta.$$

$$(21)\quad \int_{-1}^{1} v_n(x)v_m(x)\,dx = 2\,{}_3F_4\left[\begin{matrix} -n, -m, \tfrac{1}{2}; & \\ & 1 \\ 1, 1, 1, \tfrac{3}{2}; & \end{matrix}\right].$$

EXERCISES

In Exs. 1–8, $H_n(x)$, $P_n(x)$, $L_n(x)$ denote the Hermite, Legendre, and simple Laguerre polynomials, respectively. Derive each of the stated symbolic relations.

1. $H_n(x+y) \doteq [H(x)+2y]^n$.

2. $[H(x)+H(y)]^n \doteq 2^{\frac{1}{2}n} H_n(2^{-\frac{1}{2}}(x+y))$.

3. $P_n(x) \doteq n!\left[L\left(\frac{1-x}{2}\right) - L\left(\frac{1+x}{2}\right)\right]^n$

4. $H_n(\tfrac{1}{2}P(x)) \doteq [H(x)-P(x)]^n$.

5. $[H(x)-2P(x)]^{2n+1} \doteq 0$,

   $[H(x)-2P(x)]^{2n} \doteq (-1)^n 2^{2n} (\tfrac{1}{2})_n L_n(x^2-1)$.

6. $[H(x)-P(2x)]^{2n+1} \doteq 0$,

   $[H(x)-P(2x)]^{2n} \doteq (-1)^n 2^{2n} (\tfrac{1}{2})_n L_n(x^2-\tfrac{1}{4})$.

7. $H_n(\frac{1}{2}H(x)) \doteqdot 2^{\frac{1}{2}n}H_n(2^{-\frac{1}{2}}x)$.

8. $H_n(xy) \doteqdot [H(x) + 2x(y - 1)]^n$.

9. Use Laplace's first integral for $P_n(x)$ to derive equation (20), preceding these exercises.

10. For the $v_n(x)$ of equation (15), page 251, evaluate

$$\int_{-1}^{1} v_m(x)P_k(x)\,dx,$$

and use your result to establish equation (21), preceding these exercises.

11. Show that

$$(y + x)^n P_n\left(\frac{y - x}{y + x}\right) \doteqdot n!\{L(x) - L(y)\}^n.$$

12. Define polynomials $\varphi_n(x, y)$ by

$$\varphi_n(x, y) \doteqdot H_n(xL(y)).$$

Show that

$$\sum_{n=0}^{\infty} \frac{\varphi_n(x, y)t^n}{n!} = \exp(2xt - t^2)\,{}_0F_1(-; 1; -2xyt).$$

CHAPTER 16

Jacobi Polynomials

132. The Jacobi polynomials. The Jacobi polynomial $P_n^{(\alpha,\beta)}(x)$ may be defined by

$$(1)\quad P_n^{(\alpha,\beta)}(x) = \frac{(1+\alpha)_n}{n!}\,{}_2F_1\left[\begin{matrix}-n, 1+\alpha+\beta+n; \\ 1+\alpha;\end{matrix}\quad \frac{1-x}{2}\right].$$

When $\alpha = \beta = 0$, the polynomial in (1) becomes the Legendre polynomial. From (1) it follows that $P_n^{(\alpha,\beta)}(x)$ is a polynomial of degree precisely n and that

$$P_n^{(\alpha,\beta)}(1) = \frac{(1+\alpha)_n}{n!}.$$

In dealing with the Jacobi polynomials, it is natural to make much use of our knowledge of the ${}_2F_1$ of Chapter 4. An application of Theorem 20, page 60, to (1) yields

$$(2)\quad P_n^{(\alpha,\beta)}(x) = \frac{(1+\alpha)_n}{n!}\left(\frac{x+1}{2}\right)^n {}_2F_1\left[\begin{matrix}-n, -\beta-n; \\ 1+\alpha;\end{matrix}\quad \frac{x-1}{x+1}\right].$$

In the next section we shall obtain still another ${}_2F_1$ form for $P_n^{(\alpha,\beta)}(x)$, namely,

$$(3)\quad P_n^{(\alpha,\beta)}(x) = \frac{(-1)^n(1+\beta)_n}{n!}\,{}_2F_1\left[\begin{matrix}-n, 1+\alpha+\beta+n; \\ 1+\beta;\end{matrix}\quad \frac{1+x}{2}\right].$$

Each of (1)–(3) yields a finite series form for $P_n^{(\alpha,\beta)}(x)$:

$$(4)\quad P_n^{(\alpha,\beta)}(x) = \sum_{k=0}^{n} \frac{(1+\alpha)_n(1+\alpha+\beta)_{n+k}}{k!(n-k)!(1+\alpha)_k(1+\alpha+\beta)_n}\left(\frac{x-1}{2}\right)^k,$$

$$(5)\quad P_n^{(\alpha,\beta)}(x) = \sum_{k=0}^{n} \frac{(1+\alpha)_n(1+\beta)_n}{k!(n-k)!(1+\alpha)_k(1+\beta)_{n-k}}\left(\frac{x-1}{2}\right)^k\left(\frac{x+1}{2}\right)^{n-k},$$

$$(6)\quad P_n^{(\alpha,\beta)}(x) = \sum_{k=0}^{n} \frac{(-1)^{n-k}(1+\beta)_n(1+\alpha+\beta)_{n+k}}{k!(n-k)!(1+\beta)_k(1+\alpha+\beta)_n}\left(\frac{x+1}{2}\right)^k.$$

Equations (4), (5), (6) are expanded forms of (1), (2), (3), respectively. We must not use (3) or (6) until we derive them in the next section.

By reversing the order of summation in (4), (5), (6), respectively, we obtain

$$(7)\quad P_n^{(\alpha,\beta)}(x) = \frac{(1+\alpha+\beta)_{2n}}{n!(1+\alpha+\beta)_n}\left(\frac{x-1}{2}\right)^n {}_2F_1\left[\begin{matrix}-n, -\alpha-n; \\ -\alpha-\beta-2n;\end{matrix}\ \frac{2}{1-x}\right],$$

$$(8)\quad P_n^{(\alpha,\beta)}(x) = \frac{(1+\beta)_n}{n!}\left(\frac{x-1}{2}\right)^n {}_2F_1\left[\begin{matrix}-n, -\alpha-n; \\ 1+\beta;\end{matrix}\ \frac{x+1}{x-1}\right],$$

$$(9)\quad P_n^{(\alpha,\beta)}(x) = \frac{(1+\alpha+\beta)_{2n}}{n!(1+\alpha+\beta)_n}\left(\frac{x+1}{2}\right)^n {}_2F_1\left[\begin{matrix}-n, -\beta-n; \\ -\alpha-\beta-2n;\end{matrix}\ \frac{2}{x+1}\right],$$

of which (9) is not yet available for use.

A generating function for the Jacobi polynomials follows readily from (4). Consider the series

$$\psi = \sum_{n=0}^{\infty} \frac{(\alpha+\beta+1)_n P_n^{(\alpha,\beta)}(x)t^n}{(1+\alpha)_n}$$

$$= \sum_{n=0}^{\infty}\sum_{k=0}^{n} \frac{(1+\alpha+\beta)_{n+k}(\frac{1}{2}x-\frac{1}{2})^k t^n}{k!(n-k)!(1+\alpha)_k}$$

$$= \sum_{n,k=0}^{\infty} \frac{(1+\alpha+\beta)_{n+2k}(x-1)^k t^{n+k}}{k!n!(1+\alpha)_k 2^k},$$

$$\psi = \sum_{k=0}^{\infty}\sum_{n=0}^{\infty}\frac{(1+\alpha+\beta+2k)_n t^n}{n!}\cdot\frac{(1+\alpha+\beta)_{2k}(x-1)^k t^k}{k!(1+\alpha)_k 2^k}$$

$$= \sum_{k=0}^{\infty}\frac{(1+\alpha+\beta)_{2k}(x-1)^k t^k}{2^k k!(1+\alpha)_k(1-t)^{1+\alpha+\beta+2k}}.$$

Since $(a)_{2k} = 2^{2k}(\frac{1}{2}a)_k(\frac{1}{2}a+\frac{1}{2})_k$, it follows that

$$(10)\quad (1-t)^{-1-\alpha-\beta}\,{}_2F_1\left[\begin{matrix}\frac{1}{2}(1+\alpha+\beta),\ \frac{1}{2}(2+\alpha+\beta); & \dfrac{2t(x-1)}{(1-t)^2}\\ 1+\alpha; & \end{matrix}\right]$$

$$= \sum_{n=0}^{\infty}\frac{(\alpha+\beta+1)_n P_n^{(\alpha,\beta)}(x)t^n}{(1+\alpha)_n},$$

a generating relation for the Jacobi polynomials.

133. Bateman's generating function. Equation (5) of the preceding section, which came from (2) of the same section, leads us to what seems the most beautiful of the known generating functions for the Jacobi polynomials. From (5) we obtain

$$\sum_{n=0}^{\infty}\frac{P_n^{(\alpha,\beta)}(x)t^n}{(1+\alpha)_n(1+\beta)_n} = \sum_{n=0}^{\infty}\sum_{k=0}^{n}\frac{(\frac{1}{2}x-\frac{1}{2})^k(\frac{1}{2}x+\frac{1}{2})^{n-k}t^n}{k!(n-k)!(1+\alpha)_k(1+\beta)_{n-k}}$$

$$= \left[\sum_{n=0}^{\infty}\frac{(\frac{1}{2}x-\frac{1}{2})^n t^n}{n!(1+\alpha)_n}\right]\left[\sum_{n=0}^{\infty}\frac{(\frac{1}{2}x+\frac{1}{2})^n t^n}{n!(1+\beta)_n}\right].$$

We have thus derived Bateman's generating function:

$$(1)\quad {}_0F_1\left[\begin{matrix}-; & \dfrac{t(x-1)}{2}\\ 1+\alpha; & \end{matrix}\right]{}_0F_1\left[\begin{matrix}-; & \dfrac{t(x+1)}{2}\\ 1+\beta; & \end{matrix}\right] = \sum_{n=0}^{\infty}\frac{P_n^{(\alpha,\beta)}(x)t^n}{(1+\alpha)_n(1+\beta)_n}.$$

Equation (1) was first published by Bateman [1]. Bateman's methods and the form in which he put his result bear no resemblance to what was done here.

If in (1) we replace x by $(-x)$ and t by $(-t)$, the two ${}_0F_1$'s change roles. Hence from (1) we obtain

$$(2)\qquad P_n^{(\alpha,\beta)}(-x) = (-1)^n P_n^{(\beta,\alpha)}(x).$$

Applying (2) to equation (1) of Section 132 yields equation (3) of Section 132. Hence equations (3), (6), and (9) of that section are now available for use. Equation (2) also leads to

$$P_n^{(\alpha,\beta)}(-1) = \frac{(-1)^n(1+\beta)_n}{n!}.$$

Two other generating functions, one of them containing an arbitrary parameter, will be derived in Sections 140 and 141.

134. The Rodrigues formula. Equation (5), page 255, can be written

$$(1) \qquad P_n^{(\alpha,\beta)}(x) = \sum_{k=0}^{n} \frac{(1+\alpha)_n(1+\beta)_n(x-1)^k(x+1)^{n-k}}{2^n k!(n-k)!(1+\alpha)_k(1+\beta)_{n-k}}.$$

If $D = d/dx$, then for non-negative integral s and m,

$$D^s x^{m+\alpha} = (m+\alpha)(m+\alpha-1)\cdots(m+\alpha-s+1)x^{m-s+\alpha}$$

or

$$(2) \qquad D^s x^{m+\alpha} = \frac{(1+\alpha)_m x^{m-s+\alpha}}{(1+\alpha)_{m-s}}.$$

From (2) we obtain

$$(3) \qquad D^k(x+1)^{n+\beta} = \frac{(1+\beta)_n(x+1)^{n-k+\beta}}{(1+\beta)_{n-k}}$$

and

$$(4) \qquad D^{n-k}(x-1)^{n+\alpha} = \frac{(1+\alpha)_n(x-1)^{k+\alpha}}{(1+\alpha)_k}.$$

Therefore (1) can be put in the form

$$(5) \qquad P_n^{(\alpha,\beta)}(x) = \frac{(x-1)^{-\alpha}(x+1)^{-\beta}}{2^n n!} \cdot \sum_{k=0}^{n} \frac{n!}{k!(n-k)!}[D^{n-k}(x-1)^{n+\alpha}][D^k(x+1)^{n+\beta}].$$

In view of Leibnitz' rule for the derivative of a product, equation (5) yields the Rodrigues formula

$$(6) \quad P_n^{(\alpha,\beta)}(x) = \frac{(x-1)^{-\alpha}(x+1)^{-\beta}}{2^n n!} D^n[(x-1)^{n+\alpha}(x+1)^{n+\beta}],$$

or

$$(7) \quad P_n^{(\alpha,\beta)}(x) = \frac{(-1)^n(1-x)^{-\alpha}(1+x)^{-\beta}}{2^n n!} D^n[(1-x)^{n+\alpha}(1+x)^{n+\beta}].$$

Equation (7) is more desirable than (6) when we work in the interval $-1 < x < 1$.

135. Orthogonality. By using (1), Section 132, and the differential equation for ${}_2F_1\,(a, b; c; y)$, we obtain at once a differential equation satisfied by the Jacobi polynomial:

$$(1)\qquad (1-x^2)D^2\,P_n^{(\alpha,\beta)}(x)+[\beta-\alpha-(2+\alpha+\beta)x]D\,P_n^{(\alpha,\beta)}(x) + n(1+\alpha+\beta+n)P_n^{(\alpha,\beta)}(x) = 0.$$

Since $\beta-\alpha-(2+\alpha+\beta)x=(1+\beta)(1-x)-(1+\alpha)(1+x)$, we may put (1) in the form

$$(1-x)^{1+\alpha}(1+x)^{1+\beta}D^2P_n^{(\alpha,\beta)}(x) +[(1+\beta)(1-x)-(1+\alpha)(1+x)](1-x)^{\alpha}(1+x)^{\beta}DP_n^{(\alpha,\beta)}(x) +n(1+\alpha+\beta+n)(1-x)^{\alpha}(1+x)^{\beta}P_n^{(\alpha,\beta)}(x) = 0,$$

which yields

$$(2)\qquad D[(1-x)^{1+\alpha}(1+x)^{1+\beta}DP_n^{(\alpha,\beta)}(x)] + n(1+\alpha+\beta+n)(1-x)^{\alpha}(1+x)^{\beta}P_n^{(\alpha,\beta)}(x) = 0.$$

From (2) and the same equation with n replaced by m, it follows (see page 173 for the corresponding work on Legendre polynomials) that

$$[n(1+\alpha+\beta+n)-m(1+\alpha+\beta+m)](1-x)^{\alpha}(1+x)^{\beta}P_n^{(\alpha,\beta)}(x)P_m^{(\alpha,\beta)}(x) =D[(1-x)^{1+\alpha}(1+x)^{1+\beta}\{P_n^{(\alpha,\beta)}(x)DP_m^{(\alpha,\beta)}(x)-P_m^{(\alpha,\beta)}(x)DP_n^{(\alpha,\beta)}(x)\}].$$

Therefore we may conclude that

$$(3)\qquad (n-m)(1+\alpha+\beta+n+m)\cdot \int_a^b(1-x)^{\alpha}(1+x)^{\beta}P_n^{(\alpha,\beta)}(x)P_m^{(\alpha,\beta)}(x)\,dx = \Big[(1-x)^{1+\alpha}(1+x)^{1+\beta}\{P_n^{(\alpha,\beta)}(x)\,D\,P_m^{(\alpha,\beta)}(x) - P_m^{(\alpha,\beta)}(x)D\,P_n^{(\alpha,\beta)}(x)\}\Big]_a^b.$$

In particular, if $\mathrm{Re}(\alpha) > -1$ and $\mathrm{Re}(\beta) > -1$, then (3) leads us to the orthogonality property

$$(4)\qquad \int_{-1}^1(1-x)^{\alpha}(1+x)^{\beta}P_n^{(\alpha,\beta)}(x)P_m^{(\alpha,\beta)}(x)\,dx = 0, \qquad m\neq n.$$

That is, if $\mathrm{Re}(\alpha) > -1$ and $\mathrm{Re}(\beta) > -1$, the Jacobi polynomials form an orthogonal set over $(-1, 1)$ with respect to the weight function $(1-x)^{\alpha}\,(1+x)^{\beta}$. For real α and β, the Jacobi polynomials are real, so we can then apply the conclusions of Chapter 9 on real orthogonal sets of polynomials.

In order to evaluate

$$(5)\qquad g_n = \int_{-1}^{1} (1-x)^{\alpha}(1+x)^{\beta}[P_n^{(\alpha,\beta)}(x)]^2\,dx$$

we shall employ the Rodrigues formula and integration by parts. This method incidentally furnishes a second derivation of the orthogonality property (4).

From (7), page 257, we obtain

$$(6)\quad (1-x)^{\alpha}(1+x)^{\beta}P_n^{(\alpha,\beta)}(x) = \frac{(-1)^n}{2^n n!}D^n[(1-x)^{n+\alpha}(1+x)^{n+\beta}].$$

Therefore, if $\text{Re}(\alpha) > -1$ and $\text{Re}(\beta) > -1$,

$$(7)\qquad \int_{-1}^{1} (1-x)^{\alpha}(1+x)^{\beta}P_n^{(\alpha,\beta)}(x)P_m^{(\alpha,\beta)}(x)\,dx$$
$$= \frac{(-1)^n}{2^n n!}\int_{-1}^{1}\{D^n[(1-x)^{n+\alpha}(1+x)^{n+\beta}]\}P_m^{(\alpha,\beta)}(x)\,dx.$$

On the right in (7), integrate by parts n times, each time differentiating $P_m^{(\alpha,\beta)}(x)$ and integrating the quantity in curly brackets. At the kth stage the integrated part,

$$\{D^{n-k}[(1-x)^{n+\alpha}(1+x)^{n+\beta}]\}D^{k-1}P_m^{(\alpha,\beta)}(x)$$

is zero at both limits because of factors $(1-x)^{k+\alpha}(1+x)^{k+\beta}$ with $\text{Re}(\alpha) > -1$ and $\text{Re}(\beta) > -1$. After n such integrations by parts, we have

$$(8)\qquad \int_{-1}^{1} (1-x)^{\alpha}(1+x)^{\beta}P_n^{(\alpha,\beta)}(x)P_m^{(\alpha,\beta)}(x)\,dx$$
$$= \frac{(-1)^{2n}}{2^n n!}\int_{-1}^{1}(1-x)^{n+\alpha}(1+x)^{n+\beta}[D^nP_m^{(\alpha,\beta)}(x)]\,dx.$$

If $m \neq n$ we may choose n to be the larger (or interchange m and n in the preceding discussion) and therefore conclude that

$$(9)\qquad \int_{-1}^{1} (1-x)^{\alpha}(1+x)^{\beta}P_n^{(\alpha,\beta)}(x)P_m^{(\alpha,\beta)}(x)\,dx = 0, \qquad m \neq n,$$

because $P_m^{(\alpha,\beta)}(x)$ on the right in (8) is a polynomial of degree m.

In (8) we have a tool for the evaluation of the g_n of (5), but we need $D^n P_n^{(\alpha,\beta)}(x)$. From

$$P_n^{(\alpha,\beta)}(x) = \frac{(1+\alpha)_n}{n!}\,{}_2F_1\left[\begin{matrix}-n, 1+\alpha+\beta+n; & \dfrac{1-x}{2}\\ 1+\alpha; & \end{matrix}\right]$$

and repeated application of the formula for the derivative of a ${}_2F_1$ (Ex. 1, page 69), we obtain

$$D^n P_n^{(\alpha,\beta)}(x) = \frac{(-\frac{1}{2})^n(1+\alpha)_n(-n)_n(1+\alpha+\beta+n)_n}{n!(1+\alpha)_n}\cdot$$

$${}_2F_1\left[\begin{matrix}0, 1+\alpha+\beta+2n; & \dfrac{1-x}{2}\\ 1+\alpha+n; & \end{matrix}\right],$$

from which

$$D^n P_n^{(\alpha,\beta)}(x) = \frac{(1+\alpha+\beta)_{2n}}{2^n(1+\alpha+\beta)_n}. \tag{10}$$

Now (8) with $m = n$ yields

$$g_n = \int_{-1}^{1}(1-x)^\alpha(1+x)^\beta[P_n^{(\alpha,\beta)}(x)]^2\,dx$$

$$= \frac{(1+\alpha+\beta)_{2n}}{2^{2n}n!(1+\alpha+\beta)_n}\int_{-1}^{1}(1-x)^{n+\alpha}(1+x)^{n+\beta}\,dx.$$

But, by Ex. 7, page 31,

$$\int_{-1}^{1}(1-x)^{n+\alpha}(1+x)^{n+\beta}\,dx = 2^{2n+\alpha+\beta+1}B(1+\alpha+n, 1+\beta+n)$$

$$= \frac{2^{2n+\alpha+\beta+1}\Gamma(1+\alpha+n)\Gamma(1+\beta+n)}{\Gamma(2+\alpha+\beta+2n)}.$$

Hence

$$g_n = \frac{(1+\alpha+\beta)_{2n}2^{1+\alpha+\beta}\Gamma(1+\alpha+n)\Gamma(1+\beta+n)}{n!(1+\alpha+\beta)_n\Gamma(2+\alpha+\beta+2n)},$$

or

$$g_n = \frac{2^{1+\alpha+\beta}\Gamma(1+\alpha+n)\Gamma(1+\beta+n)}{n!(1+\alpha+\beta+2n)\Gamma(1+\alpha+\beta+n)}. \tag{11}$$

From (10) we conclude that

$$(12) \qquad P_n^{(\alpha,\beta)}(x) = \frac{(1+\alpha+\beta)_{2n}x^n}{2^n n!(1+\alpha+\beta)_n} + \pi_{n-1}(x),$$

in which $\pi_{n-1}(x)$ is a polynomial of degree $(n-1)$. In the notation of Theorem 58, page 154, we have

$$h_n = \frac{(1+\alpha+\beta)_{2n}}{2^n n!(1+\alpha+\beta)_n}.$$

Then

$$(13) \qquad \frac{h_n}{g_n h_{n+1}} = \frac{(n+1)!\,\Gamma(2+\alpha+\beta+n)}{2^{\alpha+\beta}(2+\alpha+\beta+2n)\Gamma(1+\alpha+n)\Gamma(1+\beta+n)}.$$

With (11) and (13) at hand, the Christoffel-Darboux formula of Theorem 58, page 154, may now be written explicitly, if desired.

By combining Theorem 54, page 148, with equations (11) and (12), we conclude that

$$(14) \qquad \int_{-1}^{1}(1-x)^\alpha(1+x)^\beta x^k P_n^{(\alpha,\beta)}(x)\,dx = 0, \qquad k = 0,1,2,\cdots,(n-1),$$

and

$$(15) \qquad \int_{-1}^{1}(1-x)^\alpha(1+x)^\beta x^n P_n^{(\alpha,\beta)}(x)\,dx = \frac{2^{1+\alpha+\beta+n}\Gamma(1+\alpha+n)\Gamma(1+\beta+n)}{\Gamma(2+\alpha+\beta+2n)}.$$

For $\alpha > -1$, $\beta > -1$, the zeros of $P_n^{(\alpha,\beta)}(x)$ are distinct and lie on the open interval $-1 < x < 1$. We may also apply Theorem 59, page 156, if we wish.

136. Differential recurrence relations. In the generating relation (10), page 256, put $x = 1 - 2v$ to get

$$(1) \qquad (1-t)^{-1-\alpha-\beta}\,{}_2F_1\left[\begin{matrix} \tfrac{1}{2}(1+\alpha+\beta),\ \tfrac{1}{2}(2+\alpha+\beta); \\ 1+\alpha; \end{matrix} \ \frac{-4tv}{(1-t)^2}\right] = \sum_{n=0}^{\infty}\frac{(1+\alpha+\beta)_n P_n^{(\alpha,\beta)}(1-2v)t^n}{(1+\alpha)_n}.$$

Now we see that Theorem 48, page 137, with

$$c = 1+\alpha+\beta, \qquad \gamma_n = \frac{(1+\alpha+\beta)_{2n}}{2^{2n}n!(1+\alpha)_n},$$

applies to the polynomial

$$f_n(v) = \frac{(1+\alpha+\beta)_n P_n^{(\alpha,\beta)}(1-2v)}{(1+\alpha)_n}.$$

The five results in Theorem 48, when put in terms of x rather than v, yield our definition, page 254, of $P_n^{(\alpha,\beta)}(x)$ and the four other properties

(2) $(1-x)^n =$

$$2^n(1+\alpha)_n \sum_{k=0}^{n} \frac{(-n)_k(1+\alpha+\beta+2k)(1+\alpha+\beta)_k P_k^{(\alpha,\beta)}(x)}{(1+\alpha+\beta)_{n+1+k}(1+\alpha)_k},$$

(3) $$(x-1)[(\alpha+\beta+n)\, D\, P_n^{(\alpha,\beta)}(x) + (\alpha+n)\, D\, P_{n-1}^{(\alpha,\beta)}(x)]$$
$$= (\alpha+\beta+n)[nP_n^{(\alpha,\beta)}(x) - (\alpha+n)P_{n-1}^{(\alpha,\beta)}(x)], \quad D = \frac{d}{dx},$$

(4) $$(x-1)\, D\, P_n^{(\alpha,\beta)}(x) - nP_n^{(\alpha,\beta)}(x)$$
$$= \frac{-(1+\alpha)_n}{(1+\alpha+\beta)_n} \sum_{k=0}^{n-1} \frac{(1+\alpha+\beta)_k}{(1+\alpha)_k} \cdot$$
$$[(1+\alpha+\beta)P_k^{(\alpha,\beta)}(x) + 2(x-1)\, D\, P_k^{(\alpha,\beta)}(x)],$$

(5) $$(x-1)\, D\, P_n^{(\alpha,\beta)}(x) - nP_n^{(\alpha,\beta)}(x) = \frac{(1+\alpha)_n}{(1+\alpha+\beta)_n} \cdot$$
$$\sum_{k=0}^{n-1} \frac{(-1)^{n-k}(1+\alpha+\beta+2k)(1+\alpha+\beta)_k P_k^{(\alpha,\beta)}(x)}{(1+\alpha)_k}.$$

The relation (3) is an ordinary differential recurrence relation, with derivatives and shift of index involved but no shift in the parameters α and β. We seek two such relations to furnish the groundwork for others and for the pure recurrence relation. We know

$$P_n^{(\beta,\alpha)}(-x) = (-1)^n P_n^{(\alpha,\beta)}(x).$$

Therefore in (3) we interchange α and β and replace x by $(-x)$ to obtain

(6) $$(x+1)[(\alpha+\beta+n)\, D\, P_n^{(\alpha,\beta)}(x) - (\beta+n)\, D\, P_{n-1}^{(\alpha,\beta)}(x)]$$
$$= (\alpha+\beta+n)[nP_n^{(\alpha,\beta)}(x) + (\beta+n)P_{n-1}^{(\alpha,\beta)}(x)].$$

Let us eliminate $D\, P_{n-1}^{(\alpha,\beta)}(x)$ from (3) and (6). The result is

(7) $(\alpha+\beta+2n)(x^2-1)\, D\, P_n^{(\alpha,\beta)}(x)$
$= n[\beta-\alpha+(\alpha+\beta+2n)x]P_n^{(\alpha,\beta)}(x) - 2(\alpha+n)(\beta+n)P_{n-1}^{(\alpha,\beta)}(x).$

137. The pure recurrence relation. If we use (7) above twice in (6), we arrive at a relation involving only shifts in indices, the pure recurrence relation

(1) $2n(\alpha+\beta+n)(\alpha+\beta+2n-2)P_n^{(\alpha,\beta)}(x)$
$= (\alpha+\beta+2n-1)[\alpha^2-\beta^2+x(\alpha+\beta+2n)(\alpha+\beta+2n-2)]P_{n-1}^{(\alpha,\beta)}(x)$
$-2(\alpha+n-1)(\beta+n-1)(\alpha+\beta+2n)P_{n-2}^{(\alpha,\beta)}(x).$

138. Mixed relations. From the definition

(1) $$P_n^{(\alpha,\beta)}(x) = \frac{(1+\alpha)_n}{n!}\, {}_2F_1\left[\begin{matrix} -n,\ 1+\alpha+\beta+n; \\ 1+\alpha; \end{matrix}\ \frac{1-x}{2}\right]$$

it follows, with $D = d/dx$ as usual, that

$$D\, P_n^{(\alpha,\beta)}(x) = \frac{n(1+\alpha)_n(1+\alpha+\beta+n)}{2(1+\alpha)n!}\cdot {}_2F_1\left[\begin{matrix} -n+1,\ 2+\alpha+\beta+n; \\ 2+\alpha; \end{matrix}\ \frac{1-x}{2}\right]$$

$$= \frac{(2+\alpha)_{n-1}(1+\alpha+\beta+n)}{2(n-1)!}\cdot {}_2F_1\left[\begin{matrix} -(n-1),\ 1+(\alpha+1)+(\beta+1)+n-1; \\ 1+(\alpha+1); \end{matrix}\ \frac{1-x}{2}\right],$$

so that

(2) $$D\, P_n^{(\alpha,\beta)}(x) = \tfrac{1}{2}(1+\alpha+\beta+n)P_{n-1}^{(\alpha+1,\beta+1)}(x).$$

Iteration of (2) yields, for $0 < k \leqq n$,

(3) $$D^k\, P_n^{(\alpha,\beta)}(x) = 2^{-k}(1+\alpha+\beta+n)_k P_{n-k}^{(\alpha+k,\beta+k)}(x).$$

In Section 132 we found that

(4) $$P_n^{(\alpha,\beta)}(x) = \frac{(1+\alpha)_n}{n!}\left(\frac{x+1}{2}\right)^n {}_2F_1\left[\begin{matrix} -n,\ -\beta-n; \\ 1+\alpha; \end{matrix}\ \frac{x-1}{x+1}\right].$$

Differentiation of both members of (4) yields

$$D\,P_n^{(\alpha,\beta)}(x) = n(x+1)^{-1}P_n^{(\alpha,\beta)}(x)$$
$$+\frac{(2+\alpha)_{n-1}(\beta+n)}{(n-1)!(x+1)}\left(\frac{x+1}{2}\right)^{n-1} {}_2F_1\left[\begin{matrix} -n+1,\ -\beta-n+1; \\ 1+(\alpha+1); \end{matrix}\ \frac{x-1}{x+1}\right],$$

from which it follows that

(5) $(x+1)\,D\,P_n^{(\alpha,\beta)}(x) = nP_n^{(\alpha,\beta)}(x) + (\beta+n)P_{n-1}^{(\alpha+1,\beta)}(x).$

In the same manner the known relation [(8) of Section 132]

(6) $$P_n^{(\alpha,\beta)}(x) = \frac{(1+\beta)_n}{n!}\left(\frac{x-1}{2}\right)^n {}_2F_1\left[\begin{matrix} -n,\ -\alpha-n; \\ 1+\beta; \end{matrix}\ \frac{x+1}{x-1}\right]$$

leads us to

(7) $(x-1)\,D\,P_n^{(\alpha,\beta)}(x) = nP_n^{(\alpha,\beta)}(x) - (\alpha+n)P_{n-1}^{(\alpha,\beta+1)}(x).$

From (5) and (7) it follows that

(8) $2D\,P_n^{(\alpha,\beta)}(x) = (\beta+n)P_{n-1}^{(\alpha+1,\beta)}(x) + (\alpha+n)P_{n-1}^{(\alpha,\beta+1)}(x).$

If we use (2) in (8) and then shift n to $(n+1)$, α to $(\alpha-1)$ and β to $(\beta-1)$, we obtain

(9) $(\alpha+\beta+n)P_n^{(\alpha,\beta)}(x) = (\beta+n)P_n^{(\alpha,\beta-1)}(x) + (\alpha+n)P_n^{(\alpha-1,\beta)}(x).$

We next seek to express each contiguous (α or β increased or decreased by unity) Jacobi polynomial in terms of the original polynomial.

In Ex. 22, page 72, we found that

(10) $$F(a, b; c; z)$$
$$= F(a-1, b+1; c; z) + c^{-1}(b+1-a)zF(a, b+1; c+1; z).$$

In (10) put $a = -n$, $b = 1+\alpha+\beta+n$, $c = 1+\alpha$, $z = \frac{1}{2}(1-x)$, and simplify the result to obtain

(11) $$\tfrac{1}{2}(2+\alpha+\beta+2n)(x-1)P_n^{(\alpha+1,\beta)}(x)$$
$$= (n+1)P_{n+1}^{(\alpha,\beta)}(x) - (1+\alpha+n)P_n^{(\alpha,\beta)}(x).$$

In Ex. 23, page 72, we found that

(12) $$(c-1-b)F(a, b; c; z)$$
$$= (c-a)F(a-1, b+1; c; z) + (a-1-b)(1-z)F(a, b+1; c; z).$$

In (12) put $a = -n, b = 1 + \alpha + \beta + n, c = 1 + \alpha, z = \frac{1}{2}(1 - x)$ to obtain

(13) $$\tfrac{1}{2}(2+ \alpha + \beta + 2n)(x + 1)P_n^{(\alpha,\beta+1)}(x) = (n + 1)P_{n+1}^{(\alpha,\beta)}(x) + (1 + \beta + n)P_n^{(\alpha,\beta)}(x).$$

The contiguous function relation

$$(a - b)F(a, b; c; z) = aF(a + 1, b; c; z) - bF(a, b + 1; c; z)$$

leads in the same way, after a shift from β to $(\beta - 1)$ to

(14) $$(\alpha + \beta + 2n)P_n^{(\alpha,\beta-1)}(x) = (\alpha + \beta + n)P_n^{(\alpha,\beta)}(x) + (\alpha + n)P_{n-1}^{(\alpha,\beta)}(x).$$

From (9) and (14) it follows that

(15) $$(\alpha + \beta + 2n)P_n^{(\alpha-1,\beta)}(x) = (\alpha + \beta + n)P_n^{(\alpha,\beta)}(x) - (\beta + n)P_{n-1}^{(\alpha,\beta)}(x).$$

Finally, (11) and (13) combine to yield

(16) $$(1 + x)P_n^{(\alpha,\beta+1)}(x) + (1 - x)P_n^{(\alpha+1,\beta)}(x) = 2P_n^{(\alpha,\beta)}(x)$$

and (14) and (15) combine to yield

(17) $$P_n^{(\alpha,\beta-1)}(x) - P_n^{(\alpha-1,\beta)}(x) = P_{n-1}^{(\alpha,\beta)}(x).$$

139. Appell's functions of two variables. One of various ways in which the ${}_2F_1$ function has been generalized is to functions of more than one variable. Of these a fruitful set is that of Appell's four functions*

$$F_1(a; b, b'; c; x, y) = \sum_{n,k=0}^{\infty} \frac{(a)_{n+k}(b)_k(b')_n x^k y^n}{k!n!(c)_{n+k}},$$

$$F_2(a; b, b'; c, c'; x, y) = \sum_{n,k=0}^{\infty} \frac{(a)_{n+k}(b)_k(b')_n x^k y^n}{k!n!(c)_k(c')_n},$$

$$F_3(a, a'; b, b'; c; x, y) = \sum_{n,k=0}^{\infty} \frac{(a)_k(a')_n(b)_k(b')_n x^k y^n}{k!n!(c)_{n+k}},$$

$$F_4(a, b; c, c'; x, y) = \sum_{n,k=0}^{\infty} \frac{(a)_{n+k}(b)_{n+k} x^k y^n}{k!n!(c)_k(c')_n}.$$

*See Appell and Kampé de Fériet [1].

The four functions above have many interesting properties, for which see the reference in the footnote or Bailey [1] or Erdélyi [1]. Here we need only two theorems on F_4, and our discussion is limited to what we need.

Consider

$$(1)\quad \psi=(1-x)^{-a}(1-y)^{-b}F_4\left(a, b; c, c'; \frac{-x}{(1-x)(1-y)}, \frac{-y}{(1-x)(1-y)}\right).$$

From the definition of F_4 we obtain

$$\psi = \sum_{n,k=0}^{\infty} \frac{(a)_{n+k}(b)_{n+k}(-1)^{n+k}x^k y^n}{k!n!(c)_k(c')_n(1-x)^{n+k+a}(1-y)^{n+k+b}}.$$

Now

$$(1-x)^{-a-n-k} = \sum_{i=0}^{\infty} \frac{(a+n+k)_i x^i}{i!}$$

and

$$(1-y)^{-b-n-k} = \sum_{s=0}^{\infty} \frac{(b+n+k)_s y^s}{s!}.$$

Hence

$$\psi = \sum_{n,k,s,i=0}^{\infty} \frac{(-1)^{n+k}(a)_{n+k+i}(b)_{n+k+s}x^{k+i}y^{n+s}}{s!i!k!n!(c)_k(c')_n}$$

so that

$$\psi = \sum_{n,k,s=0}^{\infty} \sum_{i=0}^{k} \frac{(-1)^{n+k-i}(a)_{n+k}(b)_{n+k+s-i}x^k y^{n+s}}{i!(k-i)!s!n!(c)_{k-i}(c')_n}.$$

Let us reverse the order of the inner summation and write

$$\psi = \sum_{n,k,s=0}^{\infty} \sum_{i=0}^{k} \frac{(-1)^i(b)_{n+s+i}}{i!(k-i)!(c)_i} \cdot \frac{(-1)^n(a)_{n+k}x^k y^{n+s}}{s!n!(c')_n}$$

$$= \sum_{n,k,s=0}^{\infty} {}_2F_1\left[\begin{matrix} -k, b+n+s; \\ c; \end{matrix}\ 1\right] \frac{(-1)^n(a)_{n+k}(b)_{n+s}x^k y^{n+s}}{k!s!n!(c')_n}.$$

Then

$$\psi = \sum_{n,k=0}^{\infty} \sum_{s=0}^{n} {}_2F_1\left[\begin{matrix} -k, b+n; \\ c; \end{matrix}\ 1\right] \frac{(-1)^{n-s}(a)_{n-s+k}(b)_n x^k y^n}{k!s!(n-s)!(c')_{n-s}}$$

and we reverse order in the inner summation to obtain

$$\psi = \sum_{n,k=0}^{\infty} \sum_{s=0}^{n} \frac{(-1)^s (a)_{s+k}}{s!(n-s)!(c')_s} \, {}_2F_1\left[\begin{matrix} -k, b+n; \\ c; \end{matrix} \; 1\right] \frac{(b)_n x^k y^n}{k!},$$

or

$$(2) \quad \psi = \sum_{n,k=0}^{\infty} {}_2F_1\left[\begin{matrix} -n, a+k; \\ c'; \end{matrix} \; 1\right] {}_2F_1\left[\begin{matrix} -k, b+n; \\ c; \end{matrix} \; 1\right] \frac{(a)_k (b)_n x^k y^n}{k!n!}.$$

Now recall that in Ex. 6, page 128, we found, with different notation, that if m and γ are non-negative integers,

$$(3) \qquad {}_2F_1\left[\begin{matrix} -m, d+\gamma; \\ d; \end{matrix} \; 1\right] = 0, \qquad \text{for } m > \gamma$$

$$= \frac{(-\gamma)_m}{(d)_m}, \qquad \text{for } 0 \leqq m \leqq \gamma.$$

With (3) in mind we turn to (2) and put $c' = a$, $c = b$.

Since

$$F\left[\begin{matrix} -n, a+k; \\ a; \end{matrix} \; 1\right] = 0 \qquad \text{for } n > k.$$

and

$$F\left[\begin{matrix} -k, b+n; \\ b; \end{matrix} \; 1\right] = 0 \qquad \text{for } k > n,$$

the only terms remaining in the double summation (2) will be those for which $k = n$. Recall the definition (1) of ψ. We thus arrive at

$$(1-x)^{-a}(1-y)^{-b} F_4\left(a, b; b, a; \frac{-x}{(1-x)(1-y)}, \frac{-y}{(1-x)(1-y)}\right)$$

$$= \sum_{n=0}^{\infty} \frac{(-n)_n}{(a)_n} \cdot \frac{(-n)_n}{(b)_n} \cdot \frac{(a)_n (b)_n x^n y^n}{n!n!}$$

$$= \sum_{n=0}^{\infty} x^n y^n$$

$$= (1-xy)^{-1}.$$

THEOREM 83. *If neither a nor b is zero or a negative integer,*

$$(4)\quad F_4\left(a, b; b, a; \frac{-x}{(1-x)(1-y)}, \frac{-y}{(1-x)(1-y)}\right) = (1-xy)^{-1}(1-x)^a(1-y)^b.$$

Next let us return to (2). Since the two ${}_2F_1$'s with unit argument terminate, we may write

$$\psi = \sum_{n,k=0}^{\infty} \frac{\Gamma(c')\Gamma(c'-a+n-k)\Gamma(c)\Gamma(c-b+k-n)(a)_k(b)_n x^k y^n}{\Gamma(c'+n)\Gamma(c'-a-k)\Gamma(c+k)\Gamma(c-b-n)k!n!}.$$

We know that

$$\frac{\Gamma(d+n)}{\Gamma(d)} = (d)_n, \qquad \frac{\Gamma(1-d-n)}{\Gamma(1-d)} = \frac{(-1)^n}{(d)_n},$$

so we may write

$$(5)\quad \psi = \sum_{n,k=0}^{\infty} \frac{\Gamma(c'-a+n-k)\Gamma(c-b+k-n)}{\Gamma(c'-a)\Gamma(c-b)} \cdot \frac{(a)_k(b)_n(1-c'+a)_k(1-c+b)_n(-1)^{n+k}x^k y^n}{k!n!(c')_n(c)_k}.$$

If we so choose c' that the arguments in the Gamma products add to unity, we may use the formula

$$\Gamma(z)\Gamma(1-z) = \frac{\pi}{\sin \pi z}.$$

Hence we choose

$$c' = 1 - c + a + b.$$

Then

$$\frac{\Gamma(c'-a+n-k)\Gamma(c-b+k-n)}{\Gamma(c'-a)\Gamma(c-b)} = \frac{\Gamma(1-c+b+n-k)\Gamma(c-b-n+k)}{\Gamma(1-c+b)\Gamma(c-b)}$$

$$= \frac{\sin \pi(c-b)}{\sin \pi(c-b-n+k)}$$

$$= \frac{\sin \pi(c-b)}{(-1)^{n-k}\sin \pi(c-b)} = (-1)^{n+k}.$$

Therefore (5) yields, for $c' = 1 - c + a + b$,

$$\psi = \sum_{n,k=0}^{\infty} \frac{(a)_k(b)_n(c-b)_k(1-c+b)_n x^k y^n}{k!n!(1-c+a+b)_n(c)_k},$$

$$\psi = \left[\sum_{k=0}^{\infty} \frac{(a)_k(c-b)_k x^k}{k!(c)_k}\right]\left[\sum_{n=0}^{\infty} \frac{(b)_n(1-c+b)_n y^n}{n!(1-c+a+b)_n}\right]$$

$$= {}_2F_1\left[\begin{matrix} a, c-b; \\ c; \end{matrix} \; x\right] {}_2F_1\left[\begin{matrix} 1-c+b, b; \\ 1-c+a+b; \end{matrix} \; y\right]$$

$$= (1-x)^{-a}(1-y)^{-b}\, {}_2F_1\left[\begin{matrix} a, b; \\ c; \end{matrix} \; \frac{-x}{1-x}\right] {}_2F_1\left[\begin{matrix} a, b; \\ 1-c+a+b; \end{matrix} \; \frac{-y}{1-y}\right].$$

THEOREM 84. *If neither* c *nor* $(1 - c + a + b)$ *is zero or a negative integer,*

(6) $$F_4\left(a, b; c, 1-c+a+b; \frac{-x}{(1-x)(1-y)}, \frac{-y}{(1-x)(1-y)}\right)$$

$$= {}_2F_1\left[\begin{matrix} a, b; \\ c; \end{matrix} \; \frac{-x}{1-x}\right] {}_2F_1\left[\begin{matrix} a, b; \\ 1-c+a+b; \end{matrix} \; \frac{-y}{1-y}\right].$$

In (6) replace x by $-x/(1 - x)$ and y by $-y/(1 - y)$ to obtain the equivalent result

(7) $$F_4(a, b; c, 1+a+b-c; x(1-y), y(1-x))$$

$$= {}_2F_1\left[\begin{matrix} a, b; \\ c; \end{matrix} \; x\right] {}_2F_1\left[\begin{matrix} a, b; \\ 1+a+b-c; \end{matrix} \; y\right].$$

Theorems 83 and 84 furnish the tools we need in the next two sections.

140. An elementary generating function. On page 255 we obtained the result

(1) $$P_n^{(\alpha,\beta)}(x) = \sum_{k=0}^{n} \frac{(1+\alpha)_n(1+\beta)_n}{k!(n-k)!(1+\alpha)_k(1+\beta)_{n-k}} \left(\frac{x-1}{2}\right)^k \left(\frac{x+1}{2}\right)^{n-k}.$$

With the aid of (1) we find that

$$\sum_{n=0}^{\infty} P_n^{(\alpha,\beta)}(x)t^n = \sum_{n,k=0}^{\infty} \frac{(1+\alpha)_{n+k}(1+\beta)_{n+k}[\frac{1}{2}(x-1)]^k[\frac{1}{2}(x+1)]^n t^{n+k}}{k!n!(1+\alpha)_k(1+\beta)_n},$$

of which the right member is an Appell F_4 as defined on page 265. In fact,

$$(2)\ \sum_{n=0}^{\infty} P_n^{(\alpha,\beta)}(x)t^n = F_4\left(1+\beta, 1+\alpha; 1+\alpha, 1+\beta; \tfrac{1}{2}t(x-1), \tfrac{1}{2}t(x+1)\right).$$

By Theorem 83, page 268,

$$(3)\qquad F_4\left(a, b; b, a; \frac{-u}{(1-u)(1-v)}, \frac{-v}{(1-u)(1-v)}\right) = (1-uv)^{-1}(1-u)^a(1-v)^b,$$

and we put $a = 1 + \beta$, $b = 1 + \alpha$,

$$(4)\quad \frac{-u}{(1-u)(1-v)} = \frac{t(x-1)}{2}, \qquad \frac{-v}{(1-u)(1-v)} = \frac{t(x+1)}{2}.$$

As usual, let $\rho = (1 - 2xt + t^2)^{\frac{1}{2}}$. Consider

$$(5)\qquad u = 1 - \frac{2}{1+t+\rho}, \qquad v = 1 - \frac{2}{1-t+\rho}.$$

From (5)

$$\frac{-u}{(1-u)(1-v)} = \frac{1}{1-v}\left(1 - \frac{1}{1-u}\right) = \frac{1-t+\rho}{2}\left(1 - \frac{1+t+\rho}{2}\right)$$

$$= \frac{(1-t+\rho)(1-t-\rho)}{4} = \frac{(1-t)^2-\rho^2}{4} = \frac{t(x-1)}{2},$$

and in the same way,

$$\frac{-v}{(1-u)(1-v)} = \frac{(1+t)^2-\rho^2}{4} = \frac{t(x+1)}{2},$$

as desired. Also

$$\frac{1}{1-u} = \frac{1}{2}(1+t+\rho), \qquad \frac{1}{1-v} = \frac{1}{2}(1-t+\rho),$$

from which

$$\rho = \frac{1}{1-u} + \frac{1}{1-v} - 1 = \frac{1-uv}{(1-u)(1-v)}.$$

Hence

$$(1-uv)^{-1}(1-u)^a(1-v)^b = \rho^{-1}(1-u)^{a-1}(1-v)^{b-1},$$

so that (2) and (3) yield

$$\sum_{n=0}^{\infty} P_n^{(\alpha,\beta)}(x)t^n = \rho^{-1}\left(\frac{2}{1+t+\rho}\right)^{\beta}\left(\frac{2}{1-t+\rho}\right)^{\alpha},$$

or

$$(6) \qquad \sum_{n=0}^{\infty} P_n^{(\alpha,\beta)}(x)t^n = 2^{\alpha+\beta}\rho^{-1}(1+t+\rho)^{-\beta}(1-t+\rho)^{-\alpha},$$

in which $\rho = (1 - 2xt + t^2)^{\frac{1}{2}}$.

141. Brafman's generating functions. Consider the sum

$$(1) \qquad \sum_{n=0}^{\infty} \frac{(\gamma)_n(\delta)_n P_n^{(\alpha,\beta)}(x)t^n}{(1+\alpha)_n(1+\beta)_n}$$

$$= \sum_{n=0}^{\infty}\sum_{k=0}^{n} \frac{(\gamma)_n(\delta)_n[\frac{1}{2}(x-1)]^k[\frac{1}{2}(x+1)]^{n-k}t^n}{k!(n-k)!(1+\alpha)_k(1+\beta)_{n-k}}$$

$$= \sum_{n,k=0}^{\infty} \frac{(\gamma)_{n+k}(\delta)_{n+k}[\frac{1}{2}t(x-1)]^k[\frac{1}{2}t(x+1)]^n}{k!n!(1+\alpha)_k(1+\beta)_n}$$

$$= F_4\left(\gamma, \delta; 1+\alpha, 1+\beta; \tfrac{1}{2}t(x-1), \tfrac{1}{2}t(x+1)\right).$$

Theorem 84, page 269, yields

$$F_4\left(a, b; c, 1-c+a+b; \frac{-u}{(1-u)(1-v)}, \frac{-v}{(1-u)(1-v)}\right)$$

$$= {}_2F_1\left[\begin{matrix} a, b; \\ c; \end{matrix} \quad \frac{-u}{1-u}\right] {}_2F_1\left[\begin{matrix} a, b; \\ 1-c+a+b; \end{matrix} \quad \frac{-v}{1-v}\right].$$

The F_4 in (1) will fit into Theorem 84 if we choose

$$1 + \beta = 1 - (1+\alpha) + \gamma + \delta,$$

or $\delta = 1 + \alpha + \beta - \gamma$, and

$$\frac{-u}{(1-u)(1-v)} = \frac{t(x-1)}{2}, \qquad \frac{-v}{(1-u)(1-v)} = \frac{t(x+1)}{2}.$$

These are the u and v of the preceding section and

$$\frac{-u}{1-u} = 1 - \frac{1}{1-u} = 1 - \tfrac{1}{2}(1+t+\rho) = \tfrac{1}{2}(1-t-\rho),$$

$$\frac{-v}{1-v} = 1 - \tfrac{1}{2}(1-t+\rho) = \tfrac{1}{2}(1+t-\rho).$$

Hence (1) becomes

$$(2) \quad \sum_{n=0}^{\infty} \frac{(\gamma)_n(1+\alpha+\beta-\gamma)_n P_n^{(\alpha,\beta)}(x)t^n}{(1+\alpha)_n(1+\beta)_n}$$

$$= {}_2F_1\left[\begin{matrix}\gamma, 1+\alpha+\beta-\gamma; \\ 1+\alpha;\end{matrix} \quad \frac{1-t-\rho}{2}\right].$$

$${}_2F_1\left[\begin{matrix}\gamma, 1+\alpha+\beta-\gamma; \\ 1+\beta;\end{matrix} \quad \frac{1+t-\rho}{2}\right],$$

in which $\rho = (1 - 2xt + t^2)^{\frac{1}{2}}$ and γ is arbitrary.

The set (arbitrary γ) of generating functions in (2) was discovered by Brafman [1].

142. Expansion in series of polynomials. It is at times desirable to expand the general Jacobi polynomial in a series of simpler polynomials. We shall now exhibit a technique often useful for this purpose and always available when the simpler set is one of special Jacobi polynomials.

In the next chapter we discuss briefly the ultraspherical polynomial which is merely the special case $\beta = \alpha$ of the Jacobi polynomial. As an illustration of the method under discussion we propose to obtain in explicit form the expansion

$$(1) \qquad P_n^{(\alpha,\beta)}(x) = \sum_{k=0}^{n} A_1(k, n) P_k^{(\alpha,\alpha)}(x).$$

Expansions of the Jacobi polynomial in series of other special Jacobi polynomials will be found in the exercises at the end of this chapter. An expansion of $P_n^{(\alpha,\beta)}(x)$ in a bizarre combination of elements of different sets (varying α) of ultraspherical polynomials will be found in Brafman [3].

Naturally we do not use the orthogonality property for the explicit determination of $A_1(k, n)$, since we have generating functions available.

From equation (4), page 255, we get

$$(2) \quad \frac{(1+\alpha+\beta)_n P_n^{(\alpha,\beta)}(x)}{(1+\alpha)_n} = \sum_{s=0}^{n} \frac{(-1)^s(1+\alpha+\beta)_{n+s}}{s!(n-s)!(1+\alpha)_s}\left(\frac{1-x}{2}\right)^s.$$

The use of $\beta = \alpha$ in equation (2), page 262, yields

$$(3) \quad \left(\frac{1-x}{2}\right)^s = (1+\alpha)_s s! \sum_{k=0}^{s} \frac{(-1)^k (1+2\alpha)_k (1+2\alpha+2k) P_k^{(\alpha,\alpha)}(x)}{(s-k)!(1+2\alpha)_{k+s+1}(1+\alpha)_k}.$$

Consider the series

$$(4) \qquad \psi(x, t) = \sum_{n=0}^{\infty} \frac{(1+\alpha+\beta)_n P_n^{(\alpha,\beta)}(x) t^n}{(1+\alpha)_n}.$$

Using (2) and (3) in the right member of (4), we obtain

$$\psi(x, t)$$

$$= \sum_{n=0}^{\infty} \sum_{s=0}^{n} \frac{(-1)^s (1+\alpha+\beta)_{n+s} \left(\frac{1-x}{2}\right)^s t^n}{s!(n-s)!(1+\alpha)_s}$$

$$= \sum_{n,s=0}^{\infty} \frac{(-1)^s (1+\alpha+\beta)_{n+2s} \left(\frac{1-x}{2}\right)^s t^{n+s}}{s! n! (1+\alpha)_s}$$

$$= \sum_{n,s=0}^{\infty} \sum_{k=0}^{s} \frac{(-1)^{s+k} (1+\alpha+\beta)_{n+2s} (1+2\alpha)_k (1+2\alpha+2k) P_k^{(\alpha,\alpha)}(x) t^{n+s}}{n!(s-k)!(1+2\alpha)_{s+k+1}(1+\alpha)_k}$$

$$= \sum_{n,k,s=0}^{\infty} \frac{(-1)^s (1+\alpha+\beta)_{n+2s+2k} (1+2\alpha)_k (1+2\alpha+2k) P_k^{(\alpha,\alpha)}(x) t^{n+k+s}}{n! s! (1+2\alpha)_{s+2k+1}(1+\alpha)_k}.$$

We now rearrange terms in the above to obtain

$$\psi(x, t)$$

$$= \sum_{n,k=0}^{\infty} \sum_{s=0}^{n} \frac{(-1)^s (1+\alpha+\beta)_{n+s+2k} (1+2\alpha)_k (1+2\alpha+2k) P_k^{(\alpha,\alpha)}(x) t^{n+k}}{s!(n-s)!(1+2\alpha)_{s+2k+1}(1+\alpha)_k}$$

$$= \sum_{n,k=0}^{\infty} {}_2F_1\left[\begin{matrix} -n, 1+\alpha+\beta+n+2k; \\ 2+2\alpha+2k; \end{matrix} \; 1\right] \cdot$$

$$\frac{(1+\alpha+\beta)_{n+2k} (1+2\alpha)_k (1+2\alpha+2k) P_k^{(\alpha,\alpha)}(x) t^{n+k}}{n!(1+2\alpha)_{2k+1}(1+\alpha)_k}.$$

The above ${}_2F_1$ with unit argument has the value

$$\frac{(-1)^n (\beta-\alpha)_n}{(2+2\alpha+2k)_n},$$

for which see Ex. 5, page 69. Therefore,

$$\psi(x, t)$$

$$= \sum_{n,k=0}^{\infty} \frac{(-1)^n(\beta-\alpha)_n(1+\alpha+\beta)_{n+2k}(1+2\alpha)_k(1+2\alpha+2k)P_k^{(\alpha,\alpha)}(x)t^{n+k}}{n!(1+2\alpha)_{n+2k+1}(1+\alpha)_k}$$

$$= \sum_{n=0}^{\infty}\sum_{k=0}^{n} A(k,n)P_k^{(\alpha,\alpha)}(x)t^n$$

in which

$$A(k,n) = \frac{(-1)^{n-k}(\beta-\alpha)_{n-k}(1+\alpha+\beta)_{n+k}(1+2\alpha)_k(1+2\alpha+2k)}{(n-k)!(1+2\alpha)_{n+k+1}(1+\alpha)_k}.$$

Using (4), we may now conclude that

$$(5)\quad P_n^{(\alpha,\beta)}(x) = \frac{(1+\alpha)_n}{(1+\alpha+\beta)_n}\sum_{k=0}^{n} A(k,n)P_k^{(\alpha,\alpha)}(x)$$

with

$$A(k,n) = \frac{(-1)^{n-k}(\beta-\alpha)_{n-k}(1+\alpha+\beta)_{n+k}(1+2\alpha)_k(1+2\alpha+2k)}{(n-k)!(1+2\alpha)_{n+k+1}(1+\alpha)_k},$$

which is the result desired.

EXERCISES

1. Let

$$g_n(x) = \frac{(1+x)^n P_n^{(\alpha,\beta)}\left(\dfrac{1-x}{1+x}\right)}{(1+\alpha)_n(1+\beta)_n}.$$

Use Bateman's generating function, page 256, to see that

$$\sum_{n=0}^{\infty} g_n(x)t^n = {}_0F_1(-;1+\alpha;-xt)\,{}_0F_1(-;1+\beta;t)$$

and thus show that, in the sense of Section 126, $g_n(x)$ is of σ-type zero with $\sigma = D(\theta+\alpha)$. Show also that $g_n(x)$ is of Sheffer A-type unity.

2. Show that

$$2x(\alpha+\beta+n)\,D\,P_n^{(\alpha,\beta)}(x) + [x(\alpha-\beta)-(\alpha+\beta+2n)]\,D\,P_{n-1}^{(\alpha,\beta)}(x)$$
$$= (\alpha+\beta+n)[2nP_n^{(\alpha,\beta)}(x) - (\alpha-\beta)P_{n-1}^{(\alpha,\beta)}(x)],$$

which reduces to equation (2), page 159, for $\alpha = \beta = 0$.

3. Show that

$$2(\alpha+\beta+n)\,D\,P_n^{(\alpha,\beta)}(x) + [\alpha-\beta-x(\alpha+\beta+2n)]\,D\,P_{n-1}^{(\alpha,\beta)}(x)$$
$$= (\alpha+\beta+n)(\alpha+\beta+2n)P_{n-1}^{(\alpha,\beta)}(x),$$

which reduces to equation (6), page 159, for $\alpha = \beta = 0$.

4. Show that

$$\begin{aligned}
&2n(\alpha+\beta+n+1)\, D\, P_{n+1}^{(\alpha,\beta)}(x) \\
&\quad + [(\alpha+\beta)(n+2)x + n(\alpha-\beta)]\, D\, P_n^{(\alpha,\beta)}(x) \\
&\quad + (n+1)[(\alpha-\beta)x - (\alpha+\beta+2n)]\, D\, P_{n-1}^{(\alpha,\beta)}(x) \\
&\quad = n[2(n+1)(\alpha+\beta+n) + (\alpha+\beta+n+1)(\alpha+\beta+2n+2)]P_n^{(\alpha,\beta)}(x) \\
&\quad - (\alpha-\beta)(n+1)(\alpha+\beta+n)P_{n-1}^{(\alpha,\beta)}(x),
\end{aligned}$$

which reduces to equation (5), page 159, for $\alpha = \beta = 0$.

5. Use the method of Section 142 to show that

$$P_n^{(\alpha,\beta)}(x) = \frac{(1+\alpha)_n}{(1+\alpha+\beta)_n} \sum_{k=0}^{n} \frac{(-1)^{n-k}(\beta)_{n-k}(1+\alpha+\beta)_{n+k}(1+\alpha+2k)P_k^{(\alpha,0)}(x)}{(n-k)!(1+\alpha)_{n+k+1}}.$$

6. Use the result obtained in Section 142 to evaluate

$$\int_{-1}^{1} (1-x^2)^\alpha P_n^{(\alpha,\beta)}(x) P_k^{(\alpha,\alpha)}(x)\, dx.$$

Ans. For $k > n$ the integral is zero; for $0 \leqq k \leqq n$, it has the value

$$\frac{(-1)^{n-k}(\beta-\alpha)_{n-k}2^{1+2\alpha}\Gamma(1+\alpha+n)\Gamma(1+\alpha+k)\Gamma(1+\alpha+\beta+n+k)}{k!(n-k)!\Gamma(2+2\alpha+n+k)\Gamma(1+\alpha+\beta+n)}.$$

7. Use the result in Ex. 5 above to evaluate

$$\int_{-1}^{1} (1-x)^\alpha P_n^{(\alpha,\beta)}(x) P_k^{(\alpha,0)}(x)\, dx.$$

8. Use Theorem 84, page 269, with $y = x = -v/(1-v)$ to conclude that

$$\begin{aligned}
&{}_2F_1\left[\begin{matrix} a, b; \\ c; \end{matrix}\ v\right] {}_2F_1\left[\begin{matrix} a, b; \\ 1+a+b-c; \end{matrix}\ v\right] \\
&\qquad = {}_4F_3\left[\begin{matrix} a, b, \tfrac{1}{2}(a+b), \tfrac{1}{2}(a+b+1); \\ a+b, c, 1+a+b-c; \end{matrix}\ 4v(1-v)\right].
\end{aligned}$$

9. Use the result in Ex. 8, above, and Theorem 25, page 67, to show that

$$\left\{ {}_2F_1\left[\begin{matrix} a, b; \\ a+b+\tfrac{1}{2}; \end{matrix}\ y\right] \right\}^2 = {}_3F_2\left[\begin{matrix} 2a, 2b, a+b; \\ 2a+2b, a+b+\tfrac{1}{2}; \end{matrix}\ y\right].$$

CHAPTER 17

Ultraspherical and Gegenbauer Polynomials

143. Definitions. The special case $\beta = \alpha$ of the Jacobi polynomial of Chapter 16 is called the ultraspherical polynomial and is denoted $P_n^{(\alpha,\alpha)}(x)$. The Gegenbauer polynomial $C_n^\nu(x)$ is a generalization of the Legendre polynomial and is defined by the generating relation

$$(1 - 2xt + t^2)^{-\nu} = \sum_{n=0}^{\infty} C_n^\nu(x) t^n. \tag{1}$$

If we put $\beta = \alpha$ in equation (10), page 256, we obtain

$$(1 - t)^{-1-2\alpha} {}_2F_1\left[\begin{matrix} \tfrac{1}{2} + \alpha, 1 + \alpha; \\ 1 + \alpha; \end{matrix} \quad \frac{2t(x - 1)}{(1 - t)^2}\right] = \sum_{n=0}^{\infty} \frac{(1 + 2\alpha)_n P_n^{(\alpha,\alpha)}(x) t^n}{(1 + \alpha)_n}, \tag{2}$$

in which the ${}_2F_1$ degenerates into a binomial. Indeed, the left member of (2) is

$$(1 - t)^{-1-2\alpha}\left[1 - \frac{2t(x - 1)}{(1 - t)^2}\right]^{-\frac{1}{2}-\alpha} = (1 - 2xt + t^2)^{-\frac{1}{2}-\alpha}.$$

Hence

$$(1 - 2xt + t^2)^{-\frac{1}{2}-\alpha} = \sum_{n=0}^{\infty} \frac{(1 + 2\alpha)_n P_n^{(\alpha,\alpha)}(x) t^n}{(1 + \alpha)_n}. \tag{3}$$

From (1) and (3) it follows that the Gegenbauer and ultraspherical polynomials are essentially equivalent:

$$C_n^\nu(x) = \frac{(2\nu)_n P_n^{(\nu-\frac{1}{2},\ \nu-\frac{1}{2})}(x)}{(\nu+\frac{1}{2})_n}, \tag{4}$$

$$P_n^{(\alpha,\alpha)}(x) = \frac{(1+\alpha)_n C_n^{\alpha+\frac{1}{2}}(x)}{(1+2\alpha)_n}. \tag{5}$$

For some purposes the $C_n^\nu(x)$ is a more convenient form; for others, the $P_n^{(\alpha,\alpha)}(x)$ leads to neater looking results.

144. The Gegenbauer polynomials. The polynomial $C_n^\nu(x)$ defined by

$$(1-2xt+t^2)^{-\nu} = \sum_{n=0}^{\infty} C_n^\nu(x)t^n \tag{1}$$

retains most of the properties of the Legendre polynomial of Chapter 10. Note that

$$P_n(x) = C_n^{\frac{1}{2}}(x).$$

We here derive a few of the results for the Gegenbauer polynomial $C_n^\nu(x)$. Many other properties of $C_n^\nu(x)$ are then listed, and it is left to the reader to obtain them either by using the methods of Chapter 10 or otherwise.

From (1) it follows that

$$\sum_{n=0}^{\infty} C_n^\nu(x)t^n = \sum_{n=0}^{\infty} \frac{(\nu)_n t^n(2x-t)^n}{n!}$$

$$= \sum_{n=0}^{\infty}\sum_{k=0}^{n} \frac{(-1)^k(\nu)_n(2x)^{n-k}t^{n+k}}{k!(n-k)!}$$

$$= \sum_{n=0}^{\infty}\sum_{k=0}^{[n/2]} \frac{(-1)^k(\nu)_{n-k}(2x)^{n-2k}t^n}{k!(n-2k)!}.$$

Hence

$$C_n^\nu(x) = \sum_{k=0}^{[n/2]} \frac{(-1)^k(\nu)_{n-k}(2x)^{n-2k}}{k!(n-2k)!}. \tag{2}$$

If ν is neither zero nor a negative integer, the $C_n^\nu(x)$ form a simple set of polynomials and

$$C_n^\nu(x) = \frac{2^n(\nu)_n x^n}{n!} + \pi_{n-2}.$$

It follows readily from (1) and (2) that

(3) $$C_n^\nu(-x) = (-1)^n C_n^\nu(x),$$

(4) $$C_n^\nu(1) = \frac{(2\nu)_n}{n!},$$

(5) $$C_{2n}^\nu(0) = \frac{(-1)^n(\nu)_n}{n!}, \qquad C_{2n+1}^\nu(0) = 0.$$

Since, by (1),

$$\begin{aligned}
\sum_{n=0}^{\infty} C_n^\nu(x)t^n &= [(1-xt)^2 - t^2(x^2-1)]^{-\nu} \\
&= (1-xt)^{-2\nu}\left[1 - \frac{t^2(x^2-1)}{(1-xt)^2}\right]^{-\nu} \\
&= \sum_{k=0}^{\infty} \frac{(\nu)_k(x^2-1)^k t^{2k}}{k!(1-xt)^{2\nu+2k}} \\
&= \sum_{n,k=0}^{\infty} \frac{(\nu)_k(2\nu)_{n+2k}x^n(x^2-1)^k t^{n+2k}}{k!n!(2\nu)_{2k}} \\
&= \sum_{n,k=0}^{\infty} \frac{(2\nu)_{n+2k}x^n(x^2-1)^k t^{n+2k}}{k!n!2^{2k}(\nu+\frac{1}{2})_k},
\end{aligned}$$

it follows that

(6) $$C_n^\nu(x) = \sum_{k=0}^{[n/2]} \frac{(2\nu)_n x^{n-2k}(x^2-1)^k}{2^{2k}k!(\nu+\frac{1}{2})_k(n-2k)!}.$$

Equation (6) yields

$$\sum_{n=0}^{\infty} \frac{C_n^\nu(x)t^n}{(2\nu)_n} = \sum_{n,k=0}^{\infty} \frac{x^n(x^2-1)^k t^{n+2k}}{2^{2k}k!(\nu+\frac{1}{2})_k n!}$$

so that

(7) $$e^{xt}\,{}_0F_1\left[\begin{matrix} -; \\ \nu+\frac{1}{2}; \end{matrix} \quad \frac{t^2(x^2-1)}{4}\right] = \sum_{n=0}^{\infty} \frac{C_n^\nu(x)t^n}{(2\nu)_n}.$$

For arbitrary γ, (6) yields

$$\begin{aligned}
\sum_{n=0}^{\infty} \frac{(\gamma)_n C_n^\nu(x)t^n}{(2\nu)_n} &= \sum_{n,k=0}^{\infty} \frac{(\gamma)_{n+2k}x^n(x^2-1)^k t^{n+2k}}{2^{2k}k!(\nu+\frac{1}{2})_k n!} \\
&= \sum_{k=0}^{\infty} \frac{(\gamma)_{2k}(x^2-1)^k t^{2k}}{2^{2k}k!(\nu+\frac{1}{2})_k(1-xt)^{\gamma+2k}}.
\end{aligned}$$

Hence

$$(8)\quad (1 - xt)^{-\gamma}\, {}_2F_1\left[\begin{matrix} \tfrac{1}{2}\gamma,\ \tfrac{1}{2}\gamma + \tfrac{1}{2}; \\ \nu + \tfrac{1}{2}; \end{matrix}\ \frac{t^2(x^2 - 1)}{(1 - xt)^2}\right] = \sum_{n=0}^{\infty} \frac{(\gamma)_n C_n^{\nu}(x) t^n}{(2\nu)_n}.$$

The methods of Chapter 10 also yield the following results:

$$(9)\qquad x\, DC_n^{\nu}(x) = nC_n^{\nu}(x) + DC_{n-1}^{\nu}(x), \qquad D = \frac{d}{dx},$$

$$(10)\qquad 2(\nu + n)C_n^{\nu}(x) = DC_{n+1}^{\nu}(x) - DC_{n-1}^{\nu}(x),$$

$$(11)\qquad x\, DC_n^{\nu}(x) = DC_{n+1}^{\nu}(x) - (2\nu + n)C_n^{\nu}(x),$$

$$(12)\qquad (x^2 - 1)\, DC_n^{\nu}(x) = nxC_n^{\nu}(x) - (2\nu - 1 + n)C_{n-1}^{\nu}(x),$$

$$(13)\qquad nC_n^{\nu}(x) = 2x(\nu + n - 1)C_{n-1}^{\nu}(x) - (2\nu + n - 2)C_{n-2}^{\nu}(x),$$

$$(14)\qquad (1-x^2)\, D^2 C_n^{\nu}(x) - (2\nu+1)x\, DC_n^{\nu}(x) + n(2\nu+n)C_n^{\nu}(x) = 0,$$

$$(15)\qquad C_n^{\nu}(x) = \frac{(2\nu)_n}{n!}\, {}_2F_1\left[\begin{matrix} -n,\ 2\nu + n; \\ \nu + \tfrac{1}{2}; \end{matrix}\ \frac{1 - x}{2}\right]$$

$$= \sum_{k=0}^{n} \frac{(2\nu)_{n+k}\left(\dfrac{x - 1}{2}\right)^k}{k!(n - k)!(\nu + \tfrac{1}{2})_k}.$$

Reversal of order of summation in (15) yields

$$(16)\quad C_n^{\nu}(x) = \frac{2^{2n}(\nu)_n}{n!}\left(\frac{x - 1}{2}\right)^n {}_2F_1\left[\begin{matrix} -n,\ \tfrac{1}{2} - \nu - n; \\ 1 - 2\nu - 2n; \end{matrix}\ \frac{2}{1 - x}\right].$$

The use of Theorem 20, page 60, on (15) gives us

$$(17)\quad C_n^{\nu}(x) = \frac{(2\nu)_n}{n!}\left(\frac{x + 1}{2}\right)^n {}_2F_1\left[\begin{matrix} -n,\ \tfrac{1}{2} - \nu - n; \\ \nu + \tfrac{1}{2}; \end{matrix}\ \frac{x - 1}{x + 1}\right]$$

$$= \sum_{k=0}^{n} \frac{(2\nu)_n(\nu + \tfrac{1}{2})_n}{k!(n - k)!(\nu + \tfrac{1}{2})_k(\nu + \tfrac{1}{2})_{n-k}}\left(\frac{x - 1}{2}\right)^k\left(\frac{x + 1}{2}\right)^{n-k}.$$

The property (3) of this section can be used to get other forms of $C_n^\nu(x)$ from (15), (16), and (17).

Equation (17) yields the Bateman generating relation

$$(18)\qquad {}_0F_1\left[\begin{matrix} -; \\ \nu+\frac{1}{2}; \end{matrix}\ \frac{t(x-1)}{2}\right] {}_0F_1\left[\begin{matrix} -; \\ \nu+\frac{1}{2}; \end{matrix}\ \frac{t(x+1)}{2}\right] = \sum_{n=0}^{\infty} \frac{C_n^\nu(x)t^n}{(2\nu)_n(\nu+\frac{1}{2})_n}.$$

From (2) and (6) it follows that

$$(19)\qquad C_n^\nu(x) = \frac{(\nu)_n(2x)^n}{n!}\,{}_2F_1\left[\begin{matrix} -\frac{1}{2}n,\ -\frac{1}{2}n+\frac{1}{2}; \\ 1-\nu-n; \end{matrix}\ \frac{1}{x^2}\right],$$

$$(20)\qquad C_n^\nu(x) = \frac{(2\nu)_n x^n}{n!}\,{}_2F_1\left[\begin{matrix} -\frac{1}{2}n,\ -\frac{1}{2}n+\frac{1}{2}; \\ \nu+\frac{1}{2}; \end{matrix}\ \frac{x^2-1}{x^2}\right].$$

Brafman's generating function, page 272, becomes

$$(21)\qquad {}_2F_1\left[\begin{matrix} \gamma,\ 2\nu-\gamma; \\ \nu+\frac{1}{2}; \end{matrix}\ \frac{1-t-\rho}{2}\right] {}_2F_1\left[\begin{matrix} \gamma,\ 2\nu-\gamma; \\ \nu+\frac{1}{2}; \end{matrix}\ \frac{1+t-\rho}{2}\right] = \sum_{n=0}^{\infty} \frac{(\gamma)_n(2\nu-\gamma)_n C_n^\nu(x)t^n}{(2\nu)_n(\nu+\frac{1}{2})_n}, \qquad \rho = (1-2xt+t^2)^{\frac{1}{2}}.$$

With the aid of the methods of Section 95, we conclude from the preceding (7) that

$$(22)\qquad C_n^\nu(\cos\alpha) = \left(\frac{\sin\alpha}{\sin\beta}\right)^n \sum_{k=0}^{n} \frac{(2\nu)_n}{(n-k)!(2\nu)_k} \cdot \left[\frac{\sin(\beta-\alpha)}{\sin\alpha}\right]^{n-k} C_k^\nu(\cos\beta),$$

and from (1), page 276, that

$$(23)\qquad \rho^{-n-2\nu}C_n^\nu\left(\frac{x-t}{\rho}\right) = \sum_{k=0}^{\infty} \frac{(n+k)!C_{n+k}^\nu(x)t^k}{k!n!},$$

with $\rho = (1-2xt+t^2)^{\frac{1}{2}}$.

Equations (23) and (7) of this section lead us, by the method of Section 96, to the bilateral generating relation (See Weisner [1])

$$(24)\quad \rho^{-2\nu}\exp\left[\frac{-yt(x-t)}{\rho^2}\right]{}_0F_1\left[\begin{matrix}-; & \dfrac{y^2t^2(x^2-1)}{4\rho^4}\\ \nu+\frac{1}{2}; & \end{matrix}\right] = \sum_{n=0}^{\infty}\frac{n!L_n^{(2\nu-1)}(y)C_n^{\nu}(x)t^n}{(2\nu)_n},$$

in which $L_n^{(2\nu-1)}(y)$ is the Laguerre polynomial of Chapter 12. The same procedure can be used in applying (23) to the relations (8), (18), and (21) of this section.

From (6) we obtain

$$C_n^{\nu}(x) = (2\nu)_n\sum_{k=0}^{[n/2]}\frac{(\frac{1}{2})_k x^{n-2k}(x^2-1)^k}{(2k)!(n-2k)!(\nu+\frac{1}{2})_k}$$

and thus derive

$$(25)\quad C_n^{\nu}(x) = \frac{(2\nu)_n\Gamma(\nu+\frac{1}{2})}{n!\,\Gamma(\frac{1}{2})\Gamma(\nu)}\int_0^{\pi}[x+\sqrt{x^2-1}\cos\varphi]^n\sin^{2\nu-1}\varphi\,d\varphi.$$

For $\nu > 0$ and $-1 \leqq x \leqq 1$, (25) easily yields the following bound on the Gegenbauer polynomial:

$$|C_n^{\nu}(x)| \leqq \frac{(2\nu)_n}{n!}.$$

Either the differential equation (14) or the Rodrigues formula

$$(26)\quad C_n^{\nu}(x) = \frac{(-1)^n(2\nu)_n(1-x^2)^{\frac{1}{2}-\nu}}{2^n n!(\nu+\frac{1}{2})_n}D^n(1-x^2)^{n+\nu-\frac{1}{2}};\qquad D = \frac{d}{dx},$$

leads to the orthogonality property

$$(27)\quad \int_{-1}^{1}(1-x^2)^{\nu-\frac{1}{2}}C_n^{\nu}(x)C_m^{\nu}(x)\,dx = 0,\qquad m \neq n.$$

From (26) it follows that

$$(28)\quad g_n = \int_{-1}^{1}(1-x^2)^{\nu-\frac{1}{2}}[C_n^{\nu}(x)]^2dx = \frac{(2\nu)_n\Gamma(\frac{1}{2})\Gamma(\nu+\frac{1}{2})}{n!(\nu+n)\Gamma(\nu)}.$$

In the notation of Chapter 9, the leading coefficient in $C_n^{\nu}(x)$ is $h_n = 2^n(\nu)_n/n!$, and we may now apply the theorems of Chapter 9 to the Gegenbauer polynomials, if $\nu > -\frac{1}{2}$.

The differential equation (14) leads also to the evaluation

$$(29)\quad (n-m)(n+m+2\nu)\int_a^b (1-x^2)^{\nu-\frac{1}{2}}C_n^\nu(x)C_m^\nu(x)\,dx$$

$$= \Big[(1-x^2)^{\nu+\frac{1}{2}}\{C_n^\nu(x)\,D\,C_m^\nu(x) - C_m^\nu(x)\,D\,C_n^\nu(x)\}\Big]_a^b.$$

Following the technique suggested in Exs. 1 and 2, pages 182–183, we find that

$$(30)\quad C_n^\nu(x)$$

$$= \sum_{k=0}^{n} \frac{(\nu)_k(\nu)_{n-k}(x+\sqrt{x^2-1})^{n-k}(x-\sqrt{x^2-1})^k}{k!(n-k)!}$$

$$= \frac{(\nu)_n(x+\sqrt{x^2-1})^n}{n!}\,{}_2F_1\left[\begin{matrix} -n,\ \nu; \\ 1-\nu-n; \end{matrix}\quad (x-\sqrt{x^2-1})^2\right].$$

From the generating relation (6), page 271, we obtain

$$(31)\qquad 2^{\nu-\frac{1}{2}}\rho^{-1}(1-xt+\rho)^{\frac{1}{2}-\nu} = \sum_{n=0}^{\infty} \frac{(\nu+\frac{1}{2})_n C_n^\nu(x)t^n}{(2\nu)_n},$$

with $\rho = (1-2xt+t^2)^{\frac{1}{2}}$.

Since

$$(1-2xt+t^2)^{-\nu} = (1-t)^{-2\nu}\,{}_1F_0\left[\begin{matrix} \nu; \\ -; \end{matrix}\quad \frac{-2t(1-x)}{(1-t)^2}\right],$$

Theorem 48, page 137, is applicable to $C_n^\nu(x)$ with a slight change from x to $\frac{1}{2}(1-x)$ in variable. We thus arrive at the following results:

$$(32)\qquad \left(\frac{1-x}{2}\right)^n = 2(\nu+\tfrac{1}{2})_n n! \sum_{k=0}^{n} \frac{(-1)^k(\nu+k)C_k^\nu(x)}{(n-k)!(2\nu)_{n+k+1}},$$

$$(33)\quad (x-1)\,D\,C_n^\nu(x) - nC_n^\nu(x)$$

$$= -(2\nu+n-1)C_{n-1}^\nu(x) - (x-1)\,D\,C_{n-1}^\nu(x),$$

$$(34)\quad (x-1)D\,C_n^\nu(x) - nC_n^\nu(x)$$

$$= -2\nu\sum_{k=0}^{n-1} C_k^\nu(x) - 2(x-1)\sum_{k=0}^{n-1} D\,C_k^\nu(x),$$

(35) $$(x-1)D\,C_n^{\nu}(x) - nC_n^{\nu}(x) = 2\sum_{k=0}^{n-1}(-1)^{n-k}(\nu+k)C_k^{\nu}(x).$$

The method used to derive the formula of Theorem 65, page 181, leads us also to

(36) $$\frac{(2x)^n}{n!} = \sum_{k=0}^{[n/2]}\frac{(\nu+n-2k)C_{n-2k}^{\nu}(x)}{k!(\nu)_{n+1-k}}.$$

Equations (32) and (36) simplify the task of expressing various other polynomials in series of Gegenbauer polynomials. See the exercises at the end of this chapter.

From (30) in this section it follows that

(37) $$C_n^{\nu}(\cos\theta) = \sum_{k=0}^{n}\frac{(\nu)_k(\nu)_{n-k}\cos(n-2k)\theta}{k!(n-k)!}.$$

Various mixed recurrence relations are easily derived. For instance,

(38) $$2\nu(1-x^2)C_{n-1}^{\nu+1}(x) = (n+2\nu)xC_n^{\nu}(x) - (n+1)C_{n+1}^{\nu}(x),$$

(39) $$(n+\nu)C_{n+1}^{\nu-1}(x) = (\nu-1)[C_{n+1}^{\nu}(x) - C_{n-1}^{\nu}(x)].$$

For expansions of analytic functions in series of Gegenbauer polynomials, see Boas and Buck [2; 58] for existence, uniqueness, and region of convergence; use equations (32) or (36) to obtain the explicit coefficients.

145. The ultraspherical polynomials. By means of the equation

(1) $$C_n^{\nu}(x) = \frac{(2\nu)_n P_n^{(\nu-\frac{1}{2},\nu-\frac{1}{2})}(x)}{(\nu+\frac{1}{2})_n},$$

or its equivalent,

(2) $$P_n^{(\alpha,\alpha)}(x) = \frac{(1+\alpha)_n C_n^{\alpha+\frac{1}{2}}(x)}{(1+2\alpha)_n},$$

results obtained for the Gegenbauer polynomial $C_n^{\nu}(x)$ may be transformed into results on the ultraspherical polynomial $P_n^{(\alpha,\alpha)}(x)$, or vice versa. We leave it to the reader to obtain from equations (1)–(39) of the preceding section the equivalent statements regarding $P_n^{(\alpha,\alpha)}(x)$.

Certain properties of the ultraspherical polynomial are due solely to its being a Jacobi polynomial; others are peculiar to $P_n^{(\alpha,\alpha)}(x)$ because of the equality of β and α.

EXERCISES

1. Show that the Gegenbauer polynomial $C_n^\nu(x)$ and the Hermite polynomial $H_n(x)$ are related by

$$C_n^\nu(x) = \sum_{k=0}^{[n/2]} {}_2F_0(-k, \nu + n - k; -; 1)\frac{(-1)^k(\nu)_{n-k}H_{n-2k}(x)}{k!(n-2k)!}.$$

2. Show that

$$\frac{H_n(x)}{n!} = \sum_{k=0}^{[n/2]} (-1)^k {}_1F_1(-k; 1 + \nu + n - 2k; 1)\frac{(\nu + n - 2k)C_{n-2k}^\nu(x)}{k!(\nu)_{n+1-2k}}.$$

3. Show, using the modified Bessel function of Section 65, that

$$e^{xt} = (\tfrac{1}{2}t)^{-\nu}\Gamma(\nu) \sum_{n=0}^{\infty} (\nu + n) I_{\nu+n}(t)C_n^\nu(x).$$

4. Show that

$$C_n^\nu(x) = \sum_{k=0}^{[n/2]} \frac{(\nu - \frac{1}{2})_k(\nu)_{n-k}(1 + 2n - 4k)P_{n-2k}(x)}{k!\left(\frac{3}{2}\right)_{n-k}}.$$

CHAPTER 18

Other Polynomial Sets

146. Bateman's $Z_n(x)$. One reason for interest in the polynomials $f_n(x)$ generated by

$$(1) \qquad (1-t)^{-c}\psi\left(\frac{-4xt}{(1-t)^2}\right) = \sum_{n=0}^{\infty} f_n(x)t^n,$$

which were touched upon in Section 75, is that so many special cases of the $f_n(x)$ have arisen in recent studies. In 1936 Bateman [2] was interested in constructing inverse Laplace transforms. For this purpose he introduced the polynomial

$$(2) \qquad Z_n(x) = {}_2F_2(-n, n+1; 1, 1; x),$$

for which we obtained a pure recurrence relation in Ex. 1, page 243.

By Theorem 48, page 137, with $c = 1$ and $\gamma_n = (\frac{1}{2})_n/(n!)^2$, we obtain

$$(3) \qquad (1-t)^{-1}\,{}_1F_1\left[\begin{matrix}\frac{1}{2}; \\ \\ 1;\end{matrix} \quad \frac{-4xt}{(1-t)^2}\right] = \sum_{n=0}^{\infty} Z_n(x)t^n,$$

$$(4) \qquad x^n = (n!)^2 \sum_{k=0}^{n} \frac{(-n)_k(2k+1)Z_k(x)}{(n+k+1)!},$$

$$(5) \qquad xZ_n'(x) - nZ_n(x) = -nZ_{n-1}(x) - xZ'_{n-1}(x),$$

(6) $$xZ_n'(x) - nZ_n(x) = -\sum_{k=0}^{n-1} Z_k(x) - 2x\sum_{k=0}^{n-1} Z_k'(x),$$

(7) $$xZ_n'(x) - nZ_n(x) = \sum_{k=0}^{n-1} (-1)^{n-k}(2k+1)Z_k(x).$$

Equation (3) was given by Bateman [2], (5) by Sister Celine [1], (7) by Dickinson [1], and (4) and (6) are immediate consequences of (3). Note that, because of Kummer's second formula, page 126, (3) may be written

(8) $$(1-t)^{-1}\exp\left[\frac{-2xt}{(1-t)^2}\right] {}_0F_1\left[\begin{matrix} -; & \dfrac{x^2t^2}{(1-t)^4} \\ 1; & \end{matrix}\right]$$

$$= (1-t)^{-1}\exp\left[\frac{-2xt}{(1-t)^2}\right] I_0\left[\frac{-2xt}{(1-t)^2}\right] = \sum_{n=0}^{\infty} Z_n(x)t^n,$$

which is the form Bateman used.

Bateman moved from the Z_n to the more general

(9) $${}_2F_2(-n, 2\nu+n; \nu+\tfrac{1}{2}, 1+b; t),$$

which is an instance of one of the generalizations (Rainville [6]) of the Bessel polynomials (Krall and Frink [1]) of Section 150. We shall now exhibit the relation between the polynomial (9), including Z_n as the special case $\nu = \frac{1}{2}$, $b = 0$, and the problem of inverse Laplace transforms.

With the usual notation

$$L\{F(t)\} = \int_0^\infty e^{-st}F(t)\,dt = f(s); \qquad F(t) = L^{-1}\{f(s)\},$$

we seek

(10) $$F(t) = L^{-1}\left\{\frac{\Gamma(1+b)}{(1+s)^{b+1}}\left[1 - 2z\frac{s-1}{s+1} + z^2\right]^{-\nu}\right\}.$$

In terms of the Gegenbauer polynomial of Chapter 17,

$$F(t) = \Gamma(1+b)L^{-1}\sum_{n=0}^{\infty}\frac{C_n^\nu\left(\dfrac{s-1}{s+1}\right)z^n}{(1+s)^{b+1}}$$

$$= e^{-t}\Gamma(1+b)L^{-1}\sum_{n=0}^{\infty}\frac{C_n^\nu\left(1-\dfrac{2}{s}\right)z^n}{s^{b+1}}.$$

Now

$$C_n^\nu(x) = \frac{(2\nu)_n}{n!}\,{}_2F_1(-n, 2\nu + n; \nu + \tfrac{1}{2}; \tfrac{1}{2}(1 - x)),$$

so that

$$F(t) = e^{-t}\Gamma(1 + b)L^{-1}\sum_{n=0}^{\infty}\frac{(2\nu)_n z^n}{n!\,s^{b+1}}\,{}_2F_1\left(-n, 2\nu + n; \nu + \tfrac{1}{2}; \frac{1}{s}\right).$$

Let us therefore first evaluate

$$A(t) = L^{-1}\left\{\frac{\Gamma(1 + b)}{s^{b+1}}\,{}_2F_1\left(-n, 2\nu + n; \nu + \tfrac{1}{2}; \frac{1}{s}\right)\right\}.$$

By Ex. 9, page 106, with $p = 2$, $q = 2$, $c = b$, $a_1 = -n$, $a_2 = 2\nu + n$, $b_1 = \nu + \frac{1}{2}$, $b_2 = 1 + b$, $z = 1$, we obtain

$$A(t) = t^b\,{}_2F_2(-n, 2\nu + n; \nu + \tfrac{1}{2}, b + 1; t).$$

Thus, for the $F(t)$ of (10) it follows that

$$F(t) = t^b e^{-t}\sum_{n=0}^{\infty}\frac{(2\nu)_n z^n}{n!}\,{}_2F_2(-n, 2\nu + n; \nu + \tfrac{1}{2}, b + 1; t).$$

See also Ex. 13 at the end of this chapter.

In the same paper, Bateman [2] studied another set of functions $J_n^{u,v}$ which are, except for a simple factor and changes of notation, the polynomials ${}_1F_2(-n; \beta_1, \beta_2; x)$. The $J_n^{u,v}$ occur again in Rainville [3] and Langer [1]. In the latter, Langer is studying solutions of the differential equation for

$${}_1F_2\left(\mu; \frac{1}{3}, \frac{2}{3}; \frac{-\lambda^2x^3}{9}\right)$$

and thus encounters Bateman's $J_n^{u,v}$.

147. Rice's $H_n(\zeta, p, v)$. S. O. Rice [1] made a considerable study of the polynomials defined by

$$(1) \qquad H_n(\zeta, p, v) = {}_3F_2(-n, n + 1, \zeta; 1, p; v).$$

We obtained the pure recurrence relation for H_n in Ex. 3, page 243.

A few of the simpler of Rice's results are:

For $\mathrm{Re}(p) > \mathrm{Re}(\zeta) > 0$,

$$(2) \quad H_n(\zeta, p, v) = \frac{\Gamma(p)}{\Gamma(\zeta)\Gamma(p - \zeta)}\int_0^1 t^{\zeta-1}(1 - t)^{p-\zeta-1}P_n(1 - 2vt)\,dt;$$

For $\text{Re}(\zeta) > \text{Re}(p) > 0$,

$$(3) \quad P_n(1 - 2v) = \frac{\Gamma(\zeta)}{\Gamma(p)\Gamma(\zeta - p)}\int_0^1 t^{p-1}(1 - t)^{\zeta-p-1}H_n(\zeta, p, vt)\,dt;$$

$$(4) \qquad (1 - t)^{-1}\,F\left[\begin{matrix}\zeta, \tfrac{1}{2}; \\ p;\end{matrix}\quad \frac{-4vt}{(1 - t)^2}\right] = \sum_{n=0}^{\infty} H_n(\zeta, p, v)t^n;$$

$$(5) \qquad \sum_{k=0}^{\infty} \frac{(p)_k H_n(-k, p, v)t^k}{k!} = (1 - t)^{-p}P_n\left(1 + \frac{2vt}{1 - t}\right);$$

$$(6) \quad \sum_{n=0}^{\infty} (2n + 1)Q_n(s)H_n(\zeta, p, v) = \frac{-1}{1 - s}\,F\left[\begin{matrix}\zeta, 1; \\ p;\end{matrix}\quad \frac{2v}{1 - s}\right].$$

In (2), (3), and (5), $P_n(x)$ is the Legendre polynomial. In (6) the $Q_n(s)$ is the Legendre function of the second kind defined on page 182. Sister Celine (Fasenmyer[1]) obtained the differential recurrence relation

$$(7) \quad vH_n'(\zeta, p, v) + vH_{n-1}'(\zeta, p, v) = n[H_n(\zeta, p, v) - H_{n-1}(\zeta, p, v)]$$

in which primes denote differentiation with respect to v.

Because of (4) in this section, Theorem 48, page 137, applies to Rice's H_n with the choices $c = 1$, $x = v$,

$$\gamma_n = \frac{(\tfrac{1}{2})_n(\zeta)_n}{(p)_n n!}.$$

From Theorem 48, preceding equations (1) and (7) follow, as do the results

$$(8) \qquad v^n = \frac{(p)_n(n!)^2}{(\zeta)_n}\sum_{k=0}^{n} \frac{(-1)^k(1 + 2k)H_k(\zeta, p, v)}{(n - k)!(n + k + 1)!},$$

$$(9) \qquad vH_n'(\zeta, p, v) - nH_n(\zeta, p, v) = -\sum_{k=0}^{n-1} [H_k(\zeta, p, v) + 2vH_k'(\zeta, p, v)],$$

$$(10) \quad vH_n'(\zeta, p, v) - nH_n(\zeta, p, v) = \sum_{k=0}^{n-1} (-1)^{n-k}(1 + 2k)H_k(\zeta, p, v).$$

See Rice [1] for many other properties of H_n.

148. Bateman's $F_n(z)$. Bateman [3] studied the polynomial

$$(1) \qquad F_n(z) = {}_3F_2\left(-n, n+1, \tfrac{1}{2}(1+z); 1, 1; 1\right)$$

quite extensively, and he and others kept returning to it in later papers. Note that the variable z is contained in a parameter of the ${}_3F_2$, not in the argument. That the $F_n(z)$ form a simple set of polynomials should be apparent upon consideration of the nature of the terms in a ${}_3F_2$.

Bateman obtained the generating relation

$$(2) \qquad (1-t)^{-1}\, {}_2F_1\left[\begin{matrix} \tfrac{1}{2}, \tfrac{1}{2}+\tfrac{1}{2}z; \\ 1; \end{matrix} \quad \frac{-4t}{(1-t)^2}\right] = \sum_{n=0}^{\infty} F_n(z) t^n$$

and the pure recurrence relation

$$(3) \qquad n^2 F_n(z) = -(2n-1)zF_{n-1}(z) + (n-1)^2 F_{n-2}(z)$$

together with numerous mixed relations involving a shift in argument as well as in index. Two examples are quoted here:

$$(4) \qquad \begin{aligned} &(z+1)^2[F_n(z+2) - F_n(z)] \\ &\qquad + (z-1)^2[F_n(z-2) - F_n(z)] = 4n(n+1)F_n(z), \end{aligned}$$

$$(5) \qquad \sum_{n=0}^{\infty} [F_n(z-2) - F_n(z)]t^n = \frac{2t}{(1-t)^3}\, {}_2F_1\left[\begin{matrix} \tfrac{3}{2}, \tfrac{1}{2}+\tfrac{1}{2}z; \\ 2; \end{matrix} \quad \frac{-4t}{(1-t)^2}\right].$$

Since $F_0(z) = 1$, $F_1(z) = -z$ and $F_2(z) = \frac{1}{4}(1+3z^2)$, it follows from (3) that for $|z| < 1$, $|F_n(z)| < 1$, $n \geqq 1$. Bateman also made much use of $F_n(z)$ in the study of definite integrals and certain series expansions.

In 1956 Touchard [1] introduced polynomials for which he did not give either an explicit formula or a generating relation. Later that year Wyman and Moser [1] obtained for Touchard's polynomials a finite sum formula and a generating function. Their generating function was equivalent to Bateman's (2) above, as can be seen by applying Theorems 21 and 23 of Chapter 4. In 1957 Carlitz [1] pointed out that Touchard's polynomials and Bateman's $F_n(z)$ are essentially the same, the former being

$$\frac{(-1)^n (n!)^2 F_n(1+2x)}{2^n (\frac{1}{2})_n}.$$

Also in 1957 Brafman [6] obtained two generating functions for Touchard polynomials, and one of these [his (12)] is equivalent to that of Wyman and Moser and therefore to Bateman's (2) in this section, by Theorem 20 of Chapter 4. Brafman's other generating relation is a useful contribution to the study of Bateman's $F_n(z)$; it is

$$(6) \qquad {}_1F_1(\tfrac{1}{2} - \tfrac{1}{2}z; 1; t)\, {}_1F_1(\tfrac{1}{2} + \tfrac{1}{2}z; 1; -t) = \sum_{n=0}^{\infty} \frac{F_n(z)t^n}{n!}.$$

See also Ex. 3 at the end of this chapter.

Another polynomial in which interest is concentrated on a parameter is the Mittag-Leffler polynomial

$$g_n(z) = 2z\, {}_2F_1(1 - n, 1 - z; 2; 2),$$

which was investigated to some extent in Bateman [4]. Two generating functions

$$(7) \qquad (1 + t)^z(1 - t)^{-z} = 1 + \sum_{n=1}^{\infty} g_n(z)t^n,$$

$$(8) \qquad 2ze^t\, {}_1F_1(1 - z; 2; -2t) = \sum_{n=0}^{\infty} \frac{g_{n+1}(z)t^n}{n!},$$

and several mixed recurrence relations for $g_n(z)$ were given by Bateman. He also included some discussion of a generalization of the Mittag-Leffler polynomials to

$$(9) \qquad g_n(z, r) = \frac{(-r)_n}{n!}\, {}_2F_1(-n, z; -r; 2).$$

149. Sister Celine's polynomials. Sister Celine (Fasenmyer [1]) concentrated on polynomials generated by

$$(1) \qquad (1 - t)^{-1}\, {}_pF_q\left[\begin{matrix} a_1, \cdots, a_p; \\ b_1, \cdots, b_q; \end{matrix} \; \frac{-4xt}{(1 - t)^2}\right] = \sum_{n=0}^{\infty} f_n\left[\begin{matrix} a_1, \cdots, a_p; \\ b_1, \cdots, b_q; \end{matrix} \; x\right] t^n$$

which yields

$$(2) \qquad f_n\left[\begin{matrix} a_1, \cdots, a_p; \\ b_1, \cdots, b_q; \end{matrix} \; x\right] = {}_{p+2}F_{q+2}\left[\begin{matrix} -n, n + 1, a_1, \cdots, a_p; \\ 1, \tfrac{1}{2}, b_1, \cdots, b_q; \end{matrix} \; x\right].$$

Her polynomials include as special cases Legendre polynomials

$P_n(1-2x)$, some special Jacobi polynomials, Rice's $H_n(\zeta, p, v)$, Bateman's $Z_n(x)$ and $F_n(z)$, and Pasternak's

$$F_n^m(z) = F\left[\begin{matrix}-n, n+1, \tfrac{1}{2}(1+z+m); \\ 1, m+1;\end{matrix}\; 1\right] \tag{3}$$

which is a generalization of Bateman's $F_n(z)$. The simple Bessel polynomial of Section 150 is also included.

The two major parts of Sister Celine's work are the technique for obtaining pure recurrence relations (illustrated in Chapter 14) and her extension of Rainville's work on contiguous function relations to certain terminating ${}_pF_q$'s for which $p > q + 1$. She also obtained a few results of interest for some of the simpler of her polynomials.

We quote the easily derived result

$$f_n\left[\begin{matrix}a_1, \cdots, a_p; \\ b_1, \cdots, b_q;\end{matrix}\; x\right] = \frac{1}{\sqrt{\pi}}\int_0^\infty y^{-\frac{1}{2}}e^{-y} f_n\left[\begin{matrix}a_1, \cdots, a_p; \\ \tfrac{1}{2}, b_1, \cdots, b_q;\end{matrix}\; xy\right] dy, \tag{4}$$

which includes

$$P_n(1-2x) = \frac{1}{\sqrt{\pi}}\int_0^\infty y^{-\frac{1}{2}}e^{-y} f_n(-;-;xy)\, dy \tag{5}$$

and

$$H_n(\zeta, p, v) = \frac{1}{\sqrt{\pi}}\int_0^\infty y^{-\frac{1}{2}}e^{-y} f_n(\zeta; p; vy)\, dy. \tag{6}$$

Using $p = 1$, $q = 1$, $a_1 = \frac{1}{2}$, $b_1 = 1$, we find that Sister Celine's (4) in this section becomes

$$f_n(\tfrac{1}{2}; 1; x) = \frac{1}{\sqrt{\pi}}\int_0^\infty y^{-\frac{1}{2}}e^{-y} f_n(-; 1; xy)\, dy. \tag{7}$$

As she points out, for Bateman's Z_n,

$$f_n(\tfrac{1}{2}; 1; x) = Z_n(x), \tag{8}$$

and in terms of the simple Laguerre polynomial,

$$f_n(-; 1; \tfrac{1}{4}x^2) = L_n(x)L_n(-x). \tag{9}$$

Equation (9) is the special case $\alpha = -n$, $\beta = 1$, of Ramanujan's theorem obtained in Ex. 5, page 106. By combining (7), (8), and (9), Sister Celine obtains

$$Z_n(x^2) = \frac{1}{\sqrt{\pi}}\int_0^\infty \exp(-\tfrac{1}{4}\alpha^2)L_n(\alpha x)L_n(-\alpha x)\,d\alpha. \tag{10}$$

The general polynomial of Sister Celine, (2) in this section, falls under the classification of Theorem 48, page 137, with $c = 1$. For the moment, denote the polynomial of (2) by $C_n(x)$:

$$C_n(x) = {}_{p+2}F_{q+2}\left[\begin{matrix} -n, n+1, a_1, \cdots, a_p; \\ 1, \tfrac{1}{2}, b_1, \cdots, b_q; \end{matrix}\; x\right]. \tag{11}$$

Then Theorem 48 yields

$$x^n = \frac{(\frac{1}{2})_n(n!)^2(b_1)_n\cdots(b_q)_n}{(a_1)_n\cdots(a_p)_n}\sum_{k=0}^{n}\frac{(-1)^k(2k+1)C_k(x)}{(n-k)!(n+k+1)!}, \tag{12}$$

$$xC_n'(x) - nC_n(x) = -nC_{n-1}(x) - xC'_{n-1}(x), \tag{13}$$

$$xC_n'(x) - nC_n(x) = -\sum_{k=0}^{n-1}[C_k(x) + 2xC_k'(x)], \tag{14}$$

$$xC_n'(x) - nC_n(x) = \sum_{k=0}^{n-1}(-1)^{n-k}(2k+1)C_k(x). \tag{15}$$

Next let us turn to the polynomial (2) with no a's and no b's. Put

$$f_n(x) = {}_2F_2(-n, n+1; 1, \tfrac{1}{2}; x), \tag{16}$$

a polynomial whose pure recurrence relation we derived in Chapter 14. As Sister Celine points out,

$$f_n(x^2) \doteqdot L_n(2xH(x)), \tag{17}$$

involving the Laguerre and Hermite polynomials with the symbolic notation of Chapter 15.

For the $f_n(x)$ of (16) the generating function (1) becomes

$$(1-t)^{-1}\exp\left[\frac{-4xt}{(1-t)^2}\right] = \sum_{n=0}^{\infty} f_n(x)t^n. \tag{18}$$

Therefore this $f_n(x)$ is a polynomial of Sheffer A-type zero. In the notation of Chapter 13,

$$(19) \qquad A(t) = (1-t)^{-1}, \qquad H(t) = \frac{-4t}{(1-t)^2}.$$

Then the inverse of $H(t)$ is

$$(20) \qquad J(t) = 1 - \frac{2}{1+\sqrt{1-t}} = \sum_{k=0}^{\infty} \frac{-(\frac{1}{2})_{k+1} t^{k+1}}{(k+2)!}.$$

Computation of the constants appearing in Theorems 73–76, pages 224–225, shows that

$$(21) \qquad \alpha_k = 1, \qquad h_k = -4(k+1), \qquad \epsilon_k = -4(k+1)^2,$$

$$(22) \qquad \mu_k = \frac{\frac{1}{2}(-\frac{1}{2})_{k+1}}{(k+1)!}, \qquad \nu_k = \frac{(-\frac{1}{2})_k}{k!}.$$

In particular, Theorems 75 and 76 yield the relations

$$(23) \qquad nf_n(x) = \sum_{k=0}^{n-1} [1 - 4(k+1)^2 x] f_{n-1-k}(x),$$

$$(24) \qquad f_n'(x) = -4 \sum_{k=0}^{n-1} (k+1) f_{n-1-k}(x).$$

It is interesting to note that (23) and (24) combine to yield

$$(25) \qquad xf_n'(x) - nf_n(x) = \sum_{k=0}^{n-1} [4k(k+1)x - 1] f_{n-1-k}(x)$$

$$= \sum_{k=0}^{n-1} [4(n-k)(n-1-k)x - 1] f_k(x),$$

a different result from that given by (15),

$$(26) \qquad xf_n'(x) - nf_n(x) = \sum_{k=0}^{n-1} (-1)^{n-k}(2k+1) f_k(x),$$

which came from Theorem 48.

150. Bessel polynomials. In 1949 Krall and Frink [1] initiated serious study of what they called Bessel polynomials. In their terminology the simple Bessel polynomial is

$$(1) \qquad y_n(x) = {}_2F_0\left(-n, 1+n; -; -\frac{1}{2}x\right)$$

and the generalized one is

(2) $$y_n(a, b, x) = {}_2F_0\left(-n, a - 1 + n; -; -\frac{x}{b}\right).$$

The simple Bessel polynomial $y_n(x)$ is related to the Bessel functions in the following manner:

$$y_n\left(\frac{1}{ir}\right) = \left(\frac{1}{2}\pi r\right)^{\frac{1}{2}} e^{ir}[i^{-n-1}J_{n+\frac{1}{2}}(r) + i^n J_{-n-\frac{1}{2}}(r)],$$

$$J_{n+\frac{1}{2}}(r) = (2\pi r)^{-\frac{1}{2}}\left[i^{-n-1}e^{ir}y_n\left(\frac{-1}{ir}\right) + i^{n+1}e^{-ir}y_n\left(\frac{1}{ir}\right)\right],$$

$$J_{-n-\frac{1}{2}}(r) = (2\pi r)^{-\frac{1}{2}}\left[i^n e^{ir}y_n\left(\frac{-1}{ir}\right) + i^{-n}e^{-ir}y_n\left(\frac{1}{ir}\right)\right].$$

Let us take as a standard for polynomials of this character

(3) $$\varphi_n(c, x) = \frac{(c)_n}{n!}\, {}_2F_0(-n, c + n; -; x).$$

The simple Bessel polynomial is the special case with $c = 1$ and x replaced by $(-\frac{1}{2}x)$. To get the Krall-Frink generalized Bessel polynomial, introduce the redundant parameter b by replacing x by $(-x/b)$, put $c = a - 1$ and multiply $\varphi_n(a - 1, -x/b)$ by $n!/(a - 1)_n$.

From (3) it follows that

(4) $$\varphi_n(c, x) = \sum_{k=0}^{n} \frac{(-n)_k(c + n)_k(c)_n x^k}{k!n!}$$

or

(5) $$\varphi_n(c, x) = \sum_{k=0}^{n} \frac{(-1)^k(c)_{n+k}x^k}{k!(n - k)!}.$$

Equation (4) shows that $\varphi_n(c, x)$ fits into the scheme of Theorem 48, page 137, with $\gamma_k = (\frac{1}{2}c)_k(\frac{1}{2} + \frac{1}{2}c)_k/k!$. This yields the generating relation

(6) $$(1 - t)^{-c}\, {}_2F_0\left(\tfrac{1}{2}c, \tfrac{1}{2}c + \tfrac{1}{2}; -; \frac{-4xt}{(1 - t)^2}\right) \cong \sum_{n=0}^{\infty} \varphi_n(c, x)t^n$$

and the further properties

(7) $$x^n = n! \sum_{k=0}^{n} \frac{(-1)^k(c + 2k)\varphi_k(c, x)}{(n - k)!(c)_{n+k+1}},$$

$$(8) \quad x\varphi_n'(c, x) - n\varphi_n(c, x) = -(c + n - 1)\varphi_{n-1}(c, x) - x\varphi_{n-1}'(c, x)$$

$$= - \sum_{k=0}^{n-1} [c\varphi_k(c, x) + 2x\varphi_k'(c, x)]$$

$$= \sum_{k=0}^{n-1} (-1)^{n-k}(c + 2k)\varphi_k(c, x).$$

For $c = 1$ the generating relation (6) appears in Sister Celine's work, Fasenmyer [1]. Essentially the generating relation of Theorem 48 for ψ a ${}_pF_q$ appears in Rainville [6] with minor variations in notation. Sister Celine in her Michigan thesis also obtained contiguous function relations applicable to ${}_2F_0$'s of the type of (2) or (3) in this section. Equation (7) appears in Dickinson [1].

It is a simple application of Sister Celine's technique (see Chapter 14, or Fasenmyer [2]) to obtain such relations as

$$(9) \quad (c + 2n - 1)x^2\varphi_n'(c, x) \\ = n[1 + (c + 2n - 1)x]\varphi_n(c, x) - (c + n - 1)\varphi_{n-1}(c, x),$$

$$(10) \quad c\varphi_n(c + 1, x) = x\varphi_n'(c, x) + (c + n)\varphi_n(c, x),$$

$$(11) \quad (c+2n-1)\varphi_n(c-1, x) = (c-1)[\varphi_n(c, x) + \varphi_{n-1}(c, x)],$$

$$(12) \quad n(c + 2n - 3)\varphi_n(c, x) \\ = (c + 2n - 2)[c - 1 - (c + 2n - 1)(c + 2n - 3)x]\varphi_{n-1}(c, x) \\ + (c + n - 2)(c + 2n - 1)\varphi_{n-2}(c, x).$$

For the moment let

$$(13) \quad f_n(x) = \sum_{k=0}^{n} \frac{(-1)^k(c)_{n+k}\delta_{n-k}x^k}{k!(n - k)!},$$

of which $\varphi_n(c, x)$ is the special case $\delta_n = 1$. From (13) it follows that

$$\sum_{n=0}^{\infty} \frac{f_n(x)t^n}{(c)_n} = \sum_{n,k=0}^{\infty} \frac{(-1)^k(c)_{n+2k}\delta_n x^k t^{n+k}}{k!n!(c)_{n+k}}$$

$$= \sum_{n=0}^{\infty} {}_2F_1\left[\begin{matrix} \tfrac{1}{2}c + \tfrac{1}{2}n, \tfrac{1}{2}c + \tfrac{1}{2}n + \tfrac{1}{2}; \\ c + n; \end{matrix} -4xt\right] \frac{\delta_n t^n}{n!}$$

$$= (1 + 4xt)^{-\frac{1}{2}} \sum_{n=0}^{\infty} \left(\frac{2}{1 + \sqrt{1 + 4xt}}\right)^{c+n-1} \frac{\delta_n t^n}{n!}$$

by application of the result in Ex. 10, page 70. Now let

$$F(u) = \sum_{n=0}^{\infty} \frac{\delta_n u^n}{n!}.$$

It is then possible to write

$$(14)\quad (1+4xt)^{-\frac{1}{2}}\left(\frac{2}{1+\sqrt{1+4xt}}\right)^{c-1} F\left(\frac{2t}{1+\sqrt{1+4xt}}\right) = \sum_{n=0}^{\infty} \frac{f_n(x)t^n}{(c)_n}.$$

For F a ${}_pF_q$, what is essentially the preceding (14) appears in Rainville [6] with minor changes in notation. By using $\delta_n = 1$, we obtain Burchnall's generating relation

$$(15)\quad (1+4xt)^{-\frac{1}{2}}\left(\frac{2}{1+\sqrt{1+4xt}}\right)^{c-1} \exp\left(\frac{2t}{1+\sqrt{1+4xt}}\right) = \sum_{n=0}^{\infty} \frac{\varphi_n(c,x)t^n}{(c)_n},$$

which appeared in Burchnall [1] with different notations. For $c = 1$, equation (15) appears in Krall and Frink [1]. It is evident from (15) that the reversed Bessel polynomials $x^n\varphi_n(c, 1/x)/(c)_n$ are of Sheffer A-type zero, from which several properties of Bessel polynomials arise. If the F of equation (14) is chosen to be a ${}_0F_q$, the reversed polynomials are seen to be of σ-type zero, as discussed in Chapter 13.

Brafman [4] obtained a whole class of generating relations which in our notation become

$$(16)\quad {}_2F_0\left[\begin{matrix}\alpha, c-\alpha; \\ -;\end{matrix}\ \frac{t-\sqrt{t^2-4xt}}{2}\right] {}_2F_0\left[\begin{matrix}\alpha, c-\alpha; \\ -;\end{matrix}\ \frac{t+\sqrt{t^2-4xt}}{2}\right] \cong \sum_{n=0}^{\infty} \frac{(\alpha)_n(c-\alpha)_n\varphi_n(c,x)t^n}{(c)_n},$$

for arbitrary α. Brafman, of course, discovered (16) by using a limiting process on his class of generating relations for Jacobi polynomials, for which see Section 141 or Brafman [1]. The proof in Brafman [4] is based on Whipple's result, Theorem 31, page 88.

Al-Salam [1] repeats much of the known material, with proper credits, and obtains many new results. He makes free use of pseudo-Jacobi, pseudo-Laguerre, and pseudo-Bessel polynomials. His concept of generating functions is not that defined in Chapter 8.

In addition to the various references given in this section, see Agarwal [1], Carlitz [2], and Grosswald [1].

151. Bedient's polynomials. Bedient [1], in his study of some polynomials associated with Appell's F_2 and F_3, introduced

$$(1)\quad R_n(\beta, \gamma; x) = \frac{(\beta)_n(2x)^n}{n!}\,{}_3F_2\left[\begin{matrix} -\tfrac{1}{2}n, -\tfrac{1}{2}n + \tfrac{1}{2}, \gamma - \beta; \\ \gamma, 1 - \beta - n; \end{matrix}\quad \frac{1}{x^2}\right]$$

and

$$(2)\quad G_n(\alpha, \beta; x) =$$

$$\frac{(\alpha)_n(\beta)_n(2x)^n}{n!(\alpha + \beta)_n}\,{}_3F_2\left[\begin{matrix} -\tfrac{1}{2}n, -\tfrac{1}{2}n + \tfrac{1}{2}, 1 - \alpha - \beta - n; \\ 1 - \alpha - n, 1 - \beta - n; \end{matrix}\quad \frac{1}{x^2}\right].$$

One of several reasons for interest in R_n and G_n is their connection with the Gegenbauer polynomials of Chapter 17. Indeed,

$$(3)\quad \begin{aligned} \lim_{\gamma\to\infty} R_n(\beta, \gamma; x) &= C_n{}^{\beta}(x), \\ \lim_{\alpha\to\infty} G_n(\alpha, \beta; x) &= C_n{}^{\beta}(x), \\ \lim_{\beta\to\infty} G_n(\alpha, \beta; x) &= C_n{}^{\alpha}(x). \end{aligned}$$

For the R_n, Bedient obtains the generating relations

$$(4)\quad (1 - 2xt)^{-\beta}\,{}_2F_1\left[\begin{matrix} \beta, \gamma - \beta; \\ \gamma; \end{matrix}\quad \frac{-t^2}{1 - 2xt}\right] = \sum_{n=0}^{\infty} R_n(\beta, \gamma; x)t^n$$

and

$$(5)\quad {}_1F_1\big(\beta; \gamma; t(x - \sqrt{x^2 - 1})\big)\,{}_1F_1\big(\beta; \gamma; t(x + \sqrt{x^2 - 1})\big) = \sum_{n=0}^{\infty} \frac{R_n(\beta, \gamma; x)t^n}{(\gamma)_n}.$$

For the G_n he finds that

(6) $${}_2F_1(\alpha, \beta; \alpha + \beta; 2xt - t^2) = \sum_{n=0}^{\infty} G_n(\alpha, \beta; x)t^n$$

and

(7) $${}_2F_0\big(\alpha, \beta; -; t(x - \sqrt{x^2 - 1})\big)\, {}_2F_0\big(\alpha, \beta; -; t(x + \sqrt{x^2 - 1})\big) \cong \sum_{n=0}^{\infty} (\alpha + \beta)_n G_n(\alpha, \beta; x)t^n.$$

Bedient [1] contains numerous other properties of the R_n and G_n.

152. Shively's pseudo-Laguerre and other polynomials. Shively [1] studied the pseudo-Laguerre set

(1) $$R_n(a, x) = \frac{(a)_{2n}}{n!(a)_n}\, {}_1F_1(-n; a + n; x),$$

which are related to the proper simple Laguerre polynomial

$$L_n(x) = {}_1F_1(-n; 1; x)$$

by

(2) $$R_n(a, x) = \frac{1}{(a - 1)_n} \sum_{k=0}^{n} \frac{(a - 1)_{n+k} L_{n-k}(x)}{k!}.$$

Toscano [1] had already shown that

(3) $$(1 - 4t)^{-\frac{1}{2}}\left(\frac{2}{1 + \sqrt{1 - 4t}}\right)^{a-1} \exp\left(\frac{-4xt}{(1 + \sqrt{1 - 4t})^2}\right) = \sum_{n=0}^{\infty} R_n(a, x)t^n.$$

Shively obtained Toscano's other generating relation

(4) $$e^{2t}\, {}_0F_1(-; \tfrac{1}{2} + \tfrac{1}{2}a; t^2 - xt) = \sum_{n=0}^{\infty} \frac{R_n(a, x)t^n}{(\frac{1}{2} + \frac{1}{2}a)_n}$$

and extended Toscano's (3) in this section to

(5) $$(1 - 4t)^{-\frac{1}{2}}\left(\frac{2}{1 + \sqrt{1 - 4t}}\right)^{a-1} {}_pF_q\left[\begin{matrix} \alpha_1, \cdots, \alpha_p; \\ \beta_1, \cdots, \beta_q; \end{matrix} \quad \frac{-4xt}{(1 + \sqrt{1 - 4t})^2}\right] = \sum_{n=0}^{\infty} S_n(x)t^n,$$

in which

(6) $$S_n(x) = \frac{(a)_{2n}}{n!(a)_n}\, {}_{p+1}F_{q+1}\left[\begin{matrix} -n, \alpha_1, \cdots, \alpha_p; \\ a+n, \beta_1, \cdots, \beta_q; \end{matrix} \quad x\right].$$

For the particular choice $p = 0$, $q = 1$, $\beta_1 = 1$, $a = 1$, the $S_n(x)$ becomes

(7) $$\sigma_n(x) = \frac{(2n)!}{(n!)^2}\, {}_1F_2(-n; 1+n, 1; x)$$

for which Shively has the additional generating relation

(8) $${}_0F_1\left[\begin{matrix} -; \\ 1; \end{matrix} \quad \frac{t - \sqrt{4xt + t^2}}{2}\right] {}_0F_1\left[\begin{matrix} -; \\ 1; \end{matrix} \quad \frac{t + \sqrt{4xt + t^2}}{2}\right] = \sum_{n=0}^{\infty} \frac{\sigma_n(x) t^n}{(2n)!}.$$

The $R_n(a, x)$ of (1) is of Sheffer A-type zero, as pointed out by Shively. He obtains many other properties of $R_n(a, x)$.

153. Bernoulli polynomials. Much good has come from the study of $B_n(x)$ defined by

(1) $$\frac{te^{xt}}{e^t - 1} = \sum_{n=0}^{\infty} \frac{B_n(x) t^n}{n!},$$

particularly in the Theory of Numbers. The $B_n(x)$ of (1) are the Bernoulli polynomials, which have been generalized in numerous directions. See Erdélyi [1] and [3]. The $B_n(x)/n!$ are of Sheffer A-type zero, from which fact various interesting properties may be obtained. It is also a simple matter to show that

(2) $$B_n(x+1) - B_n(x) = nx^{n-1},$$

(3) $$B_n(x+1) \doteqdot \{1 + B(x)\}^n,$$

(4) $$B_n(1-x) = (-1)^n B_n(x).$$

One definition of Bernoulli numbers B_n is

(5) $$B_n = B_n(0).$$

It follows that

(6) $$B_n(x) \doteqdot \{B + x\}^n$$

and

(7) $$B_n(1) = B_n, \quad n \geqq 2, \quad B_1(1) = 1 + B_1.$$

Since

$$\frac{t}{e^t - 1} = -t + \frac{-t}{e^{-t} - 1},$$

B_1 is the only nonzero Bernoulli number with odd index. Thus the generating relation

(8) $$\frac{t}{e^t - 1} = \sum_{n=0}^{\infty} \frac{B_n t^n}{n!}$$

may also be written

(9) $$\frac{t}{e^t - 1} = B_0 + B_1 t + \sum_{n=1}^{\infty} \frac{B_{2n} t^{2n}}{(2n)!},$$

in which $B_0 = 1$, $B_1 = -\frac{1}{2}$, etc.

For us one natural extension of $B_n(x)$ is to replace the factor e^{xt} in the generating relation (1) by any ${}_0F_q$ with argument xt. The resultant polynomials are of σ-type zero in the sense of Chapter 13.

154. Euler polynomials. The polynomials $E_n(x)$ defined by

(1) $$\frac{2e^{xt}}{e^t + 1} = \sum_{n=0}^{\infty} \frac{E_n(x) t^n}{n!}$$

are called Euler polynomials, and the numbers

(2) $$E_n = 2^n E_n(\tfrac{1}{2})$$

are called Euler numbers. The polynomials $E_n(x)/n!$ are of Sheffer A-type zero.

It is not difficult to obtain such results as

(3) $$E_n(x + 1) + E_n(x) = 2x^n,$$

(4) $$E_n(1 - x) = (-1)^n E_n(x),$$

(5) $$E_n(x + 1) \doteqdot \{1 + E(x)\}^n,$$

(6) $$nE_{n-1}(x) = 2B_n(x) - 2^{n+1} B_n(\tfrac{1}{2}x).$$

Since the Euler numbers defined in (2) have the generating relation

(7) $$\sum_{n=0}^{\infty} \frac{E_n t^n}{n!} = \frac{2e^t}{e^{2t} + 1} = \frac{2e^{-t}}{1 + e^{-2t}},$$

it follows that $E_{2n+1} = 0$.

The Euler polynomials have been generalized in various ways. See Erdélyi [1] and [3].

155. Tchebicheff polynomials. The Tchebicheff polynomials $T_n(x)$ and $U_n(x)$ of the first and second kinds, respectively, are special ultraspherical polynomials. See Szegö [1] and Chapter 17 above. In detail,

$$T_n(x) = \frac{n!}{(\frac{1}{2})_n} P_n^{(-\frac{1}{2},-\frac{1}{2})}(x), \tag{1}$$

$$U_n(x) = \frac{(n+1)!}{(\frac{3}{2})_n} P_n^{(\frac{1}{2},\frac{1}{2})}(x). \tag{2}$$

These are often introduced by the relations

$$T_n(\cos\theta) = \cos n\theta, \qquad U_n(\cos\theta) = \frac{\sin(n+1)\theta}{\sin\theta}.$$

From (1) and (2) many of the formulas on Jacobi polynomials, Chapter 16, can be converted into results on Tchebicheff polynomials by choosing $\alpha = \beta = \frac{1}{2}$ or $\alpha = \beta = -\frac{1}{2}$.

Some of the generating functions to be found are

$$(1 - 2xt + t^2)^{-1} = \sum_{n=0}^{\infty} U_n(x)t^n, \tag{3}$$

$$(1 - xt)(1 - 2xt + t^2)^{-1} = \sum_{n=0}^{\infty} T_n(x)t^n, \tag{4}$$

$$\sum_{n=0}^{\infty} \frac{T_n(x)t^n}{n!} = e^{xt}\cosh(t\sqrt{x^2 - 1}), \tag{5}$$

$$\sum_{n=0}^{\infty} \frac{U_n(x)t^{n+1}}{(n+1)!} = \frac{e^{xt}\sinh(t\sqrt{x^2 - 1})}{\sqrt{x^2 - 1}}. \tag{6}$$

Relations (3) and (4) at once yield

$$T_n(x) = U_n(x) - xU_{n-1}(x), \qquad n \geqq 1, \tag{7}$$

$$(1 - x^2)U_n(x) = xT_{n+1}(x) - T_{n+2}(x). \tag{8}$$

Of the several explicit formulas for T_n and U_n (see Chapter 17 with appropriate α and β), we note

$$T_n(x) = \sum_{k=0}^{[n/2]} \frac{n!x^{n-2k}(x^2 - 1)^k}{(2k)!(n - 2k)!}, \tag{9}$$

$$T_n(x) = \tfrac{1}{2}\left[\left(x + \sqrt{x^2 - 1}\right)^n + \left(x - \sqrt{x^2 - 1}\right)^n\right], \tag{10}$$

(11) $$U_n(x) = \sum_{k=0}^{[n/2]} \frac{(n+1)!x^{n-2k}(x^2-1)^k}{(2k+1)!(n-2k)!}.$$

Equation (10) relates $T_n(x)$ to the Neumann polynomials of Chapter 6. See equations (4) and (6), pages 116–117.

EXERCISES

1. For the Bernoulli polynomial of Section 153 show that

$$x^n = \sum_{k=0}^{n} \frac{n!B_k(x)}{k!(n-k+1)!}.$$

2. Let $B_n(x)$ and $B_n = B_n(0)$ denote the Bernoulli polynomials and numbers as treated in Section 153. Define the differential operator $A(x, D)$ by

$$A(x, D) = (x - \tfrac{1}{2})D - \sum_{k=1}^{\infty} \frac{B_{2k}D^{2k}}{(2k)!}.$$

Prove that $A(x, D)B_n(x) = nB_n(x)$.

3. Consider the polynomials

$$\psi_n(c, x, y) = \frac{(-1)^n(\tfrac{1}{2} + \tfrac{1}{2}x)_n}{(c)_n} {}_3F_2\left[\begin{matrix} -n, \tfrac{1}{2} - \tfrac{1}{2}x, 1 - c - n; \\ c, \tfrac{1}{2} - \tfrac{1}{2}x - n; \end{matrix} \quad y\right].$$

Show that

$${}_1F_1(\tfrac{1}{2} - \tfrac{1}{2}x; c; yt)\, {}_1F_1(\tfrac{1}{2} + \tfrac{1}{2}x; c; -t) = \sum_{n=0}^{\infty} \frac{\psi_n(c, x, y)t^n}{n!},$$

$$(1 - yt)^{-\frac{1}{2}+\frac{1}{2}x}(1 + t)^{-\frac{1}{2}-\frac{1}{2}x}\, {}_2F_1\left[\begin{matrix} \tfrac{1}{2} - \tfrac{1}{2}x, \tfrac{1}{2} + \tfrac{1}{2}x; \\ c; \end{matrix} \quad \frac{-yt^2}{(1 - yt)(1 + t)}\right]$$

$$= \sum_{n=0}^{\infty} \frac{(c)_n\psi_n(c, x, y)t^n}{n!},$$

and that $\psi_n(1, x, 1) = F_n(x)$, where F_n is Bateman's polynomial of Section 148.

4. Sylvester (1879) studied polynomials (see page 255 of Erdélyi [3]),

$$\varphi_n(x) = \frac{x^n}{n!} {}_2F_0\left(-n, x; -; -\frac{1}{x}\right).$$

Show that

$$(1 - t)^{-x}e^{xt} = \sum_{n=0}^{\infty} \varphi_n(x)t^n,$$

$$(1 - xt)^{-c}\, {}_2F_0\left(c, x; -; \frac{t}{1 - xt}\right) \cong \sum_{n=0}^{\infty} (c)_n\varphi_n(x)t^n.$$

5. For Sylvester's polynomials of Ex. 4 find what properties you can from the fact that $\varphi_n(x)$ is of Sheffer A-type zero.

6. Show that Bateman's $Z_n(x)$, the Legendre polynomial $P_n(x)$, and the Laguerre polynomial $L_n(x)$ are related symbolically by

$$Z_n(x) \doteqdot P_n(2L(x) - 1).$$

7. Show that Sister Celine's polynomial

$$f_n(x) = {}_2F_2(-n, n+1; 1, \tfrac{1}{2}; x)$$

of equation (16), page 292, is such that

$$\int_0^\infty e^{-x} f_n(x)\, dx = (-1)^n (2n+1).$$

8. For Sister Celine's $f_n(x)$ of Ex. 7, show that

$$\int_0^\infty e^{-x} f_n(x) f_m(x)\, dx$$
$$= (-1)^{n+m}(2n+1)(2m+1)\,{}_4F_3\left[\begin{matrix} -n, n+1, -m, m+1; & \\ & 1 \\ 1, \tfrac{3}{2}\ \tfrac{3}{2}; & \end{matrix}\right],$$

$$\int_0^\infty e^{-x} L_k(x) f_n(x)\, dx = \frac{(-1)^n(2n+1)(-n)_k(n+1)_k}{k!\left(\dfrac{3}{2}\right)_k}; \qquad 0 \leqq k \leqq n,$$
$$= 0; \qquad k > n,$$

$$\int_0^\infty e^{-x} f_n(x) Z_k(x)\, dx = (-1)^{n+k}(2n+1)\,{}_4F_3\left[\begin{matrix} -n, n+1, -k, k+1; & \\ & 1 \\ 1, 1\ \tfrac{3}{2}; & \end{matrix}\right].$$

9. Show that

$$\int_0^\infty e^{-x} Z_n(x) Z_k(x)\, dx = (-1)^{n+k}{}_4F_3\left[\begin{matrix} -n, n+1, -k, k+1; & \\ & 1 \\ 1, 1, 1; & \end{matrix}\right].$$

10. Gottlieb [1] introduced the polynomials

$$\varphi_n(x; \lambda) = e^{-n\lambda}\,{}_2F_1(-n, -x; 1; 1 - e^\lambda).$$

Show that

$$\sum_{n=0}^\infty \frac{\varphi_n(x;\lambda) t^n}{n!} = e^t\,{}_1F_1\bigl(1 + x; 1; -t(1 - e^{-\lambda})\bigr),$$

$$\varphi_n(x, \lambda + \mu) = \frac{(e^\mu - 1)^n}{e^{\mu n}(1 - e^\lambda)^n} \sum_{k=0}^n \frac{n!(1 - e^{\lambda+\mu})^k \varphi_k(x; \lambda)}{k!(n-k)!(e^\mu - 1)^k},$$

$$\int_0^\infty e^{-st} \varphi_n(x; -t)\, dt = \frac{\Gamma(s-n)\Gamma(s+x+1)}{\Gamma(s+1)\Gamma(s+x-n+1)},$$

$$(x - n)\varphi_n(x; \lambda) = x\varphi_n(x-1; \lambda) - ne^{-\lambda}\varphi_{n-1}(x; \lambda),$$

$$n[\varphi_n(x;\lambda) - \varphi_{n-1}(x;\lambda)] = (x+1)[\varphi_n(x+1;\lambda) - \varphi_n(x;\lambda)],$$

$$(x + n + 1)\varphi_n(x;\lambda) = xe^\lambda \varphi_n(x-1;\lambda) + (n+1)e^\lambda \varphi_{n+1}(x;\lambda),$$

and see Gottlieb [1] for many other results on $\varphi_n(x;\lambda)$.

11. With f_n denoting Sister Celine's polynomials of Section 149, show that

$$\int_0^t x^{\frac{1}{2}}(t-x)^{\frac{1}{2}} f_n(-;-;x)\,dx = \pi Z_n(t),$$

$$\int_0^t x^{\frac{1}{2}}(t-x)^{\frac{1}{2}} f_n\big(-;-;x(t-x)\big)\,dx = \pi Z_n(\tfrac{1}{4}t^2),$$

and

$$\int_0^t x^{\frac{1}{2}}(t-x)^{\frac{1}{2}} f_n\big(-;\tfrac{1}{2};x(t-x)\big)\,dx = \pi L_n(t)L_n(-t).$$

12. Define polynomials $\varphi_n(x)$ by

$$\varphi_n(x) = \frac{(c)_n}{n!}\,{}_2F_2(-n, c+n; 1, 1; x).$$

Show that

$$\varphi_n(x) = \sum_{k=0}^{n} \frac{(-1)^{n-k}(c-1)_{n-k}(c)_{n+k}(2k+1)Z_k(x)}{(n-k)!(n+k+1)!}.$$

13. Show that the $F(t)$ of equation (10), page 286, can be put in the form

$$F(t) = t^b e^{-t}(1-z)^{-2\nu}\,{}_1F_1\left[\begin{matrix} \nu; \\ b+1; \end{matrix}\quad \frac{-4zt}{(1-z)^2}\right].$$

CHAPTER 19

Elliptic Functions

156. Doubly periodic functions. Let the function $f(z)$ have two distinct periods $2\omega_1$ and $2\omega_2$:

(1) $f(z+2\omega_1)=f(z)$,

(2) $f(z+2\omega_2)=f(z)$.

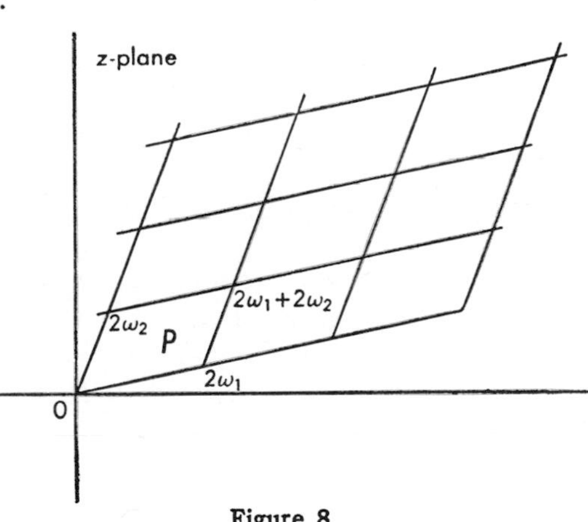

Figure 8

Further, let the periods have a nonreal ratio, ω_1/ω_2 not real, so that there exists a network of period parallelograms, or meshes, as shown in Figure 8. The period parallelograms are called *fundamental* if they are the smallest such meshes for the function $f(z)$.

The values of $f(z)$ in any parallelogram P in Figure 8 are repeated in each of the other parallelograms, as indicated in (1) and (2) above.

Doubly periodic functions are a natural extension of such simply periodic functions as the familiar trigonometric functions of elementary mathematics.

157. Elliptic functions. If a doubly periodic analytic function $f(z)$ has in the finite plane no singular points other than poles, then $f(z)$ is called an *elliptic function.*

Consider a fundamental period parallelogram P of an elliptic function $f(z)$. There are only a finite number of poles and zeros of $f(z)$ in the closed region P because any point of condensation of poles or of zeros of $f(z)$ would be an essential singularity of the function, which is contrary to the definition. Therefore there exists a point t in P such that the parallelogram with vertices at t, $t + 2\omega_1$, $t + 2\omega_1 + 2\omega_2$, $t + 2\omega_2$ has on its boundary no poles and no zeros of $f(z)$. That parallelogram is called a *cell,* shown in Figure 9.

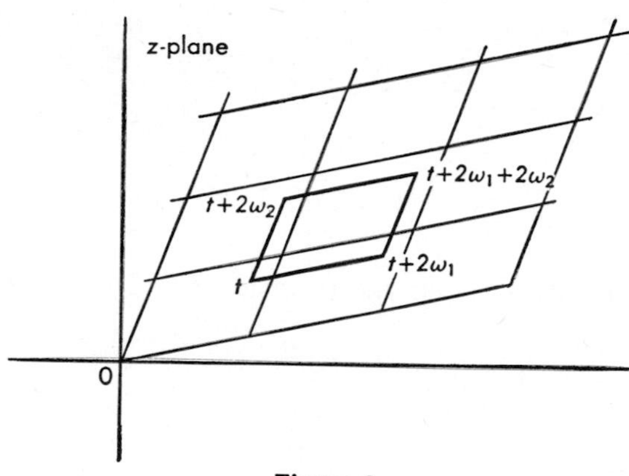

Figure 9

We replace the network of fundamental period parallelograms of Figure 8 with a network of cells, each of which contains no pole and no zero on its boundary. One reason for this is that we wish to perform integrations along the boundary of a cell; we could not integrate $f(z)$ around the boundary of P for fear of encountering a singular point of the integrand.

158. Elementary properties. Consider an elliptic function $f(z)$ with periodicity properties

$$f(z + 2\omega_1) = f(z), \tag{1}$$

$$f(z + 2\omega_2) = f(z), \tag{2}$$

and a network of cells as defined in the preceding section.

Certain simple properties of $f(z)$ are at once available.

THEOREM 85. *The number of poles of $f(z)$ in any cell is finite, and the number of zeros of $f(z)$ in any cell is finite.*

THEOREM 86. *An elliptic function with no poles in a cell is a constant.*

We have already seen that Theorem 85 is a consequence of the fact that $f(z)$ is permitted no essential singularity in the finite plane.

For Theorem 86, note that if $f(z)$ has no poles in a cell, it is analytic and bounded in the closed cell and also throughout the finite plane because of its periodicity. Then by Liouville's theorem,* it follows that $f(z)$ is a constant.

THEOREM 87. *The sum of the residues at the poles in a cell of an elliptic function is zero.*

Proof: Let S_R be the sum of the residues at the poles of the elliptic function $f(z)$ in the cell shown in Figure 9. Then $2\pi i S_R$ equals the integral of $f(z)$ around the cell:

$$(3)\quad 2\pi i S_R = \int_t^{t+2\omega_1} f(z)\,dz + \int_{t+2\omega_1}^{t+2\omega_1+2\omega_2} f(z)\,dz + \int_{t+2\omega_1+2\omega_2}^{t+2\omega_2} f(z)\,dz + \int_{t+2\omega_2}^{t} f(z)\,dz,$$

in which each integration takes place along a straight line path.

Let us change variable of integration in the second integral on the right in (3) by replacing z by $z + 2\omega_1$, and in the third integral on the right in (3) by replacing z by $z + 2\omega_2$. We thus obtain

$$(4)\quad 2\pi i S_R = \int_t^{t+2\omega_1} f(z)\,dz + \int_t^{t+2\omega_2} f(z+2\omega_1)\,dz + \int_{t+2\omega_1}^{t} f(z+2\omega_2)\,dz + \int_{t+2\omega_2}^{t} f(z)\,dz.$$

But $f(z + 2\omega_2) = f(z)$, so that the first and third integrals differ only in sign (direction of integration), and because $f(z + 2\omega_1) = f(z)$, the second and fourth integrals also have the sum zero. Then $2\pi i S_R = 0$; so, $S_R = 0$, and the proof is complete.

THEOREM 88. *For an elliptic function $f(z)$, the number of zeros in a cell equals the number of poles in a cell.*

Proof: Let N_z be the number of zeros and N_p the number of poles of $f(z)$ in a cell. Here the multiplicity is to be counted; a pole of order two counts as two poles; a zero of order three counts as

*See any book on complex variables or function theory; for instance, Churchill [3; 96].

three zeros, etc. Then from the study of functions of a complex variable we know that

$$(5) \qquad 2\pi i(N_z - N_p) = \int_B \frac{f'(z)\,dz}{f(z)},$$

in which the integration is to be carried out around the boundary B of the cell.

From

$$(6) \qquad f(z + 2\omega_1) = f(z) \qquad \text{and} \qquad f(z + 2\omega_2) = f(z),$$

it follows that

$$(7) \qquad f'(z + 2\omega_1) = f'(z) \qquad \text{and} \qquad f'(z + 2\omega_2) = f'(z).$$

Therefore $f'(z)$ and also $f'(z)/f(z)$ have the same periods as does $f(z)$. Hence the device used in showing that the integral $\int_B f(z)\,dz$ is zero may be used to show that

$$\int_B \frac{f'(z)\,dz}{f(z)} = 0.$$

Therefore $N_z - N_p = 0$, which we wished to prove.

The actual situation is that Theorem 88 is merely a particular instance of the fact that the number of times an elliptic function $f(z)$ assumes any particular value c in a cell is independent of c. For zeros of $f(z)$, $c = 0$; for poles of $f(z)$, $c = \infty$.

THEOREM 89. *For an elliptic function $f(z)$ and for any finite c, the number of zeros of $f(z) - c$ is independent of c.*

Proof: The function $f(z) - c$ has the same periods as $f(z)$ and so does $f'(z)$. The poles of $f(z) - c$ are the poles of $f(z)$. Then the difference between the number of poles of $f(z)$ and the number of zeros of $f(z) - c$ is

$$\frac{1}{2\pi i}\int_B \frac{f'(z)\,dz}{f(z) - c},$$

which can be shown to be zero by the same device as was used in the proofs of Theorems 87 and 88.

159. Order of an elliptic function. For an elliptic function $f(z)$ we define as its *order* the number (counting multiplicity) of poles of $f(z)$ in a cell. We already know that the order cannot be zero unless $f(z)$ is constant. If $f(z)$ were of order one, then in a cell there would be a single, and simple, pole. By Theorem 87, page 307, the residue at that pole must be zero, so the point is not a pole but a point of analyticity of $f(z)$. Hence $f(z)$ has no poles in the cell, and $f(z)$ must be constant. We have proved the following result which will be used over and over in Chapter 20.

THEOREM 90. *An elliptic function of order less than two is a constant.*

From the standpoint of its order, the simplest elliptic function will be one of order two. There are essentially two types of elliptic functions of order two. If there is but one pole in each cell, that pole must be of order two and the residue at the pole must be zero. If there are two simple poles in each cell, the residues at those poles must be negatives of each other. We shall encounter specific instances of each of these types of elliptic functions of order two.

160. The Weierstrass function P(z). Let m and n be integers and define

$$\Omega_{m,n} = 2m\omega_1 + 2n\omega_2, \tag{1}$$

where $2\omega_1$ and $2\omega_2$ are to be the fundamental periods of our elliptic function and ω_1/ω_2 is not real. We shall show that the function in (2), below, devised by Weierstrass, is an elliptic function of order two, with each cell containing a second order pole.

Let a prime attached to the summation symbol indicate that the indices of summation are not both to be zero at once. Then define the Weierstrass function $P(z)$ by

$$P(z) = \frac{1}{z^2} + \sum_{m,n=-\infty}^{\infty}{}' \left\{ \frac{1}{(z - \Omega_{m,n})^2} - \frac{1}{\Omega_{m,n}^2} \right\}. \tag{2}$$

The deletion of the $m = n = 0$ term from the summation is forced upon us by the last term in (2). Since the general term in the series in (2) is $O(\Omega_{m,n}^{-3})$, the series is absolutely and uniformly convergent in any closed region which excludes the origin and all the other points $z = \Omega_{m,n}$. Then $P(z)$ is analytic throughout the finite plane except for a double pole at each of the points $z = \Omega_{m,n}$ for integral m and n.

Now

$$(3) \qquad P'(z) = \sum_{m,n=-\infty}^{\infty} \frac{-2}{(z - \Omega_{m,n})^3}$$

in which the $m = n = 0$ term need not be deleted and the differentiation is justified by the uniform convergence of the series in (3). Now $P'(z)$ is an odd function of z, since

$$(4) \qquad P'(-z) = \sum_{m,n=-\infty}^{\infty} \frac{2}{(z + \Omega_{m,n})^3} = \sum_{m,n=-\infty}^{\infty} \frac{2}{(z - \Omega_{m,n})^3},$$

where the second series is obtained by replacing m by $(-m)$ and n by $(-n)$ in the first series. By (4),

$$(5) \qquad P'(-z) = -P'(z).$$

The same device can be used on the series in (2) to show that $P(z)$ is an even function of z,

$$(6) \qquad P(-z) = P(z).$$

We next show that $P'(z)$ has the periods $2\omega_1$ and $2\omega_2$. Consider

$$P'(z + 2\omega_1) = \sum_{m,n=-\infty}^{\infty} \frac{-2}{(z + 2\omega_1 - \Omega_{m,n})^3}.$$

Since $\Omega_{m,n} = 2m\omega_1 + 2n\omega_2$, we have

$$P'(z + 2\omega_1) = \sum_{m,n=-\infty}^{\infty} \frac{-2}{(z - \Omega_{m-1,n})^3}$$

and a shift of index from m to $(m + 1)$ shows that

$$(7) \qquad P'(z + 2\omega_1) = P'(z).$$

It follows in the same way that

$$(8) \qquad P'(z + 2\omega_2) = P'(z),$$

and we now see that $P'(z)$ is an elliptic function.

Integration of each member of (7) yields the following equation:

$$(9) \qquad P(z + 2\omega_1) = P(z) + c,$$

with c constant. In (9) use $z = -\omega_1$ to obtain

$$P(\omega_1) = P(-\omega_1) + c.$$

But by (6), $P(\omega_1) = P(-\omega_1)$, so $c = 0$ and (9) yields

$$(10) \qquad P(z + 2\omega_1) = P(z).$$

The same device shows that

$$\text{(11)} \qquad P(z + 2\omega_2) = P(z).$$

Hence $P(z)$ is an elliptic function with poles of order two at each of the points $z = \Omega_{m,n}$, m and n integral.

161. Other elliptic functions. From the Weierstrass function $P(z)$ we easily obtain an elliptic function with any specified set of zeros and poles in a cell.

First consider an even elliptic function which, in a cell, is to have zeros at $a_1, a_2 \cdots, a_n$ and (automatically) at the points congruent to $-a_1, -a_2, \cdots, -a_n$. Let multiplicity be accounted for by repetition, that is, by equality of a's. Let the desired poles in the cell be at $b_1, b_2, \cdots, b_n$ and at the points congruent to $-b_1, -b_2, \cdots, -b_n$, with multiplicity treated as with the zeros. Then

$$\text{(1)} \qquad f(z) = \prod_{i=1}^{n} \frac{P(z) - P(a_i)}{P(z) - P(b_i)}$$

has the desired zeros and poles in a cell and is an even elliptic function. Odd elliptic functions may be constructed in the same way, using $P'(z)$, or may be treated by using the product of $P'(z)$ and a rational function of $P(z)$. Since any function may be split into its even and odd parts, the procedure sketched here permits construction of elliptic functions with any desired permissible distribution of zeros and poles in a cell. See Whittaker and Watson [1; 448 ff].

162. A differential equation for P(z). We know that $P(z)$ is analytic in a deleted neighborhood of its second-order pole at $z = 0$. Indeed, from the definition

$$\text{(1)} \qquad P(z) = \frac{1}{z^2} + \sum_{m,n=-\infty}^{\infty}{}' \left\{ \frac{1}{(z - \Omega_{m,n})^2} - \frac{1}{\Omega_{m,n}^2} \right\},$$

we find that $P(z) - z^{-2}$ is analytic at the origin and is zero there. Also $P(z) - z^{-2}$ is an even function of z, so there exists a Taylor expansion of the form

$$\text{(2)} \qquad P(z) - z^{-2} = \sum_{n=1}^{\infty} a_n z^{2n}$$

convergent in a region around $z = 0$.

From (2) we need only a few terms. We write

(3) $$P(z) = z^{-2} + a_1z^2 + a_2z^4 + o(z^5), \qquad \text{as } z \to 0.$$

We keep in mind that the terms lumped together in $o(z^5)$ are actually as exhibited in (2), so that differentiations with respect to z are legitimate.

From (3) we obtain

(4) $$P'(z) = -2z^{-3} + 2a_1z + 4a_2z^3 + o(z^4).$$

The series (3) for $P(z)$ contains one negative power, z^{-2}; the series (4) for $P'(z)$ contains one negative power, $-2z^{-3}$. In order to eliminate negative powers of z, we are thus led to consider $[P(z)]^3$ and $[P'(z)]^2$. In detail,

(5) $$[P'(z)]^2 = 4z^{-6} - 8a_1z^{-2} - 16a_2 + o(z)$$

and

(6) $$[P(z)]^3 = z^{-6} + 3a_1z^{-2} + 3a_2 + o(z).$$

Then

$$[P'(z)]^2 - 4[P(z)]^3 = -20a_1z^{-2} - 28a_2 + o(z).$$

But

$$20a_1P(z) = 20a_1z^{-2} + o(z),$$

and we arrive at

(7) $$[P'(z)]^2 - 4[P(z)]^3 + 20a_1P(z) + 28a_2 = o(z).$$

The left member of (7) is an elliptic function with periods $2\omega_1$ and $2\omega_2$. That function is analytic at $z = 0$, and therefore also at $z = \Omega_{m,n}$. The function is therefore constant, and that constant is zero, as can be seen by letting $z \to 0$ and noting the right member of (7).

We have shown that $P(z)$ satisfies the first-order nonlinear differential equation

(8) $$[P'(z)]^2 = 4[P(z)]^3 - 20a_1P(z) - 28a_2,$$

in which a_1 and a_2 are the specific coefficients stipulated in (2). It is customary to put $20a_1 = g_2$ and $28a_2 = g_3$. We then express our result by stating that $y = P(z)$ is a solution of the differential equation

(9) $$\left(\frac{dy}{dz}\right)^2 = 4y^3 - g_2y - g_3.$$

Since equation (9) is invariant under a translation in the z-plane, the general solution of (9) is $y = P(z + \alpha)$, with α an arbitrary constant. It can be shown* that every solution of (9) is of the form $y = P(z + \alpha)$.

163. Connection with elliptic integrals. By elementary means it follows from the differential equation

$$\left(\frac{dy}{dz}\right)^2 = 4y^3 - g_2 y - g_3 \tag{1}$$

that one solution is

$$z = \int_y^\infty \frac{dt}{\sqrt{4t^3 - g_2 t - g_3}}. \tag{2}$$

But $y = P(z + \alpha)$, and if we let $y \to \infty$, $z \to 0$, so that $P(z)$ has a pole at $z = \alpha$. Then $\alpha = \Omega_{m,n}$ for some integers m and n. Then

$$y = P(z + \Omega_{m,n}) = P(z).$$

That is, the inverse of the integral in (2) is the Weierstrass function,

$$y = P(z). \tag{3}$$

Any nondegenerate integral whose integrand contains the square root of a cubic or quartic polynomial in the variable of integration together with rational functions is called an *elliptic integral.* The name arose from the occurrence of one such integral in the problem of finding the length of arc on an ellipse. As the foregoing discussion shows in relating (2) and (3), an elliptic function is the inverse of an elliptic integral. A corresponding fact in elementary singly periodic functions is that the integral

$$z = \int_0^y \frac{dt}{\sqrt{1 - t^2}} \tag{4}$$

has as its inverse the simply periodic trigonometric function

$$y = \sin z. \tag{5}$$

---

*See pages 437 and 484–485 of Whittaker and Watson [1]. The major problem is to show that ω_1 and ω_2 exist for given g_2 and g_3.

CHAPTER 20

Theta Functions

164. Definitions. In this chapter we study in some detail four functions first treated extensively by Jacobi. These are the four theta functions:

(1) $$\theta_1(z, q) = 2\sum_{n=0}^{\infty} (-1)^n q^{(n+\frac{1}{2})^2} \sin(2n+1)z,$$

(2) $$\theta_2(z, q) = 2\sum_{n=0}^{\infty} q^{(n+\frac{1}{2})^2} \cos(2n+1)z,$$

(3) $$\theta_3(z, q) = 1 + 2\sum_{n=1}^{\infty} q^{n^2} \cos 2nz,$$

(4) $$\theta_4(z, q) = 1 + 2\sum_{n=1}^{\infty} (-1)^n q^{n^2} \cos 2nz.$$

It seems safe to say that no topic in mathematics is more replete with beautiful formulas than that on which we now embark.

In the functions defined by equations (1)–(4), we ordinarily think of each θ as a function of z, with q playing the role of a parameter. When only the z is to be emphasized, the q will be suppressed, and we shall write

(5) $$\theta_i(z) = \theta_i(z, q), \qquad i = 1, 2, 3, 4.$$

In the definitions (1)–(4) we need only to require that $|q| < 1$ to get absolute convergence for all finite z. At times the role of the parameter q becomes important, and we use the additional notation

(6) $$q = \exp(\pi i \tau).$$

The requirment $|q| < 1$ implies that the coefficient in the imaginary part of τ be positive, $\operatorname{Im}(\tau) > 0$. When the dependence of $\theta_i(z, q)$ upon τ is to be emphasized, we write $\theta_i(z) = \theta_i(z|\tau)$ for $i = 1, 2, 3, 4$.

One last set of notations is needed. For $i = 2, 3, 4$, the value of $\theta_i(z, q)$ at $z = 0$ plays an important role, as does the value of $\dfrac{d}{dz}\theta_1(z, q)$ at $z = 0$. For these functions of q alone we use the simplified notations

(7) $$\theta_i = \theta_i(0, q), \qquad i = 2, 3, 4,$$

(8) $$\theta_1' = \left[\frac{d}{dz}\theta_1(z, q)\right]_{z=0}.$$

165. Elementary properties. Some properties of the theta functions follow at once from the definitions (1)–(4) of the preceding section. The function $\theta_1(z)$ is an odd function of z; the functions $\theta_2(z)$, $\theta_3(z)$, and $\theta_4(z)$ are all even functions of z.

Since

$$\begin{aligned}\sin[(2n + 1)(z + \tfrac{1}{2}\pi)] &= \cos(2n + 1)z \sin[\tfrac{1}{2}(2n + 1)\pi] \\ &= (-1)^n \cos(2n + 1)z,\end{aligned}$$

$$\begin{aligned}\cos[(2n+ 1)(z + \tfrac{1}{2}\pi)] &= -\sin(2n + 1)z \sin[\tfrac{1}{2}(2n + 1)\pi] \\ &= (-1)^{n+1}\sin(2n + 1)z,\end{aligned}$$

and

$$\cos[2n(z + \tfrac{1}{2}\pi)] = \cos 2nz \cos n\pi = (-1)^n \cos 2nz,$$

it follows readily that

(1) $$\theta_1(z + \tfrac{1}{2}\pi) = \theta_2(z),$$

(2) $$\theta_2(z + \tfrac{1}{2}\pi) = -\theta_1(z),$$

(3) $$\theta_3(z + \tfrac{1}{2}\pi) = \theta_4(z),$$

(4) $$\theta_4(z + \tfrac{1}{2}\pi) = \theta_3(z).$$

From (1) and (2) above we obtain

$$\theta_1(z + \pi) = \theta_2(z + \tfrac{1}{2}\pi) = -\theta_1(z),$$

and we perform similar operations on the other theta functions to arrive at the results

(5) $$\theta_1(z+\pi) = -\theta_1(z),$$

(6) $$\theta_2(z+\pi) = -\theta_2(z),$$

(7) $$\theta_3(z+\pi) = \theta_3(z),$$

(8) $$\theta_4(z+\pi) = \theta_4(z).$$

We now know that $\theta_3(z)$ and $\theta_4(z)$ have the common period π and that $\theta_1(z)$ and $\theta_2(z)$ have the common period 2π. This is, of course, a reflection of the periodicity of the sine and cosine functions. But the exponential function is also periodic,

(9) $$\exp(u + 2\pi i) = \exp(u),$$

and the theta functions involve an exponential in that $q = \exp(\pi i\tau)$. Hence we are led to investigate the possibility of another period for the theta functions. It turns out that these functions are not doubly periodic but that they miss that property by so small a margin (a multiplicative factor) that it is a simple matter to use the theta functions to construct functions which are doubly periodic.

166. The basic property table. The discussion in the last paragraph of the preceding section leads us to investigate for each $\theta_i(z)$ the value of $\theta_i(z + \frac{1}{2}\pi\tau)$.

We defined the theta functions by the equations

(1) $$\theta_1(z) = 2\sum_{n=0}^{\infty}(-1)^n q^{(n+\frac{1}{2})^2}\sin(2n+1)z,$$

(2) $$\theta_2(z) = 2\sum_{n=0}^{\infty} q^{(n+\frac{1}{2})^2}\cos(2n+1)z,$$

(3) $$\theta_3(z) = 1 + 2\sum_{n=1}^{\infty} q^{n^2}\cos 2nz,$$

(4) $$\theta_4(z) = 1 + 2\sum_{n=1}^{\infty}(-1)^n q^{n^2}\cos 2nz.$$

Let us now put the trigonometric functions involved into exponential form. We may write, for instance,

$$\theta_1(z) = i^{-1}\sum_{n=0}^{\infty}(-1)^n q^{(n+\frac{1}{2})^2}\left[e^{(2n+1)iz} - e^{-(2n+1)iz}\right]$$

$$= -i\sum_{n=0}^{\infty}(-1)^n q^{(n+\frac{1}{2})^2} e^{(2n+1)iz} + i\sum_{n=0}^{\infty}(-1)^n q^{(n+\frac{1}{2})^2} e^{-(2n+1)iz}.$$

In the last summation, replace n by $(-n-1)$ to obtain

$$\theta_1(z) = -i\sum_{n=0}^{\infty}(-1)^n q^{(n+\frac{1}{2})^2}e^{(2n+1)iz} + i\sum_{n=-1}^{-\infty}(-1)^{n+1}q^{(-n-\frac{1}{2})^2}e^{(2n+1)iz},$$

or

$$(5) \qquad \theta_1(z) = -i\sum_{n=-\infty}^{\infty}(-1)^n q^{(n+\frac{1}{2})^2}e^{(2n+1)iz}.$$

The same procedure yields also, and even more easily,

$$(6) \qquad \theta_2(z) = \sum_{n=-\infty}^{\infty} q^{(n+\frac{1}{2})^2}e^{(2n+1)iz},$$

$$(7) \qquad \theta_3(z) = \sum_{n=-\infty}^{\infty} q^{n^2}e^{2niz},$$

$$(8) \qquad \theta_4(z) = \sum_{n=-\infty}^{\infty} (-1)^n q^{n^2}e^{2niz}.$$

We are now in a position to obtain $\theta_i(z+\frac{1}{2}\pi\tau)$ for $i = 1, 2, 3, 4$. Since $e^{\pi i\tau} = q$, it follows that

$$\exp[(2n+1)i(z+\tfrac{1}{2}\pi\tau)] = \exp[(2n+1)iz]\exp[(n+\tfrac{1}{2})i\pi\tau]$$

or

$$(9) \qquad \exp[(2n+1)i(z+\tfrac{1}{2}\pi\tau)] = e^{(2n+1)iz}q^{n+\frac{1}{2}}.$$

We may now write from (5) and (9),

$$\theta_1(z+\tfrac{1}{2}\pi\tau) = -i\sum_{n=-\infty}^{\infty}(-1)^n q^{n^2+n+\frac{1}{4}}e^{(2n+1)iz}q^{n+\frac{1}{2}}$$

$$= -iq^{-\frac{1}{4}}\sum_{n=-\infty}^{\infty}(-1)^n q^{(n+1)^2}e^{(2n+1)iz}.$$

In the last summation replace n by $(n-1)$ to obtain

$$\theta_1(z+\tfrac{1}{2}\pi\tau) = -iq^{-\frac{1}{4}}\sum_{n=-\infty}^{\infty}(-1)^{n-1}q^{n^2}e^{(2n-1)iz}$$

$$= iq^{-\frac{1}{4}}e^{-iz}\sum_{n=-\infty}^{\infty}(-1)^n q^{n^2}e^{2niz}.$$

We have thus derived the result

$$(10) \qquad \theta_1(z+\tfrac{1}{2}\pi\tau) = iq^{-\frac{1}{4}}e^{-iz}\theta_4(z).$$

Equations (6) and (9) yield

$$\theta_2(z + \tfrac{1}{2}\pi\tau) = \sum_{n=-\infty}^{\infty} q^{n^2+n+\frac{1}{4}} e^{(2n+1)iz} q^{n+\frac{1}{2}}$$

$$= q^{-\frac{1}{4}} \sum_{n=-\infty}^{\infty} q^{(n+1)^2} e^{(2n+1)iz}$$

$$= q^{-\frac{1}{4}} \sum_{n=-\infty}^{\infty} q^{n^2} e^{(2n-1)iz},$$

from which we get

(11) $$\theta_2(z + \tfrac{1}{2}\pi\tau) = q^{-\frac{1}{4}} e^{-iz} \theta_3(z).$$

Since $\exp[2ni(z + \frac{1}{2}\pi\tau)] = e^{2niz}q^n$, we may obtain from (7)

$$\theta_3(z + \tfrac{1}{2}\pi\tau) = \sum_{n=-\infty}^{\infty} q^{n^2} e^{2niz} q^n = q^{-\frac{1}{4}} \sum_{n=-\infty}^{\infty} q^{(n+\frac{1}{2})^2} e^{2niz},$$

so that we find

(12) $$\theta_3(z + \tfrac{1}{2}\pi\tau) = q^{-\frac{1}{4}} e^{-iz} \theta_2(z).$$

In the same way we derive

(13) $$\theta_4(z + \tfrac{1}{2}\pi\tau) = iq^{-\frac{1}{4}} e^{-iz} \theta_1(z).$$

We now use equations (10)–(13) to obtain the functions of argument $(z + \pi\tau)$. Using first (10) and then (13), we obtain

$$\begin{aligned}\theta_1(z + \pi\tau) &= iq^{-\frac{1}{4}} e^{-i(z+\frac{1}{2}\pi\tau)} \theta_4(z + \tfrac{1}{2}\pi\tau) \\ &= iq^{-\frac{1}{4}} e^{-iz} q^{-\frac{1}{2}} iq^{-\frac{1}{4}} e^{-iz} \theta_1(z),\end{aligned}$$

which yields

(14) $$\theta_1(z + \pi\tau) = -q^{-1} e^{-2iz} \theta_1(z).$$

From (11) and (12) it follows that

$$\begin{aligned}\theta_2(z + \pi\tau) &= q^{-\frac{1}{4}} e^{-i(z+\frac{1}{2}\pi\tau)} \theta_3(z + \tfrac{1}{2}\pi\tau) \\ &= q^{-\frac{1}{4}} e^{-iz} q^{-\frac{1}{2}} \cdot q^{-\frac{1}{4}} e^{-iz} \theta_2(z),\end{aligned}$$

or

(15) $$\theta_2(z + \pi\tau) = q^{-1} e^{-2iz} \theta_2(z).$$

In the same manner we obtain

(16) $$\theta_3(z + \pi\tau) = q^{-1} e^{-2iz} \theta_3(z),$$

(17) $$\theta_4(z + \pi\tau) = -q^{-1} e^{-2iz} \theta_4(z).$$

Examination of equations (14)–(17) show that the theta functions barely missed having another period $\pi\tau$. Indeed certain ratios of theta functions do have that period. We shall eventually build doubly periodic functions from the four theta functions by means of well chosen ratios.

The information contained in equations (1)–(8) of Section 165 and in equations (10)–(17) of this section is tabulated below. This table is one to which we refer almost constantly in this and the next chapter.

$y =$	z	$z + \frac{1}{2}\pi$	$z + \pi$	$z + \frac{1}{2}\pi\tau$	$z + \pi\tau$
$\theta_1(y) =$	$\theta_1(z)$	$\theta_2(z)$	$-\theta_1(z)$	$iq^{-\frac{1}{4}}e^{-iz}\theta_4(z)$	$-q^{-1}e^{-2iz}\theta_1(z)$
$\theta_2(y) =$	$\theta_2(z)$	$-\theta_1(z)$	$-\theta_2(z)$	$q^{-\frac{1}{4}}e^{-iz}\theta_3(z)$	$q^{-1}e^{-2iz}\theta_2(z)$
$\theta_3(y) =$	$\theta_3(z)$	$\theta_4(z)$	$\theta_3(z)$	$q^{-\frac{1}{4}}e^{-iz}\theta_2(z)$	$q^{-1}e^{-2iz}\theta_3(z)$
$\theta_4(y) =$	$\theta_4(z)$	$\theta_3(z)$	$\theta_4(z)$	$iq^{-\frac{1}{4}}e^{-iz}\theta_1(z)$	$-q^{-1}e^{-2iz}\theta_4(z)$

The basic table may be used also to depress, as well as to augment, the argument of a theta function. For instance, since $\theta_4(z)$ is an even function of z,

$$\theta_4(z - \pi\tau) = \theta_4(-z + \pi\tau) = -q^{-1}e^{2iz}\theta_4(-z) = -q^{-1}e^{2iz}\theta_4(z),$$

a result also easily obtained by reading the table in reverse order. We can in these ways obtain a set of results useful to us later:

$$\theta_1(z - \pi) = -\theta_1(z), \qquad \theta_2(z - \pi) = -\theta_2(z),$$
$$\theta_3(z - \pi) = \theta_3(z), \qquad \theta_4(z - \pi) = \theta_4(z),$$
$$\theta_1(z - \pi\tau) = -q^{-1}e^{2iz}\theta_1(z), \qquad \theta_2(z - \pi\tau) = q^{-1}e^{2iz}\theta_2(z),$$
$$\theta_3(z - \pi\tau) = q^{-1}e^{2iz}\theta_3(z), \qquad \theta_4(z - \pi\tau) = -q^{-1}e^{2iz}\theta_4(z).$$

167. Location of zeros. The basic table makes it a simple matter to obtain the zeros of the theta functions. Since

$$(1) \qquad \theta_1(z) = 2\sum_{n=0}^{\infty}(-1)^n q^{(n+\frac{1}{2})^2}\sin(2n + 1)z,$$

we see that $\theta_1(0) = 0$; $z = 0$ is a zero of $\theta_1(z)$. Again from the table it follows that

$$\theta_2(\tfrac{1}{2}\pi) = \theta_2(0 + \tfrac{1}{2}\pi) = -\theta_1(0) = 0,$$

so that $z = \frac{1}{2}\pi$ is a zero of $\theta_2(z)$. Since

$$\theta_3(z + \tfrac{1}{2}\pi\tau) = q^{-\frac{1}{4}}e^{-iz}\theta_2(z),$$

we find that

$$\theta_3(\tfrac{1}{2}\pi + \tfrac{1}{2}\pi\tau) = q^{-\frac{1}{4}}e^{-\frac{1}{2}i\pi}\theta_2(\tfrac{1}{2}\pi) = 0;$$

$z = \frac{1}{2}\pi + \frac{1}{2}\pi\tau$ is a zero of $\theta_3(z)$. Finally, from

$$\theta_4(z + \tfrac{1}{2}\pi\tau) = iq^{-\frac{1}{4}}e^{-iz}\theta_1(z)$$

it follows that

$$\theta_4(\tfrac{1}{2}\pi\tau) = iq^{-\frac{1}{4}}e^{0}\theta_1(0) = 0;$$

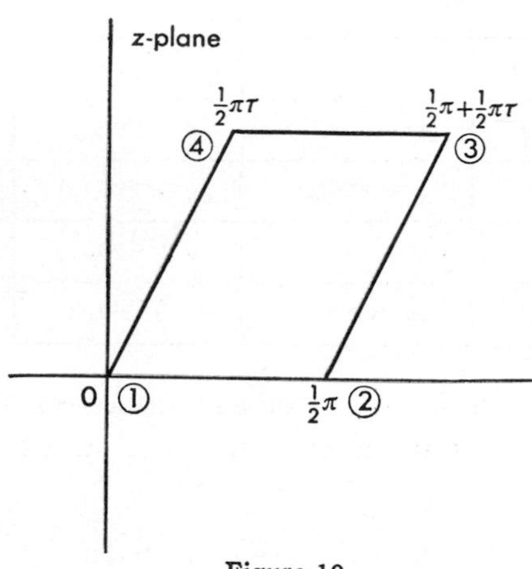

Figure 10

$z = \frac{1}{2}\pi\tau$ is a zero of $\theta_4(z)$.

The zeros obtained so far are exhibited in Figure 10. In the figure an encircled number indicates which theta function possesses that zero. For instance, ③ near $\frac{1}{2}\pi + \frac{1}{2}\pi\tau$ is a reminder that $z = \frac{1}{2}\pi + \frac{1}{2}\pi\tau$ is a zero of $\theta_3(z)$. Note the counter clockwise order of the zeros in the figure.

The fourth and sixth columns of the basic table, page 319, show that, if we add π or $\pi\tau$ to (or subtract π or $\pi\tau$ from) a zero of any theta function, we arrive again at a zero of that function. So far we have shown that if m and n are any integers,

$$\text{A}\qquad \begin{cases} \theta_1(z) = 0 & \text{at } z = m\pi + n\pi\tau, \\ \theta_2(z) = 0 & \text{at } z = \frac{1}{2}\pi + m\pi + n\pi\tau, \\ \theta_3(z) = 0 & \text{at } z = \frac{1}{2}\pi + \frac{1}{2}\pi\tau + m\pi + n\pi\tau, \\ \theta_4(z) = 0 & \text{at } z = \frac{1}{2}\pi\tau + m\pi + n\pi\tau. \end{cases}$$

Our next task is to show that these are the only zeros of the theta functions.

Directly from the basic table we obtain

$$(2)\qquad \theta_k(z + \pi) = \pm\,\theta_k(z); \qquad k = 1, 2, 3, 4,$$

in which the minus sign is to be used for $k = 1, 2$, and the plus sign for $k = 3, 4$. Logarithmic differentiation of the members of (2) yields

$$(3) \qquad \frac{\theta_k'(z+\pi)}{\theta_k(z+\pi)} = \frac{\theta_k'(z)}{\theta_k(z)}; \qquad k = 1, 2, 3, 4.$$

Note that the relation (3) is the same for all pertinent k values.

The basic table also yields

$$(4) \qquad \theta_k(z+\pi\tau) = \pm q^{-1}e^{-2iz}\theta_k(z); \qquad k = 1, 2, 3, 4,$$

in which the plus sign applies for $k = 2, 3$, and the minus sign for $k = 1, 4$. Logarithmic differentiation of the members of (4) gives

$$(5) \qquad \frac{\theta_k'(z+\pi\tau)}{\theta_k(z+\pi\tau)} = -2i + \frac{\theta_k'(z)}{\theta_k(z)}; \qquad k = 1, 2, 3, 4.$$

Let t be any point inside the parallelogram of Figure 10. Consider the parallelogram P shown in Figure 11, that with vertices at t, $t+\pi$, $t+\pi+\pi\tau$, $t+\pi\tau$. Let N_k be the number of zeros of $\theta_k(z)$ inside P. Then, since $\theta_k(z)$ has no poles,

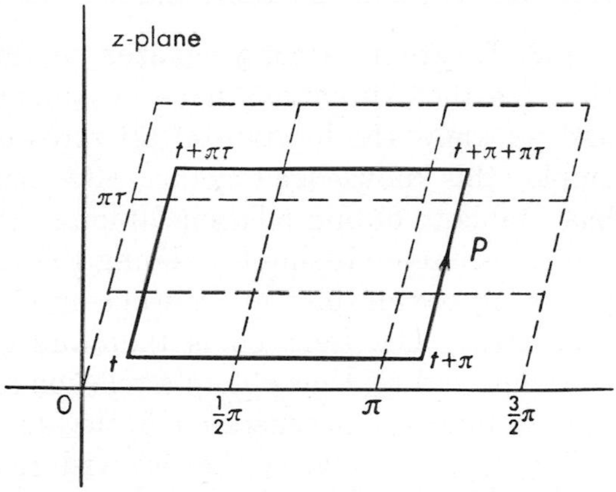

Figure 11

$$(6) \qquad 2\pi i N_k = \int_P \frac{\theta_k'(z)\,dz}{\theta_k(z)},$$

or

$$(7) \qquad 2\pi i N_k = \int_t^{t+\pi} \frac{\theta_k'(z)\,dz}{\theta_k(z)} + \int_{t+\pi}^{t+\pi+\pi\tau} \frac{\theta_k'(z)\,dz}{\theta_k(z)} + \int_{t+\pi+\pi\tau}^{t+\pi\tau} \frac{\theta_k'(z)\,dz}{\theta_k(z)} + \int_{t+\pi\tau}^{t} \frac{\theta_k'(z)\,dz}{\theta_k(z)}.$$

In the second integral in (7) replace z by $(z+\pi)$; in the third integral in (7) replace z by $(z+\pi\tau)$. The result is

$$(8) \qquad 2\pi i N_k = \int_t^{t+\pi} \frac{\theta_k'(z)\,dz}{\theta_k(z)} + \int_t^{t+\pi\tau} \frac{\theta_k'(z+\pi)\,dz}{\theta_k(z+\pi)} + \int_{t+\pi}^{t} \frac{\theta_k'(z+\pi\tau)\,dz}{\theta_k(z+\pi\tau)} + \int_{t+\pi\tau}^{t} \frac{\theta_k'(z)\,dz}{\theta_k(z)}.$$

Next use the relation (3) on the second integrand in (8) and the relation (5) on the third integrand in (8) to obtain

(9) $$2\pi i N_k = \int_t^{t+\pi} \frac{\theta_k'(z)\,dz}{\theta_k(z)} + \int_t^{t+\pi\tau} \frac{\theta_k'(z)\,dz}{\theta_k(z)} + \int_{t+\pi}^{t} \left[-2i + \frac{\theta_k'(z)}{\theta_k(z)}\right] dz + \int_{t+\pi\tau}^{t} \frac{\theta_k'(z)\,dz}{\theta_k(z)}.$$

From (9) it follows that

$$2\pi i N_k = \int_{t+\pi}^{t} -2i dz = 2\pi i,$$

so that $N_k = 1$ for each $k = 1, 2, 3, 4$.

We have thus proved that the zeros tabulated in A are the only zeros of the theta functions and that each zero is a simple zero.

168. Relations among squares of theta functions. We know that the theta functions have no singularities in the finite plane, and we know the location of all zeros of these functions. Let us employ this knowledge together with the properties exhibited in the basic table to obtain relations among the theta functions.

The technique is simple. Using the information described in the preceding paragraph we construct an elliptic function of order less than two. The function is therefore (Theorem 90, page 309) a constant, and we thus obtain a relation among the component parts used to form the degenerate elliptic function.

Suppose, for instance, that we wish to obtain a relation involving $\theta_1^2(z)$. We set up the function

(1) $$\varphi(z) = \frac{N(z)}{\theta_1^2(z)}$$

which we shall force to be an elliptic function of order less than two with period parallelogram C, as shown in Figure 12. The denominator on the right in (1) has a double zero at $z = 0$, and that is its only zero inside C. Furthermore, the basic table, page 319, yields

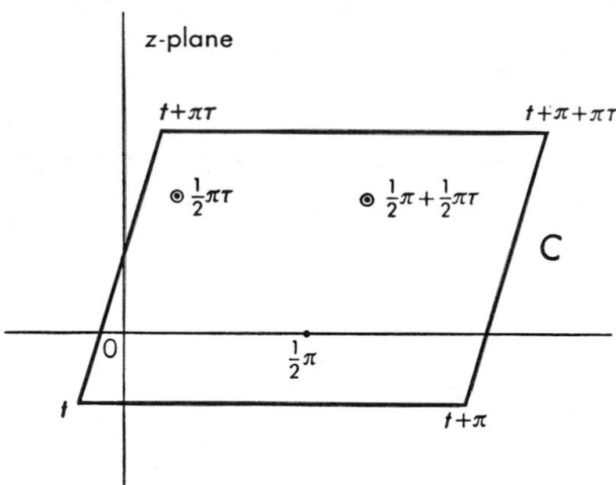

Figure 12

(2) $$\theta_1^2(z+\pi) = \theta_1^2(z),$$

(3) $$\theta_1^2(z+\pi\tau) = q^{-2}e^{-4iz}\theta_1^2(z).$$

We therefore choose the numerator $N(z)$ in (1) so that $N(z)$ will have at least a single zero at $z = 0$ and also possess the properties

$$(4) \qquad N(z + \pi) = N(z),$$

$$(5) \qquad N(z + \pi\tau) = q^{-2}e^{-4iz}N(z).$$

Then $\varphi(z)$ will have the two periods π and $\pi\tau$ and will have at most a simple pole inside the cell C. Thus $\varphi(z)$ will be an elliptic function of order less than two and will be constant.

The basic table shows us that the square of any theta function will have the properties (4) and (5). We form for $N(z)$ a linear combination such as

$$A\theta_2^2(z) + B\theta_3^2(z).$$

Recall the notation $\theta_k = \theta_k(0)$ for $k = 2, 3, 4$. If we choose

$$(6) \qquad N(z) = \theta_3^2\theta_2^2(z) - \theta_2^2\theta_3^2(z),$$

surely $N(z)$ will vanish at $z = 0$. Furthermore, $N(z)$ will have the properties (4) and (5), as we see from the basic table. Hence

$$(7) \qquad \varphi(z) = \frac{\theta_3^2\theta_2^2(z) - \theta_2^2\theta_3^2(z)}{\theta_1^2(z)} = h,$$

a constant, since $\varphi(z)$ is an elliptic function of order less than two.

To evaluate the constant h we use any convenient z. Choose $z = \frac{1}{2}\pi$. Then from the basic table, page 319,

$$\theta_1^2(\tfrac{1}{2}\pi) = \theta_2^2(0) = \theta_2^2,$$

$$\theta_2^2(\tfrac{1}{2}\pi) = 0,$$

$$\theta_3^2(\tfrac{1}{2}\pi) = \theta_4^2(0) = \theta_4^2.$$

Hence

$$h = \frac{0 - \theta_2^2\theta_4^2}{\theta_2^2} = -\theta_4^2.$$

With this value for h, equation (7) may be rewritten as

$$(8) \qquad \theta_4^2\theta_1^2(z) + \theta_3^2\theta_2^2(z) = \theta_2^2\theta_3^2(z).$$

Additional relations are

$$(9) \qquad \theta_3^2\theta_1^2(z) + \theta_4^2\theta_2^2(z) = \theta_2^2\theta_4^2(z),$$

$$(10) \qquad \theta_2^2\theta_1^2(z) + \theta_4^2\theta_3^2(z) = \theta_3^2\theta_4^2(z),$$

$$(11) \qquad \theta_2^2\theta_2^2(z) + \theta_4^2\theta_4^2(z) = \theta_3^2\theta_3^2(z),$$

each of which can be discovered by the process used to get (8). We leave (9) and (10) as exercises and now prove the validity of (11).

Consider the function

$$\varphi_1(z) = \frac{\theta_3^2(\tfrac{1}{2}\pi)\theta_4^2(z) - \theta_4^2(\tfrac{1}{2}\pi)\theta_3^2(z)}{\theta_2^2(z)}. \tag{12}$$

The basic table yields at once

$$\varphi_1(z + \pi) = \varphi_1(z), \qquad \varphi_1(z + \pi\tau) = \varphi_1(z).$$

In the cell C of Figure 12, the function $\varphi_1(z)$ can have no singular point except possibly at $z = \frac{1}{2}\pi$, the zero of $\theta_2(z)$. At $z = \frac{1}{2}\pi$, the denominator in (12) has a double zero and numerator at least a single zero. Hence $\varphi_1(z)$ is an elliptic function of order less than two, and it must be constant. But

$$\varphi_1(\tfrac{1}{2}\pi\tau) = \frac{0 - \theta_4^2(\tfrac{1}{2}\pi)\theta_3^2(\tfrac{1}{2}\pi\tau)}{\theta_2^2(\tfrac{1}{2}\pi\tau)}.$$

From the basic table, page 319, we get

$$\theta_3^2(\tfrac{1}{2}\pi\tau) = \theta_3^2(0 + \tfrac{1}{2}\pi\tau) = q^{-\frac{1}{2}}e^0\theta_2^2(0) = q^{-\frac{1}{2}}\theta_2^2,$$
$$\theta_2^2(\tfrac{1}{2}\pi\tau) = \theta_2^2(0 + \tfrac{1}{2}\pi\tau) = q^{-\frac{1}{2}}e^0\theta_3^2(0) = q^{-\frac{1}{2}}\theta_3^2,$$
$$\theta_4^2(\tfrac{1}{2}\pi) = \theta_3^2(0) = \theta_3^2,$$
$$\theta_3^2(\tfrac{1}{2}\pi) = \theta_4^2(0) = \theta_4^2.$$

Therefore the constant value of $\varphi_1(z)$ is

$$\varphi_1(z) = \varphi_1(\tfrac{1}{2}\pi\tau) = \frac{-\theta_3^2 \cdot q^{-\frac{1}{2}}\theta_2^2}{q^{-\frac{1}{2}}\theta_3^2} = -\theta_2^2$$

and (12) now yields

$$\frac{\theta_4^2\theta_4^2(z) - \theta_3^2\theta_3^2(z)}{\theta_2^2(z)} = -\theta_2^2,$$

which is merely a rearrangement of (11).

From (11) we obtain by putting $z = 0$ the identity

$$\theta_2^4 + \theta_4^4 = \theta_3^4. \tag{13}$$

Now θ_2, θ_3, and θ_4 are functions of the parameter q. From the original definitions of the theta functions we obtain

$$\theta_2 = 2\sum_{n=0}^{\infty} q^{(n+\frac{1}{2})^2} = 2q^{\frac{1}{4}}\sum_{n=0}^{\infty} q^{n^2+n},$$

$$\theta_4 = 1 + 2\sum_{n=1}^{\infty}(-1)^n q^{n^2},$$

$$\theta_3 = 1 + 2\sum_{n=1}^{\infty} q^{n^2},$$

so that (13) may be interpreted as a compact form of the statement that, for $|q| < 1$,

$$16q\left[\sum_{n=0}^{\infty} q^{n^2+n}\right]^4 + \left[1 + 2\sum_{n=1}^{\infty}(-1)^n q^{n^2}\right]^4 = \left[1 + 2\sum_{n=1}^{\infty} q^{n^2}\right]^4.$$

169. Pseudo addition theorems. We shall now derive formulas involving $\theta_k(x + y)$ and $\theta_k(x - y)$ for $k = 1, 2, 3, 4$. The technique used will be much the same as that employed in getting relations among the squares of the theta functions.

Suppose, for instance, that we wish to construct a formula for $\theta_1(x + y)\ \theta_1(x - y)$. We use that product as the denominator of a fraction which we force to become an elliptic function of order less than two when considered as a function of y with x fixed. It is desirable first to determine the periodicity properties and the zeros of the product. Let

$$D_{11}(y) = \theta_1(x + y)\theta_1(x - y).$$

Then, from the basic table, page 319, we find that

$$D_{11}(y + \pi) = \theta_1(x + y + \pi)\theta_1(x - y - \pi) = \theta_1(x + y)\theta_1(x - y),$$

or

(1) $$D_{11}(y + \pi) = D_{11}(y).$$

Also

$$\begin{aligned} D_{11}(y + \pi\tau) &= \theta_1(x + y + \pi\tau)\theta_1(x - y - \pi\tau) \\ &= -q^{-1}e^{-2i(x+y)}\theta_1(x + y)(-q^{-1})e^{2i(x-y)}\theta_1(x - y), \end{aligned}$$

or

(2) $$D_{11}(y + \pi\tau) = q^{-2}e^{-4iy}D_{11}(y).$$

Therefore our denominator $D_{11}(y)$ has periodicity factors unity for period π and $q^{-2}e^{-4iy}$ for quasi-period $\pi\tau$. The square of any individual theta function of y will also have those periodicity factors. We may then choose as our numerator such a combination as

(3) $$A\theta_1^2(y) + B\theta_2^2(y)$$

and thus obtain a fraction with periods π and $\pi\tau$. The A and B in (3) may be any functions of x.

We need next to locate the zeros of our denominator $D_{11}(y)$ and knock out one of them by an appropriate choice of A and B. Since

$$D_{11}(y) = \theta_1(x+y)\theta_1(x-y),$$

we observe that $D_{11}(y)$ has simple zeros at $x+y=0$ and at $x-y=0$. Let us force the numerator (3) to vanish at $y = x$ by choosing $A = \theta_2^2(x)$, $B = -\theta_1^2(x)$. Then the fraction

$$\text{(4)} \qquad \frac{\theta_2^2(x)\theta_1^2(y) - \theta_1^2(x)\theta_2^2(y)}{\theta_1(x+y)\theta_1(x-y)}$$

is doubly periodic with periods π and $\pi\tau$ and has in a cell no singular points except possibly a simple pole at $y = -x$. (Actually the pole at $y = -x$ was accidentally removed along with that at $y = x$.) Then (4) is an elliptic function of order less than two and is therefore constant,

$$\text{(5)} \qquad \frac{\theta_2^2(x)\theta_1^2(y) - \theta_1^2(x)\theta_2^2(y)}{\theta_1(x+y)\theta_1(x-y)} = C.$$

The constant C might depend upon x, of course. To evaluate C, let us put $y = 0$ and thus obtain

$$C = \frac{0 - \theta_1^2(x)\theta_2^2}{\theta_1(x)\theta_1(x)} = -\theta_2^2.$$

Insertion of the value of C into (5) leads us to the desired relation,

$$\text{(6)} \qquad \theta_2^2\theta_1(x+y)\theta_1(x-y) = \theta_1^2(x)\theta_2^2(y) - \theta_2^2(x)\theta_1^2(y).$$

We may move from (6) to other expressions for the product $\theta_1(x+y)\theta_1(x-y)$ by using on the right the various identities involving squares of theta functions, equations (8), (9), (10), (11) of Section 168. As an example, the known identity

$$\theta_4^2\theta_1^2(z) + \theta_3^2\theta_2^2(z) = \theta_2^2\theta_3^2(z),$$

used first with argument x, then with argument y, permits us to rewrite (6) as

$$\begin{aligned}\theta_4^2\theta_2^2\theta_1(x+y)\theta_1(x-y) &\\ = \theta_2^2(y)[\theta_2^2\theta_3^2(x) - \theta_3^2\theta_2^2(x)] &- \theta_2^2(x)[\theta_2^2\theta_3^2(y) - \theta_3^2\theta_2^2(y)]\\ = \theta_2^2\theta_2^2(y)\theta_3^2(x) &- \theta_2^2\theta_2^2(x)\theta_3^2(y).\end{aligned}$$

We thus obtain

$$\theta_4^2\theta_1(x+y)\theta_1(x-y) = \theta_2^2(y)\theta_3^2(x) - \theta_2^2(x)\theta_3^2(y), \tag{7}$$

a result also easily discovered by the method we used to get (6) in the first place.

The reader can easily show that for each $k = 1, 2, 3, 4$, the product

$$D_{kk}(y) = \theta_k(x+y)\theta_k(x-y)$$

has the properties

$$D_{kk}(y+\pi) = D_{kk}(y), \qquad D_{kk}(y+\pi\tau) = q^{-2}e^{-4iy}D_{kk}(y).$$

The derivation of the consequent formulas for the various products $D_{kk}(y)$ is left for exercises at the end of this chapter.

Let us next obtain one formula in which two different theta functions, one of argument $(x+y)$ and one of argument $(x-y)$, are involved. Let

$$D_{14}(y) = \theta_1(x+y)\theta_4(x-y).$$

Then the basic table, page 319, yields the results

$$D_{14}(y+\pi) = -D_{14}(y),$$
$$D_{14}(y+\pi\tau) = q^{-2}e^{-4iy}D_{14}(y).$$

Now the products $\theta_2(y)\ \theta_3(y)$ and $\theta_1(y)\ \theta_4(y)$ behave in that same way. That is,

$$\theta_2(y+\pi)\theta_3(y+\pi) = -\theta_2(y)\theta_3(y),$$
$$\theta_1(y+\pi)\theta_4(y+\pi) = -\theta_1(y)\theta_4(y),$$
$$\theta_2(y+\pi\tau)\theta_3(y+\pi\tau) = q^{-2}e^{-4iy}\theta_2(y)\theta_3(y),$$
$$\theta_1(y+\pi\tau)\theta_4(y+\pi\tau) = q^{-2}e^{-4iy}\theta_1(y)\theta_4(y).$$

The fraction

$$\frac{A\theta_2(y)\theta_3(y) + B\theta_1(y)\theta_4(y)}{\theta_1(x+y)\theta_4(x-y)} \tag{8}$$

has periods π and $\pi\tau$ in its argument y, as long as A and B are independent of y. The denominator in (8) has simple zeros at $x+y=0$ and at $x-y=\frac{1}{2}\pi\tau$. By choosing $A=\theta_1(x)\theta_4(x)$ and $B=\theta_2(x)\theta_3(x)$, we make the numerator of (8) vanish at $y=-x$. Then

$$\frac{\theta_1(x)\theta_4(x)\theta_2(y)\theta_3(y) + \theta_2(x)\theta_3(x)\theta_1(y)\theta_4(y)}{\theta_1(x+y)\theta_4(x-y)} = C, \tag{9}$$

where C is independent of y because the left member of (9) is an elliptic function of order less than two. The value of C in (9) may be found by using $y = 0$:

$$C = \frac{\theta_1(x)\theta_4(x)\theta_2\theta_3 + 0}{\theta_1(x)\theta_4(x)} = \theta_2\theta_3.$$

We have thus discovered the identity

$$(10)\quad \theta_2\theta_3\theta_1(x+y)\theta_4(x-y) = \theta_1(x)\theta_2(y)\theta_3(y)\theta_4(x) + \theta_1(y)\theta_2(x)\theta_3(x)\theta_4(y).$$

Similar results for other products $\theta_k(x + y)\theta_h(x - y)$ will be found in the exercises.

An important special case of (10) is obtained by choosing $y = x$. The result is

$$(11)\qquad \theta_2\theta_3\theta_4\theta_1(2x) = 2\theta_1(x)\theta_2(x)\theta_3(x)\theta_4(x),$$

which may be called a duplication formula for $\theta_1(x)$. Similar results for the other theta functions appear in the exercises.

170. Relation to the heat equation. The simple (one dimensional) heat equation, or equation of diffusion,

$$(1)\qquad \frac{\partial u}{\partial t} = h^2 \frac{\partial^2 u}{\partial x^2}$$

appears in many phases of applied mathematics. The four theta functions, when considered as functions of z and τ, satisfy an equation of the form (1).

At the start of this chapter we defined the theta functions by series. As an example, consider

$$(2)\qquad \theta_4(z, q) = 1 + 2\sum_{n=1}^{\infty} (-1)^n q^{n^2} \cos 2nz.$$

With $q = \exp(\pi i\tau)$, we write (2) in the form

$$(3)\qquad \theta_4(z|\tau) = 1 + 2\sum_{n=1}^{\infty} (-1)^n \exp(n^2\pi i\tau) \cos 2nz.$$

From (3) we obtain

$$\frac{\partial^2}{\partial z^2}\theta_4(z|\tau) = 2\sum_{n=1}^{\infty} (-1)^n(-4n^2) \exp(n^2\pi i\tau) \cos 2nz$$

and

$$\frac{\partial}{\partial \tau}\theta_4(z|\tau) = 2\sum_{n=1}^{\infty} (-1)^n(\pi i n^2) \exp(n^2\pi i\tau) \cos 2nz.$$

Hence

$$(4)\qquad \frac{\partial}{\partial\tau}\theta_4(z\,|\,\tau) = -\frac{1}{4}\pi i\frac{\partial^2}{\partial z^2}\theta_4(z\,|\,\tau).$$

Indeed, examination of the series definitions of the theta functions shows that equation (4) is still valid when the subscript 4 is replaced by 1, 2, or 3. That is,

$$(5)\qquad \frac{\partial}{\partial\tau}\theta_k(z\,|\,\tau) = -\frac{1}{4}\pi i\frac{\partial^2}{\partial z^2}\theta_k(z\,|\,\tau);\qquad k = 1, 2, 3, 4.$$

171. The relation $\theta'_1 = \theta_2\theta_3\theta_4$. The values θ_2, θ_3, θ_4 of three theta functions at $z = 0$ have already entered our work in many places. We define θ_1' as the value of $\theta_1'(z)$ at $z = 0$. The four functions of q alone, θ_1', θ_2, θ_3, θ_4 are connected by the remarkably neat relation

$$(1)\qquad \theta_1' = \theta_2\theta_3\theta_4$$

which we now proceed to prove.

Equation (11) of Section 169 gives us

$$(2)\qquad \theta_2\theta_3\theta_4\theta_1(2z) = 2\theta_1(z)\theta_2(z)\theta_3(z)\theta_4(z).$$

In (2) take logarithms of each member and differentiate throughout with respect to z to obtain

$$(3)\qquad \frac{2\theta_1'(2z)}{\theta_1(2z)} = \frac{\theta_1'(z)}{\theta_1(z)} + \sum_{k=2}^{4}\frac{\theta_k'(z)}{\theta_k(z)}.$$

Differentiate each member of (3) with respect to z, thus arriving at

$$\frac{4\theta_1''(2z)}{\theta_1(2z)} - 4\left[\frac{\theta_1'(2z)}{\theta_1(2z)}\right]^2 = \frac{\theta_1''(z)}{\theta_1(z)} - \left[\frac{\theta_1'(z)}{\theta_1(z)}\right]^2 + \sum_{k=2}^{4}\left[\frac{\theta_k''(z)}{\theta_k(z)} - \left\{\frac{\theta_k'(z)}{\theta_k(z)}\right\}^2\right],$$

or

$$(4)\qquad \frac{4\theta_1''(2z)}{\theta_1(2z)} - \frac{\theta_1''(z)}{\theta_1(z)} - \left[4\left\{\frac{\theta_1'(2z)}{\theta_1(2z)}\right\}^2 - \left\{\frac{\theta_1'(z)}{\theta_1(z)}\right\}^2\right] = \sum_{k=2}^{4}\left[\frac{\theta_k''(z)}{\theta_k(z)} - \left\{\frac{\theta_k'(z)}{\theta_k(z)}\right\}^2\right].$$

We wish to let $z \to 0$ in (4). The functions $\theta_2(z)$, $\theta_3(z)$, $\theta_4(z)$ are even functions of z; the function $\theta_1(z)$ is an odd function of z. Thus the limits of the terms on the right in (4) are easily found, but the terms on the left yield indeterminate forms. We shall attempt to simplify the proof by employing three lemmas.

Lemma 13. $$\lim_{z\to 0} \frac{\theta_1(2z)}{\theta_1(z)} = 2.$$

Lemma 14. $$\lim_{z\to 0} \frac{\theta_1''(z)}{\theta_1(z)} = \frac{\theta_1'''(0)}{\theta_1'(0)}.$$

Lemma 15.

$$\lim_{z\to 0} \left[4\left\{\frac{\theta_1'(2z)}{\theta_1(2z)}\right\}^2 - \left\{\frac{\theta_1'(z)}{\theta_1(z)}\right\}^2 \right] = \frac{2\theta_1'''(0)}{\theta_1'(0)}.$$

Lemma 13 follows at once from equation (2) of this section. Lemma 14 is a consequence of a single application of L'Hôpital's rule.

To obtain the result in Lemma 15, note first that

$$4\left\{\frac{\theta_1'(2z)}{\theta_1(2z)}\right\}^2 - \left\{\frac{\theta_1'(z)}{\theta_1(z)}\right\}^2 = \frac{4\theta_1^2(z)[\theta_1'(2z)]^2 - \theta_1^2(2z)[\theta_1'(z)]^2}{\theta_1^2(z)\theta_1^2(2z)}$$
$$= ABC,$$

where

$$A = \frac{\theta_1(z)}{\theta_1(2z)},$$

$$B = \frac{2\theta_1(z)\theta_1'(2z) + \theta_1(2z)\theta_1'(z)}{\theta_1(2z)},$$

$$C = \frac{2\theta_1(z)\theta_1'(2z) - \theta_1(2z)\theta_1'(z)}{\theta_1^3(z)}.$$

By Lemma 13, $\lim_{z\to 0} A = \frac{1}{2}$. Using Lemma 13 again we obtain

$$\lim_{z\to 0} B = 2 \cdot \tfrac{1}{2}\theta_1'(0) + \theta_1'(0) = 2\theta_1'(0).$$

Finally, with the aid of L'Hôpital's rule, we get

$$\lim_{z\to 0} C = \lim_{z\to 0} \frac{4\theta_1(z)\theta_1''(2z) - \theta_1(2z)\theta_1''(z)}{3\theta_1^2(z)\theta_1'(z)}$$

$$= \frac{1}{3\theta_1'(0)} \lim_{z\to 0} \left[\frac{4\theta_1''(2z)}{\theta_1(2z)} \cdot \frac{\theta_1(2z)}{\theta_1(z)} - \frac{\theta_1(2z)}{\theta_1(z)} \cdot \frac{\theta_1''(z)}{\theta_1(z)} \right].$$

Hence, employing Lemmas 13 and 14, we find that

$$\lim_{z\to 0} C = \frac{1}{3\theta_1'(0)}\left[\frac{4\theta_1'''(0)}{\theta_1'(0)}\cdot 2 - 2\cdot\frac{\theta_1'''(0)}{\theta_1'(0)}\right] = \frac{2\theta_1'''(0)}{[\theta_1'(0)]^2}.$$

Since the left member in Lemma 15 is the product of the limits of A, B, and C, it equals

$$\tfrac{1}{2}\cdot 2\theta_1'(0)\frac{2\theta_1'''(0)}{[\theta_1'(0)]^2} = \frac{2\theta_1'''(0)}{\theta_1'(0)},$$

as stated in the lemma.

We return now to equation (4); let $z \to 0$ and use Lemmas 14 and 15 to obtain

$$\frac{4\theta_1'''(0)}{\theta_1'(0)} - \frac{\theta_1'''(0)}{\theta_1'(0)} - \frac{2\theta_1'''(0)}{\theta_1'(0)} = \sum_{k=2}^{4}\frac{\theta_k''(0)}{\theta_k(0)},$$

or

$$\frac{\theta_1'''(0)}{\theta_1'(0)} = \sum_{k=2}^{4}\frac{\theta_k''(0)}{\theta_k(0)}. \tag{5}$$

In Section 170 we found that

$$\frac{\partial}{\partial\tau}\theta_k(z\,|\,\tau) = -\tfrac{1}{4}\pi i\frac{\partial^2}{\partial z^2}\theta_k(z\,|\,\tau); \qquad k = 1, 2, 3, 4. \tag{6}$$

From (6) we also get

$$\frac{\partial}{\partial\tau}\frac{\partial}{\partial z}\theta_1(z\,|\,\tau) = -\tfrac{1}{4}\pi i\frac{\partial^3}{\partial z^3}\theta_1(z\,|\,\tau). \tag{7}$$

Because of (6) with $k = 2, 3, 4$, and of (7), equation (5) may be put in the form

$$\frac{\frac{\partial}{\partial\tau}\theta_1'(0\,|\,\tau)}{\theta_1'(0\,|\,\tau)} = \sum_{k=2}^{4}\frac{\frac{\partial}{\partial\tau}\theta_k(0\,|\,\tau)}{\theta_k(0\,|\,\tau)}. \tag{8}$$

Now integration with respect to τ leads us to the result

$$\theta_1'(0\,|\,\tau) = E\theta_2(0\,|\,\tau)\theta_3(0\,|\,\tau)\theta_4(0\,|\,\tau),$$

or

$$\theta_1' = E\theta_2\theta_3\theta_4, \tag{9}$$

in which θ_2, θ_3, θ_4, and θ_1' are functions of $q = \exp(\pi i\tau)$, but E is constant. In (9) insert a factor $q^{-\frac{1}{4}}$ on each side to write

$$q^{-\frac{1}{4}}\theta_1' = E(q^{-\frac{1}{4}}\theta_2)\theta_3\theta_4,$$

and then let $q \to 0$. From the definitions of the theta functions we see that

$$\lim_{q \to 0} (q^{-\frac{1}{4}}\theta_1') = \lim_{q \to 0} (q^{-\frac{1}{4}}\theta_2) = 2$$

and

$$\lim_{q \to 0} \theta_3 = \lim_{q \to 0} \theta_4 = 1.$$

Hence $E = 1$, which concludes the proof that

$$(1) \qquad \theta_1' = \theta_2\theta_3\theta_4.$$

Equation (1) is interesting in itself, but it will also prove of value to us in future developments.

172. Infinite products. Since we know all the zeros of the theta functions, it is natural to seek infinite product representations for those functions. From any one such representation the others will follow by using properties from the basic table, page 319. It turns out that either $\theta_3(z)$ or $\theta_4(z)$ furnishes a simple starting point.

Consider $\theta_4(z)$. Its zeros are at

$$z = \tfrac{1}{2}\pi\tau + n\pi\tau + m\pi,$$

for integral n and m. We wish to form an infinite product which vanishes for

$$(1) \qquad e^{2iz} = \exp(\pi i\tau + 2n\pi i\tau + 2m\pi i) = q^{2n+1},$$

for every integer n. Since we wish our product index to run from 1 to ∞, we separate the zeros in (1) into

$$(2) \qquad e^{2iz} = q^{2n-1} \qquad \text{and} \qquad e^{2iz} = q^{-(2n-1)},$$

for n integral, $n \geqq 1$. With these zeros in mind we form the product

$$(3) \qquad f(z) = \prod_{n=1}^{\infty} [(1 - q^{2n-1}e^{2iz})(1 - q^{2n-1}e^{-2iz})],$$

which is absolutely convergent for $|q| < 1$. The function $f(z)$ has exactly the zeros possessed by $\theta_4(z)$.

In order to employ our customary technique, we recall that $\theta_4(z)$ has the periodicity properties

$$(4) \qquad \theta_4(z + \pi) = \theta_4(z),$$

$$(5) \qquad \theta_4(z + \pi\tau) = -q^{-1}e^{-2iz}\theta_4(z),$$

and seek corresponding information about $f(z)$.

Since $e^{2i(z+\pi)} = e^{2iz}$, we get

$$f(z+\pi) = f(z). \tag{6}$$

Now

$$f(z+\pi\tau) = \prod_{n=1}^{\infty} [(1 - q^{2n-1}e^{2iz+2i\pi\tau})(1 - q^{2n-1}e^{-2iz-2i\pi\tau})]$$

$$= \prod_{n=1}^{\infty} [(1 - q^{2n+1}e^{2iz})(1 - q^{2n-3}e^{-2iz})].$$

Hence

$$f(z+\pi\tau) = \frac{1 - q^{-1}e^{-2iz}}{1 - qe^{2iz}} f(z)$$

$$= q^{-1}e^{-2iz}\frac{qe^{2iz} - 1}{1 - qe^{2iz}} f(z),$$

so that

$$f(z+\pi\tau) = -q^{-1}e^{-2iz}f(z). \tag{7}$$

Therefore $\theta_4(z)/f(z)$ is an elliptic function of order less than two and must be constant. It is customary to call the constant G. It is a function of q but not of z. We have shown that

$$\theta_4(z) = G\prod_{n=1}^{\infty} [(1 - q^{2n-1}e^{2iz})(1 - q^{2n-1}e^{-2iz})], \tag{8}$$

which may also be written

$$\theta_4(z) = G\prod_{n=1}^{\infty} (1 - 2q^{2n-1}\cos 2z + q^{4n-2}). \tag{9}$$

The value of G will be obtained in the next section.

We now wish to find infinite products for the other theta functions. From the basic table, page 319, we have

$$\theta_3(z) = \theta_4(z + \tfrac{1}{2}\pi),$$

so that (9) leads to

$$\theta_3(z) = G\prod_{n=1}^{\infty} (1 + 2q^{2n-1}\cos 2z + q^{4n-2}). \tag{10}$$

Again from the basic table we obtain

$$\theta_1(z) = -iqe^{\frac{1}{4}iz}\theta_4(z + \tfrac{1}{2}\pi\tau),$$

which, in view of (8), leads to

$$\theta_1(z) = -iq^{\frac{1}{4}}e^{iz}G\prod_{n=1}^{\infty}(1 - q^{2n-1}e^{2iz+i\pi\tau})\prod_{n=1}^{\infty}(1 - q^{2n-1}e^{-2iz-i\pi\tau})$$

$$= -iq^{\frac{1}{4}}e^{iz}G\prod_{n=1}^{\infty}(1 - q^{2n}e^{2iz})\prod_{n=1}^{\infty}(1 - q^{2n-2}e^{-2iz}).$$

A shift of index from n to $(n + 1)$ in the last infinite product yields

$$\theta_1(z) = -iq^{\frac{1}{4}}e^{iz}G\prod_{n=1}^{\infty}(1 - q^{2n}e^{2iz})\prod_{n=0}^{\infty}(1 - q^{2n}e^{-2iz}).$$

Therefore

$$\theta_1(z) = -iq^{\frac{1}{4}}e^{iz}(1 - e^{-2iz})G\prod_{n=1}^{\infty}[(1 - q^{2n}e^{2iz})(1 - q^{2n}e^{-2iz})].$$

Now $-ie^{iz}(1 - e^{-2iz}) = -i(e^{iz} - e^{-iz}) = 2\sin z$. Hence

(11) $$\theta_1(z) = 2q^{\frac{1}{4}}G\sin z\prod_{n=1}^{\infty}[(1 - q^{2n}e^{2iz})(1 - q^{2n}e^{-2iz})],$$

which may also be written in the form

(12) $$\theta_1(z) = 2q^{\frac{1}{4}}G\sin z\prod_{n=1}^{\infty}(1 - 2q^{2n}\cos 2z + q^{4n}).$$

The basic table yields $\theta_2(z) = \theta_1(z + \frac{1}{2}\pi)$, with the aid of which (12) yields

(13) $$\theta_2(z) = 2q^{\frac{1}{4}}G\cos z\prod_{n=1}^{\infty}(1 + 2q^{2n}\cos 2z + q^{4n}).$$

Equations (9), (10), (12), and (13) are the desired infinite product forms for the theta functions except that we have yet to determine how G depends upon q.

173. The value of G. We shall find G from the known (page 332) relation

(1) $$\theta_1' = \theta_2\theta_3\theta_4$$

and the infinite products obtained in the preceding section. By differentiating with respect to z each member of the identity

$$\theta_1(z) = 2q^{\frac{1}{4}}G\sin z\prod_{n=1}^{\infty}(1 - 2q^{2n}\cos 2z + q^{4n})$$

and then using $z = 0$, we arrive at

(2) $$\theta_1' = 2q^{\frac{1}{4}}G\prod_{n=1}^{\infty}(1 - q^{2n})^2.$$

From the identities

$$\theta_2(z) = 2q^{\frac{1}{4}}G\cos z\prod_{n=1}^{\infty}(1 + 2q^{2n}\cos 2z + q^{4n}),$$

$$\theta_3(z) = G\prod_{n=1}^{\infty}(1 + 2q^{2n-1}\cos 2z + q^{4n-2}),$$

$$\theta_4(z) = G\prod_{n=1}^{\infty}(1 - 2q^{2n-1}\cos 2z + q^{4n-2}),$$

it follows that

(3) $$\theta_2 = 2q^{\frac{1}{4}}G\prod_{n=1}^{\infty}(1 + q^{2n})^2,$$

(4) $$\theta_3 = G\prod_{n=1}^{\infty}(1 + q^{2n-1})^2,$$

(5) $$\theta_4 = G\prod_{n=1}^{\infty}(1 - q^{2n-1})^2.$$

Since $\theta_1' = \theta_2\theta_3\theta_4$, we are led to the relation

(6) $$\prod_{n=1}^{\infty}(1 - q^{2n})^2 = G^2\prod_{n=1}^{\infty}(1 + q^{2n})^2\prod_{n=1}^{\infty}(1 + q^{2n-1})^2\prod_{n=1}^{\infty}(1 - q^{2n-1})^2,$$

in which each of the infinite products is absolutely convergent because of the restriction $|q| < 1$.

Since the set of even positive integers plus the set of odd positive integers is the set of all positive integers,

$$\prod_{n=1}^{\infty}(1 + q^{2n})^2\prod_{n=1}^{\infty}(1 + q^{2n-1})^2 = \prod_{n=1}^{\infty}(1 + q^{n})^2.$$

Equation (6) may now be put in the form

$$\prod_{n=1}^{\infty}(1 - q^{2n})^2 = G^2\prod_{n=1}^{\infty}(1 + q^{n})^2\prod_{n=1}^{\infty}(1 - q^{2n-1})^2,$$

from which

$$\prod_{n=1}^{\infty}(1 - q^{2n})^4 = G^2\prod_{n=1}^{\infty}(1 + q^{n})^2\prod_{n=1}^{\infty}(1 - q^{2n-1})^2\prod_{n=1}^{\infty}(1 - q^{2n})^2.$$

But

$$\prod_{n=1}^{\infty}(1 - q^{2n-1})^2\prod_{n=1}^{\infty}(1 - q^{2n})^2 = \prod_{n=1}^{\infty}(1 - q^{n})^2.$$

Hence we have

$$\prod_{n=1}^{\infty}(1-q^{2n})^4 = G^2\prod_{n=1}^{\infty}(1+q^n)^2\prod_{n=1}^{\infty}(1-q^n)^2,$$

or

$$(7)\qquad \prod_{n=1}^{\infty}(1-q^{2n})^4 = G^2\prod_{n=1}^{\infty}(1-q^{2n})^2.$$

But $|q| < 1$, so there are no zero factors in (7). Therefore

$$(8)\qquad \prod_{n=1}^{\infty}(1-q^{2n})^2 = G^2.$$

For $q = 0$, $G = +1$, as may be seen from

$$(4)\qquad \theta_3 = G\prod_{n=1}^{\infty}(1+q^{2n-1})^2$$

and the series

$$(9)\qquad \theta_3 = 1 + 2\sum_{n=1}^{\infty} q^{n^2}$$

obtained early in this chapter. We may therefore conclude from (8) in this section that

$$(10)\qquad G = \prod_{n=1}^{\infty}(1-q^{2n}).$$

EXERCISES

1. Show that $\theta_1' = 2q^{\frac{1}{4}}G^3$.

2. The following formulas are drawn from Section 168. Derive (9) and (10):

$$(8)\qquad \theta_4^2\theta_1^2(z) + \theta_3^2\theta_2^2(z) = \theta_2^2\theta_3^2(z),$$

$$(9)\qquad \theta_3^2\theta_1^2(z) + \theta_4^2\theta_2^2(z) = \theta_2^2\theta_4^2(z),$$

$$(10)\qquad \theta_2^2\theta_1^2(z) + \theta_4^2\theta_3^2(z) = \theta_3^2\theta_4^2(z),$$

$$(11)\qquad \theta_2^2\theta_2^2(z) + \theta_4^2\theta_4^2(z) = \theta_3^2\theta_3^2(z).$$

3. Use (8) and (10) of Ex. 2 and the equation $\theta_2^4 + \theta_4^4 = \theta_3^4$ to show that

$$\theta_1^4(z) + \theta_3^4(z) = \theta_2^4(z) + \theta_4^4(z).$$

4. The first of the following relations was derived in Section 169. Obtain the other three by using appropriate changes of variable and the basic table, page 319. For example, change x to $(x + \frac{1}{2}\pi)$, or x to $(x + \frac{1}{2}\pi\tau)$, etc.

$$\theta_2^2\theta_1(x+y)\theta_1(x-y) = \theta_1^2(x)\theta_2^2(y) - \theta_2^2(x)\theta_1^2(y),$$

$$\theta_2^2\theta_2(x+y)\theta_2(x-y) = \theta_2^2(x)\theta_2^2(y) - \theta_1^2(x)\theta_1^2(y),$$

$$\theta_2^2\theta_3(x+y)\theta_3(x-y) = \theta_3^2(x)\theta_2^2(y) + \theta_4^2(x)\theta_1^2(y),$$

$$\theta_2^2\theta_4(x+y)\theta_4(x-y) = \theta_4^2(x)\theta_2^2(y) + \theta_3^2(x)\theta_1^2(y).$$

5. Use the identities in Ex. 2 and the relation $\theta_2^4 + \theta_4^4 = \theta_3^4$ to transform the first relation of Ex. 4 into the first relation below. Then obtain the remaining three relations with the aid of the basic table, page 319.

$$\theta_2^2\theta_1(x + y)\theta_1(x - y) = \theta_4^2(x)\theta_3^2(y) - \theta_3^2(x)\theta_4^2(y),$$

$$\theta_2^2\theta_2(x + y)\theta_2(x - y) = \theta_3^2(x)\theta_3^2(y) - \theta_4^2(x)\theta_4^2(y),$$

$$\theta_2^2\theta_3(x + y)\theta_3(x - y) = \theta_2^2(x)\theta_3^2(y) + \theta_1^2(x)\theta_4^2(y),$$

$$\theta_2^2\theta_4(x + y)\theta_4(x - y) = \theta_1^2(x)\theta_3^2(y) + \theta_2^2(x)\theta_4^2(y).$$

6. Use equation (10) of Ex. 2 and the first relation of Ex. 5 to obtain the first relation below; then use the basic table to get the remaining three results.

$$\theta_3^2\theta_1(x + y)\theta_1(x - y) = \theta_1^2(x)\theta_3^2(y) - \theta_3^2(x)\theta_1^2(y),$$

$$\theta_3^2\theta_2(x + y)\theta_2(x - y) = \theta_2^2(x)\theta_3^2(y) - \theta_4^2(x)\theta_1^2(y),$$

$$\theta_3^2\theta_3(x + y)\theta_3(x - y) = \theta_3^2(x)\theta_3^2(y) + \theta_1^2(x)\theta_1^2(y),$$

$$\theta_3^2\theta_4(x + y)\theta_4(x - y) = \theta_4^2(x)\theta_3^2(y) + \theta_2^2(x)\theta_1^2(y).$$

7. Use the first relation of Ex. 6 and identities from Ex. 2 to obtain the first relation below. Derive the other three relations from the first one.

$$\theta_3^2\theta_1(x + y)\theta_1(x - y) = \theta_4^2(x)\theta_2^2(y) - \theta_2^2(x)\theta_4^2(y),$$

$$\theta_3^2\theta_2(x + y)\theta_2(x - y) = \theta_3^2(x)\theta_2^2(y) - \theta_1^2(x)\theta_4^2(y),$$

$$\theta_3^2\theta_3(x + y)\theta_3(x - y) = \theta_2^2(x)\theta_2^2(y) + \theta_4^2(x)\theta_4^2(y),$$

$$\theta_3^2\theta_4(x + y)\theta_4(x - y) = \theta_1^2(x)\theta_2^2(y) + \theta_3^2(x)\theta_4^2(y).$$

8. Use equation (10) of Ex. 2 and the first relation of Ex. 6 to obtain the first relation below. Derive the others with the aid of the basic table, page 319.

$$\theta_4^2\theta_1(x + y)\theta_1(x - y) = \theta_1^2(x)\theta_4^2(y) - \theta_4^2(x)\theta_1^2(y),$$

$$\theta_4^2\theta_2(x + y)\theta_2(x - y) = \theta_2^2(x)\theta_4^2(y) - \theta_3^2(x)\theta_1^2(y),$$

$$\theta_4^2\theta_3(x + y)\theta_3(x - y) = \theta_3^2(x)\theta_4^2(y) - \theta_2^2(x)\theta_1^2(y),$$

$$\theta_4^2\theta_4(x + y)\theta_4(x - y) = \theta_4^2(x)\theta_4^2(y) - \theta_1^2(x)\theta_1^2(y).$$

9. The first relation below was derived in Section 169. Obtain the other three.

$$\theta_4^2\theta_1(x + y)\theta_1(x - y) = \theta_3^2(x)\theta_2^2(y) - \theta_2^2(x)\theta_3^2(y),$$

$$\theta_4^2\theta_2(x + y)\theta_2(x - y) = \theta_4^2(x)\theta_2^2(y) - \theta_1^2(x)\theta_3^2(y),$$

$$\theta_4^2\theta_3(x + y)\theta_3(x - y) = \theta_4^2(x)\theta_3^2(y) - \theta_1^2(x)\theta_2^2(y),$$

$$\theta_4^2\theta_4(x + y)\theta_4(x - y) = \theta_3^2(x)\theta_3^2(y) - \theta_2^2(x)\theta_2^2(y).$$

10. Use the method of Section 169 together with the basic table, page 319, to derive the following results, one of which was obtained in Section 169.

$$\theta_3\theta_4\theta_1(x + y)\theta_2(x - y) = \theta_1(x)\theta_2(x)\theta_3(y)\theta_4(y) + \theta_1(y)\theta_2(y)\theta_3(x)\theta_4(x),$$

$$\theta_2\theta_4\theta_1(x + y)\theta_3(x - y) = \theta_1(x)\theta_2(y)\theta_3(x)\theta_4(y) + \theta_1(y)\theta_2(x)\theta_3(y)\theta_4(x),$$

$$\theta_2\theta_3\theta_1(x + y)\theta_4(x - y) = \theta_1(x)\theta_2(y)\theta_3(y)\theta_4(x) + \theta_1(y)\theta_2(x)\theta_3(x)\theta_4(y),$$

$$\theta_2\theta_3\theta_2(x + y)\theta_3(x - y) = \theta_2(x)\theta_3(x)\theta_2(y)\theta_3(y) - \theta_1(x)\theta_4(x)\theta_1(y)\theta_4(y),$$

$$\theta_2\theta_4\theta_2(x + y)\theta_4(x - y) = \theta_2(x)\theta_4(x)\theta_2(y)\theta_4(y) - \theta_1(x)\theta_3(x)\theta_1(y)\theta_3(y),$$

$$\theta_3\theta_4\theta_3(x + y)\theta_4(x - y) = \theta_3(x)\theta_4(x)\theta_3(y)\theta_4(y) - \theta_1(x)\theta_2(x)\theta_1(y)\theta_2(y).$$

11. Use the results in Exs. 4–10 above to show that

$$\theta_2^3\theta_2(2x) = \theta_2^4(x) - \theta_1^4(x) = \theta_3^4(x) - \theta_4^4(x),$$

$$\theta_3^3\theta_3(2x) = \theta_1^4(x) + \theta_3^4(x) = \theta_2^4(x) + \theta_4^4(x),$$

$$\theta_4^3\theta_4(2x) = \theta_4^4(x) - \theta_1^4(x) = \theta_3^4(x) - \theta_2^4(x),$$

$$\theta_3^2\theta_2\theta_2(2x) = \theta_2^2(x)\theta_3^2(x) - \theta_1^2(x)\theta_4^2(x),$$

$$\theta_4^2\theta_2\theta_2(2x) = \theta_2^2(x)\theta_4^2(x) - \theta_1^2(x)\theta_3^2(x),$$

$$\theta_2^2\theta_3\theta_3(2x) = \theta_2^2(x)\theta_3^2(x) + \theta_1^2(x)\theta_4^2(x),$$

$$\theta_4^2\theta_3\theta_3(2x) = \theta_3^2(x)\theta_4^2(x) - \theta_1^2(x)\theta_2^2(x),$$

$$\theta_2^2\theta_4\theta_4(2x) = \theta_1^2(x)\theta_3^2(x) + \theta_2^2(x)\theta_4^2(x).$$

$$\theta_3^2\theta_4\theta_4(2x) = \theta_1^2(x)\theta_2^2(x) + \theta_3^2(x)\theta_4^2(x).$$

12. Use the method of Section 168 to show that

$$\theta_3(z\,|\,\tau) = (-i\tau)^{-\frac{1}{2}}\exp\left(\frac{z^2}{\pi i\tau}\right)\theta_3\left(\frac{z}{\tau}\,\middle|\,\frac{-1}{\tau}\right).$$

From the above identity obtain corresponding ones for the other theta functions with the aid of the basic table, page 319.

CHAPTER 21

Jacobian Elliptic Functions

174. A differential equation involving theta functions. The basic table, page 319, shows that the ratio

$$\frac{\theta_1(z)}{\theta_4(z)}$$

has the periodicity factors minus one for period π and plus one for period $\pi\tau$. Then

(1) $$\frac{d}{dz}\frac{\theta_1(z)}{\theta_4(z)}$$

has the same periodicity factors. Again from the table it can be seen that

(2) $$\frac{\theta_2(z)}{\theta_4(z)} \cdot \frac{\theta_3(z)}{\theta_4(z)}$$

has the periodicity factors minus one for period π and plus one for period $\pi\tau$. Therefore the function

(3) $$\varphi(z) = \frac{\theta_4{}^2(z)}{\theta_2(z)\theta_3(z)}\frac{d}{dz}\frac{\theta_1(z)}{\theta_4(z)},$$

the ratio of the functions in (1) and (2), has π and $\pi\tau$ as periods. Actually, $\frac{1}{2}\pi\tau$ is a period of $\varphi(z)$, a fact which we now proceed to prove.

From (3) we get

(4) $$\varphi(z) = \frac{\theta_4(z)\theta_1'(z) - \theta_1(z)\theta_4'(z)}{\theta_2(z)\theta_3(z)}.$$

The basic table yields

(5) $$\theta_1(z + \tfrac{1}{2}\pi\tau) = iq^{-\frac{1}{4}}e^{-iz}\theta_4(z)$$

and

(6) $$\theta_4(z + \tfrac{1}{2}\pi\tau) = iq^{-\frac{1}{4}}e^{-iz}\theta_1(z).$$

From (5) and (6) it follows that

(7) $$\theta_1'(z + \tfrac{1}{2}\pi\tau) = iq^{-\frac{1}{4}}e^{-iz}[\theta_4'(z) - i\theta_4(z)],$$

(8) $$\theta_4'(z + \tfrac{1}{2}\pi\tau) = iq^{-\frac{1}{4}}e^{-iz}[\theta_1'(z) - i\theta_1(z)].$$

Hence

$$\theta_4(z + \tfrac{1}{2}\pi\tau)\theta_1'(z + \tfrac{1}{2}\pi\tau) - \theta_1(z + \tfrac{1}{2}\pi\tau)\theta_4'(z + \tfrac{1}{2}\pi\tau)$$
$$= -q^{-\frac{1}{2}}e^{-2iz}[\theta_1(z)\theta_4'(z) - \theta_4(z)\theta_1'(z)].$$

The basic table gives us

$$\theta_2(z + \tfrac{1}{2}\pi\tau)\theta_3(z + \tfrac{1}{2}\pi\tau) = q^{-\frac{1}{2}}e^{-2iz}\theta_2(z)\theta_3(z),$$

and we may therefore conclude that

(9) $$\varphi(z + \tfrac{1}{2}\pi\tau) = \varphi(z).$$

The function $\varphi(z)$ of (3) or (4) is now known to have the periods π and $\frac{1}{2}\pi\tau$. The only singular points which $\varphi(z)$ can have in the finite plane are the zeros of $\theta_2(z)$ and $\theta_3(z)$. As we saw on page 322, the zeros of $\theta_2(z)$ and $\theta_3(z)$ are all simple ones and are located at

$$z = \tfrac{1}{2}\pi + m\pi + n\pi\tau,$$
$$z = \tfrac{1}{2}\pi + \tfrac{1}{2}\pi\tau + m\pi + n\pi\tau,$$

for integral m and n. Of these zeros only one is contained in a cell of $\varphi(z)$. Hence $\varphi(z)$ is an elliptic function of order less than two and is therefore a constant.

From

$$\varphi(z) = \frac{\theta_4(z)\theta_1'(z) - \theta_1(z)\theta_4'(z)}{\theta_2(z)\theta_3(z)} = c,$$

we obtain c by using $z = 0$:

$$c = \frac{\theta_4\theta_1' - 0}{\theta_2\theta_3} = \frac{\theta_4\theta_2\theta_3\theta_4}{\theta_2\theta_3} = \theta_4^2,$$

with the aid of the known relation $\theta_1' = \theta_2\theta_3\theta_4$. Thus we arrive at the differential equation

$$(10) \qquad \frac{d}{dz}\frac{\theta_1(z)}{\theta_4(z)} = \frac{\theta_4{}^2\theta_2(z)\theta_3(z)}{\theta_4{}^2(z)}.$$

The equation (10) involves all four of the theta functions. We have previously (Section 168) found the relations

$$(11) \qquad \theta_3{}^2\theta_1{}^2(z) + \theta_4{}^2\theta_2{}^2(z) = \theta_2{}^2\theta_4{}^2(z),$$

$$(12) \qquad \theta_2{}^2\theta_1{}^2(z) + \theta_4{}^2\theta_3{}^2(z) = \theta_3{}^2\theta_4{}^2(z).$$

With the aid of (11) and (12) equation (10) can be written in a form involving only the dependent variable $\theta_1(z)/\theta_4(z)$. To that end we square both sides of (10) and write

$$(13) \qquad \left[\frac{d}{dz}\frac{\theta_1(z)}{\theta_4(z)}\right]^2 = \frac{\theta_4{}^2\theta_2{}^2(z)}{\theta_4{}^2(z)} \cdot \frac{\theta_4{}^2\theta_3{}^2(z)}{\theta_4{}^2(z)}.$$

Equations (11) and (12) may now be used to put (13) in the form

$$(14) \qquad \left[\frac{d}{dz}\frac{\theta_1(z)}{\theta_4(z)}\right]^2 = \left[\theta_2{}^2 - \theta_3{}^2\frac{\theta_1{}^2(z)}{\theta_4{}^2(z)}\right]\left[\theta_3{}^2 - \theta_2{}^2\frac{\theta_1{}^2(z)}{\theta_4{}^2(z)}\right].$$

With the dependent variable $w = \theta_1(z)/\theta_4(z)$ we now have

$$(15) \qquad \left(\frac{dw}{az}\right)^2 = (\theta_2{}^2 - \theta_3{}^2w^2)(\theta_3{}^2 - \theta_2{}^2w^2).$$

In an attempt to simplify the appearance of equation (15), let us introduce a new independent variable u by $z = u/\theta_3{}^2$ and a new dependent variable y by $w = \theta_2 y/\theta_3$. Then (15) becomes

$$\frac{\theta_2{}^2}{\theta_3{}^2}\,\theta_3{}^4\left(\frac{dy}{du}\right)^2 = \left(\theta_2{}^2 - \frac{\theta_3{}^2\theta_2{}^2}{\theta_3{}^2}\,y^2\right)\left(\theta_3{}^2 - \frac{\theta_2{}^2\theta_2{}^2}{\theta_3{}^2}\,y^2\right)$$

or

$$(16) \qquad \left(\frac{dy}{du}\right)^2 = (1 - y^2)(1 - k^2y^2),$$

in which

$$(17) \qquad k = \frac{\theta_2{}^2}{\theta_3{}^2}.$$

In retrospect, we have shown that the function

$$(18) \qquad y = \frac{\theta_3}{\theta_2}\frac{\theta_1(\theta_3{}^{-2}u)}{\theta_4(\theta_3{}^{-2}u)}$$

is one solution of the differential equation (16) with the constant k being defined by (17).

175. The function $sn(u)$. The y of equation (18) of the preceding section is called the Jacobian elliptic function $sn(u)$, defined by

$$sn(u) = \frac{\theta_3}{\theta_2} \frac{\theta_1(\theta_3^{-2}u)}{\theta_4(\theta_3^{-2}u)} \tag{1}$$

and also written

$$sn(u, k) = \frac{\theta_3}{\theta_2} \frac{\theta_1(\theta_3^{-2}u)}{\theta_4(\theta_3^{-2}u)}; \qquad k = \frac{\theta_2^2}{\theta_3^2}, \tag{2}$$

when it is desired to exhibit the parameter or "modulus" k. We know that $y = sn(u, k)$ is a solution of the differential equation (16) of the preceding section. We next prove that $sn(u, k)$ is an elliptic function, and we shall study some of its properties.

The basic table, page 319, shows that the ratio $\theta_1(z)/\theta_4(z)$ has the period $\pi\tau$ in z and the periodicity factor minus one for an increase of π in the argument z. Hence, as a function of z, $\theta_1(z)/\theta_4(z)$ has periods 2π and $\pi\tau$. Therefore, as a function of u,

$$\frac{\theta_1(\theta_3^{-2}u)}{\theta_4(\theta_3^{-2}u)}$$

has periods $2\pi\theta_3^2$ and $\pi\tau\theta_3^2$. That is,

$$sn(u + 2\pi\theta_3^2) = sn(u), \tag{3}$$

$$sn(u + \pi\tau\theta_3^2) = sn(u), \tag{4}$$

so that $sn(u)$ is a doubly periodic function.

Let us now examine the singular points of $sn(u)$. Since $\theta_1(z)$ and $\theta_4(z)$ are analytic for all finite z, the only singular points of $sn(u)$ are at the zeros of the denominator $\theta_4(\theta_6^{-2}u)$. Hence $sn(u)$ has simple poles where

$$\theta_3^{-2}u = \tfrac{1}{2}\pi\tau + n\pi + m\pi\tau.$$

Because of the periods $2\pi\theta_3^2$ and $\pi\tau\theta_3^2$, we see that a representative cell for $sn(u)$ has in it two simple poles, one at

$$u = \tfrac{1}{2}\pi\tau\theta_3^2$$

and one at

$$u = (\tfrac{1}{2}\pi\tau + \pi)\theta_3^2$$

Thus $sn(u)$ is an elliptic function of order two, and in contrast to the Weierstrass P(z) of Chapter 19, $sn(u)$ has two simple poles rather than one double pole in each cell.

A common notation for the periods of $sn(u, k)$ is $4K$ and $2iK'$, where

$$(5) \qquad K = \tfrac{1}{2}\pi\theta_3^2,$$

$$(6) \qquad K' = -\tfrac{1}{2}i\pi\tau\theta_3^2.$$

With reference to the notation in (6), recall that we require $\mathrm{Im}(\tau)>0$ so that $|q| = |e^{\pi i\tau}| < 1$. If τ is pure imaginary, then K' is positive.

One more notation will simplify some of the relations to be encountered later in this chapter. We have already defined in association with $sn(u)$ the modulus

$$(7) \qquad k = \frac{\theta_2^2}{\theta_3^2}.$$

We now define a complementary modulus k' by

$$(8) \qquad k' = \frac{\theta_4^2}{\theta_3^2}.$$

In view of the relation, page 324,

$$\theta_2^4 + \theta_4^4 = \theta_3^4,$$

we may conclude that

$$(9) \qquad k^2 + (k')^2 = 1.$$

176. The functions *cn(u)* and *dn(u)*. We have one Jacobian elliptic function $sn(u)$ defined by

$$(1) \qquad sn(u) = \frac{\theta_3}{\theta_2}\,\frac{\theta_1(\theta_3^{-2}u)}{\theta_4(\theta_3^{-2}u)}.$$

Closely associated with $sn(u)$ are two other functions defined by

$$(2) \qquad cn(u) = \frac{\theta_4}{\theta_2}\,\frac{\theta_2(\theta_3^{-2}u)}{\theta_4(\theta_3^{-2}u)}$$

and

$$(3) \qquad dn(u) = \frac{\theta_4}{\theta_3}\,\frac{\theta_3(\theta_3^{-2}u)}{\theta_4(\theta_3^{-2}u)}.$$

We may refer to the basic table, page 319, to see that $cn(u)$ is an elliptic function of order two with periods $2\pi\theta_3^2$ and $(\pi + \pi\tau)\theta_3^2$

and with a representative cell containing two simple poles, one at $u = \frac{1}{2}\pi\tau\theta_3^2$, one at $u = (\pi + \frac{1}{2}\pi\tau)\theta_3^2$.

In the same way it can be shown that $dn\ (u)$ is an elliptic function of order two with periods $\pi\theta_3^2$ and $2\pi\tau\theta_3^2$ and with a representative cell containing two simple poles, one at $u = \frac{1}{2}\pi\tau\theta_3^2$, one at $u = \frac{3}{2}\pi\tau\theta_3^2$.

In terms of the parameters K and K' given by

$$(4) \qquad K = \tfrac{1}{2}\pi\theta_3^2, \qquad K' = -\tfrac{1}{2}i\pi\tau\theta_3^2,$$

the periodicity properties of the Jacobian elliptic functions $sn\ (u)$, $cn\ (u)$, $dn\ (u)$ may be written as follows:

$$(5) \qquad sn(u + 4K) = sn(u + 2iK') = sn\ (u),$$

$$(6) \qquad cn(u + 4K) = cn(u + 2K + 2iK') = cn\ (u),$$

$$(7) \qquad dn(u + 2K) = dn(u + 4iK') = dn\ (u).$$

From our knowledge of the properties of the theta functions, we obtain certain elementary facts about the Jacobian elliptic functions. Recall that $\theta_m(z)$ is an even function of z for $m = 2, 3, 4$ and that $\theta_1(z)$ is an odd function of z. It follows that

$$sn(-u) = -sn(u), \qquad cn(-u) = cn(u), \qquad dn(-u) = dn(u);$$

that is, $sn\ (u)$ is an odd function of u and both $cn\ (u)$ and $dn\ (u)$ are even functions of u. Furthermore we have

$$sn(0) = 0, \qquad cn(0) = 1, \qquad dn(0) = 1.$$

177. Relations involving squares. Information on the theta functions acquired in Chapter 20 naturally reflects itself in information on the Jacobian functions. We found in Section 168 that

$$(1) \qquad \theta_3^2\theta_1^2(z) + \theta_4^2\theta_2^2(z) = \theta_2^2\theta_4^2(z),$$

$$(2) \qquad \theta_2^2\theta_1^2(z) + \theta_4^2\theta_3^2(z) = \theta_3^2\theta_4^2(z).$$

The choice $z = \theta_3^{-2}\, u$ permits (1) and (2) to be written as

$$(3) \qquad \frac{\theta_3^2}{\theta_2^2}\frac{\theta_1^2(\theta_3^{-2}u)}{\theta_4^2(\theta_3^{-2}u)} + \frac{\theta_4^2}{\theta_2^2}\frac{\theta_2^2(\theta_3^{-2}u)}{\theta_4^2(\theta_3^{-2}u)} = 1$$

and

$$(4) \qquad \frac{\theta_2^2}{\theta_3^2}\frac{\theta_1^2(\theta_3^{-2}u)}{\theta_4^2(\theta_3^{-2}u)} + \frac{\theta_4^2}{\theta_3^2}\frac{\theta_3^2(\theta_3^{-2}u)}{\theta_4^2(\theta_3^{-2}u)} = 1.$$

By the definitions of $sn\,(u)$, $cn\,(u)$, $dn\,(u)$ in the preceding section, we see that (3) and (4) yield

$$sn^2\,(u) + cn^2\,(u) = 1, \tag{5}$$

$$k^2\,sn^2\,(u) + dn^2\,(u) = 1, \tag{6}$$

in which $k = \theta_2^2/\theta_3^2$, as usual.

The elimination of $sn\,(u)$ from (5) and (6) yields

$$dn^2\,(u) - k^2\,cn^2\,(u) = 1 - k^2,$$

or

$$dn^2\,(u) - k^2\,cn^2\,(u) = (k')^2, \tag{7}$$

with $k = \theta_2^2/\theta_3^2$ and $k' = \theta_4^2/\theta_3^2$ satisfying the relation

$$k^2 + (k')^2 = 1$$

as noted earlier.

178. Relations involving derivatives. On page 341 we obtained the result

$$\frac{d}{dz}\frac{\theta_1(z)}{\theta_4(z)} = \frac{\theta_4^2\theta_2(z)\theta_3(z)}{\theta_4^2(z)}. \tag{1}$$

From the definition

$$sn(u) = \frac{\theta_3}{\theta_2}\frac{\theta_1(\theta_3^{-2}u)}{\theta_4(\theta_3^{-2}u)}, \tag{2}$$

it follows that

$$\frac{d}{du}\,sn(u) = \frac{\theta_3}{\theta_2}\frac{1}{\theta_3^2}\frac{d}{d(\theta_3^{-2}u)}\frac{\theta_1(\theta_3^{-2}u)}{\theta_4(\theta_3^{-2}u)}. \tag{3}$$

By (1), equation (3) becomes

$$\frac{d}{du}\,sn(u) = \frac{\theta_4^2}{\theta_2\theta_3}\frac{\theta_2(\theta_3^{-2}u)\theta_3(\theta_3^{-2}u)}{\theta_4^2(\theta_3^{-2}u)}. \tag{4}$$

But, by definition,

$$cn(u) = \frac{\theta_4}{\theta_2}\frac{\theta_2(\theta_3^{-2}u)}{\theta_4(\theta_3^{-2}u)} \tag{5}$$

$$dn(u) = \frac{\theta_4}{\theta_3}\frac{\theta_3(\theta_3^{-2}u)}{\theta_4(\theta_3^{-2}u)}, \tag{6}$$

so that (4) yields

$$\frac{d}{du}\,sn(u) = cn(u)\,dn(u). \tag{7}$$

We now know how to differentiate $sn\,(u)$ and we know a relation

$$sn^2\,(u) + cn^2\,(u) = 1 \tag{8}$$

between $sn\,(u)$ and $cn\,(u)$. Then we differentiate each member of (8) to obtain

$$sn(u)\;cn(u)\;dn(u) + cn(u)\frac{d}{du}\,cn(u) = 0,$$

from which it follows that

$$\frac{d}{du}\,cn(u) = -sn(u)\;dn(u). \tag{9}$$

In the same manner we conclude from (7) and the known relation (page 345)

$$k^2\,sn^2\,(u) + dn^2\,(u) = 1 \tag{10}$$

that

$$k^2\,sn(u)\;cn(u)\;dn(u) + dn(u)\frac{d}{du}\,dn(u) = 0,$$

which yields

$$\frac{d}{du}\,dn(u) = -k^2\,sn(u)\;cn(u). \tag{11}$$

We already know from equation (16), page 341, that $y_1 = sn\,(u)$ is a solution of the differential equation

$$\left(\frac{dy_1}{du}\right)^2 = (1 - y_1^2)(1 - k^2y_1^2), \tag{12}$$

in which $k = \theta_2^2/\theta_3^2$. Let us obtain corresponding differential equations satisfied by $cn\,(u)$ and $dn\,(u)$.

Put $y_2 = cn\,(u)$. Then, by (9),

$$\left(\frac{dy_2}{du}\right)^2 = sn^2(u)\;dn^2(u).$$

But $sn^2\,(u) = 1 - cn^2\,(u)$ and $dn^2\,(u) = (k')^2 + k^2\,cn^2\,(u)$, for which see Section 177. It follows that $y_2 = cn\,(u)$ is a solution of the differential equation

$$\left(\frac{dy_2}{du}\right)^2 = (1 - y_2^2)[(k')^2 + k^2y_2^2]. \tag{13}$$

Now let $y_3 = dn\,(u)$. By (11) we get

$$\left(\frac{dy_3}{du}\right)^2 = k^4\, sn^2(u)\, cn^2(u).$$

But $k^2\, sn^2\,(u) = 1 - dn^2\,(u)$ and $k^2\, cn^2\,(u) = dn^2\,(u) - (k')^2$. Hence $y_3 = dn\,(u)$ is a solution of the differential equation

$$\left(\frac{dy_3}{du}\right)^2 = (1 - y_3^2)[y_3^2 - (k')^2]. \tag{14}$$

179. Addition theorems. We have defined the Jacobian elliptic functions by

$$sn(u) = \frac{\theta_3}{\theta_2}\frac{\theta_1(\theta_3^{-2}u)}{\theta_4(\theta_3^{-2}u)}, \tag{1}$$

$$cn(u) = \frac{\theta_4}{\theta_2}\frac{\theta_2(\theta_3^{-2}u)}{\theta_4(\theta_3^{-2}u)}, \tag{2}$$

$$dn(u) = \frac{\theta_4}{\theta_3}\frac{\theta_3(\theta_3^{-2}u)}{\theta_4(\theta_3^{-2}u)}, \tag{3}$$

and we have established formulas involving the theta functions with arguments $(x + y)$ and $(x - y)$. We should therefore be able to conclude something of interest about $sn(u + v)$, $cn(u + v)$, etc.

In Section 169 and Ex. 8, p. 337, we found that

$$\theta_2\theta_3\theta_1(x+y)\theta_4(x-y) = \theta_1(x)\theta_2(y)\theta_3(y)\theta_4(x) + \theta_1(y)\theta_2(x)\theta_3(x)\theta_4(y), \tag{4}$$

$$\theta_4^2\theta_4(x + y)\theta_4(x - y) = \theta_4^2(x)\theta_4^2(y) - \theta_1^2(x)\theta_1^2(y). \tag{5}$$

Equations (4) and (5) yield

$$\frac{\theta_2\theta_3}{\theta_4^2}\frac{\theta_1(x + y)}{\theta_4(x + y)} = \frac{\theta_1(x)\theta_2(y)\theta_3(y)\theta_4(x) + \theta_1(y)\theta_2(x)\theta_3(x)\theta_4(y)}{\theta_4^2(x)\theta_4^2(y) - \theta_1^2(x)\theta_1^2(y)}$$

$$= \frac{\dfrac{\theta_1(x)}{\theta_4(x)}\dfrac{\theta_2(y)}{\theta_4(y)}\dfrac{\theta_3(y)}{\theta_4(y)} + \dfrac{\theta_1(y)}{\theta_4(y)}\dfrac{\theta_2(x)}{\theta_4(x)}\dfrac{\theta_3(x)}{\theta_4(x)}}{1 - \dfrac{\theta_1^2(x)}{\theta_4^2(x)}\dfrac{\theta_1^2(y)}{\theta_4^2(y)}}.$$

Now put $x = \theta_3^{-2}u$ and $y = \theta_3^{-2}v$ and use (1), (2), and (3) to obtain

$$\frac{\theta_2^2}{\theta_4^2}\, sn(u + v) = \frac{\dfrac{\theta_2^2}{\theta_4^2}\, sn(u)\, cn(v)\, dn(v) + \dfrac{\theta_2^2}{\theta_4^2}\, sn(v)\, cn(u)\, dn(u)}{1 - \dfrac{\theta_2^4}{\theta_3^4}\, sn^2(u)\, sn^2(v)},$$

from which it follows that

$$(6) \qquad sn(u+v) = \frac{sn(u)\,cn(v)\,dn(v) + sn(v)\,cn(u)\,dn(u)}{1 - k^2\,sn^2(u)\,sn^2(v)}.$$

Equation (6) is called an addition theorem. The companion results

$$(7) \qquad cn(u+v) = \frac{cn(u)\,cn(v) - sn(u)\,sn(v)\,dn(u)\,dn(v)}{1 - k^2\,sn^2(u)\,sn^2(v)}$$

and

$$(8) \qquad dn(u+v) = \frac{dn(u)\,dn(v) - k^2\,sn(u)\,sn(v)\,cn(u)\,cn(v)}{1 - k^2\,sn^2(u)\,sn^2(v)}$$

may be obtained in a similar manner and are left as exercises.

EXERCISES

1. Derive the preceding addition theorem (7) by the method of Section 179. You may use results from the exercises at the end of Chapter 20.

2. Derive preceding (8) by the method of Section 179, with the aid of results from the exercises at the end of Chapter 20.

3. Show that $\int cn^3(x)\,dn(x)\,dx = sn(x) - \frac{1}{3}\,sn^3(x) + c.$

4. Show that if $g(x)$ and $h(x)$ are any two different ones of the three functions $sn(x)$, $cn(x)$, $dn(x)$, and if m is a non-negative integer, you can perform the integration

$$\int g^{2m+1}(x)h(x)\,dx.$$

5. Obtain the result

$$\int sn(x)\,dx = \frac{1}{k}\,\text{Log}[dn(x) - k\,cn(x)] + c.$$

6. Show that

$$dn(2x) - k\,cn(2x) = \frac{(1-k)[1 + k\,sn^2(x)]}{1 - k\,sn^2(x)}.$$

Bibliography

Agarwal, R. P.

1. On Bessel polynomials, *Canadian Journal of Math.*, 6, 1954, pp. 410–415.

Al-Salam, W. A.

1. The Bessel polynomials, *Duke Math. Journal*, 24, 1957, pp. 529–545.

Appell, Paul

1. *Sur les Fonctions Hypergéométriques de Plusieurs Variables.* Paris: Gauthier-Villars, 1925.

Appell, Paul, and Kampé de Fériet, J.

1. *Fonctions Hypergéométriques et Hypersphériques; Polynomes d'Hermites.* Paris: Gauthier-Villars, 1926.

Bailey, W. N.

1. *Generalized Hypergeometric Series.* Cambridge: Cambridge Univ. Press, 1935.
2. Products of generalized hypergeometric series, *Proc. London Math. Soc.*, series 2, 28, 1928, pp. 242–254.
3. Some theorems concerning products of hypergeometric series, *Proc. London Math. Soc.*, series 2, 38, 1935, pp. 377–384.

Barnes, E. W.

1. A transformation of generalized hypergeometric series, *Quarterly Journal of Math.*, 41, 1910, pp. 136–140.
2. A new development of the theory of the hypergeometric function, *Proc. London Math. Soc.*, series 2, 6, 1908, pp. 141–177.

Bateman, H.

1. A generalization of the Legendre polynomial, *Proc. London Math. Soc.*, series 2, 3, 1905, pp. 111–123.
2. Two systems of polynomials for the solution of Laplace's integral equation, *Duke Math. Journal*, 2, 1936, pp. 569–577.
3. Some properties of a certain set of polynomials, *Tôhoku Math. Journal*, 37, 1933, pp. 23–38.
4. The polynomial of Mittag-Lefler, *Proc. Nat. Acad. Sci.*, 26, 1940, pp. 491–496.
5. Polynomials associated with those of Lerch, *Monatshefte für Math. und Physik*, 43, 1936, pp. 75–80.
6. The transformation of a Lagrangian series into a Newtonian series, *Proc. Nat. Acad. Sci.*, 25, 1939, pp. 262–265.
7. Relations between confluent hypergeometric functions, *Proc. Nat. Acad. Sci.*, 17, 1931, pp. 689–690.

8. An orthogonal property of the hypergeometric polynomial, *Proc. Nat. Acad. Sci.*, 28, 1942, pp. 374–377.

9. Note on the function $F(a, b; c-n; z)$, *Proc. Nat. Acad. Sci.*, 30, 1944, pp. 28–30.

10. The polynomial $F_n(x)$, *Annals of Math.*, 35, 1934, pp. 767–775.

11. The polynomial $F_n(x)$ and its relation to other functions, *Annals of Math.*, 38, 1937, pp. 303–310.

12. The k-function, a particular case of the confluent hypergeometric function, *Trans. Amer. Math. Soc.*, 33, 1931, pp. 817–831.

Bedient, P. E.

1. Polynomials related to Appell functions of two variables, Michigan thesis, 1958.

Bhonsle, B. R.

1. On a series of Rainville involving Legendre polynomials, *Proc. Amer. Math. Soc.*, 8, 1957, pp. 10–14.

Boas, R. P., Jr., and Buck, R. C.

1. Polynomials defined by generating relations, *Amer. Math. Monthly*, 63, 1956, pp. 626–632.

2. *Polynomial Expansions of Analytic Functions*. Berlin: Springer, 1958.

Brafman, Fred

1. Generating functions of Jacobi and related polynomials, *Proc. Amer. Math. Soc.*, 2, 1951, pp. 942–949.

2. Some generating functions for Laguerre and Hermite polynomials, *Canadian Journal of Math.*, 9, 1957, pp. 180–187.

3. A relation between ultraspherical and Jacobi polynomial sets, *Canadian Journal of Math.*, 5, 1953, pp. 301–305.

4. A set of generating functions for Bessel polynomials, *Proc. Amer. Math. Soc.*, 4, 1953, pp. 275–277.

5. Unusual generating functions for ultraspherical polynomials, *Michigan Math. Journal*, 1, 1952, pp. 131–138.

6. On Touchard polynomials, *Canadian Journal of Math.*, 9, 1957, pp. 191–193.

7. Series of products of Gegenbauer polynomials, *Math. Zeitschr.*, 62, 1955, pp. 438–442.

8. A generating function for associated Legendre polynomials, *Quarterly Journal of Math.*, Oxford 2nd series, 8, 1957, pp. 81–83.

9. An ultraspherical generating function, *Pacific Journal of Math.*, 7, 1957, pp. 1319–1323.

Brenke, W. C.

1. On generating functions of polynomial systems, *Amer. Math. Monthly*, 52, 1945, pp. 297-301.

Burchnall, J. L.

1. The Bessel polynomials, *Canadian Journal of Math.*, 3, 1951, pp. 62–68.
2. A relation between hypergeometric series, *Quarterly Journal of Math.*, Oxford series, 3, 1932, pp. 318–320.

Carlitz, Leonard

1. Some polynomials of Touchard connected with the Bernoulli numbers, *Canadian Journal of Math.*, 9, 1957, pp. 188–191.
2. On the Bessel polynomials, *Duke Math. Journal*, 24, 1957, pp. 151–162.
3. On Jacobi polynomials, *Boll. Unione Mat. Ital.*, series 3, 9, 1956, pp. 371–381.

Chaundy, T. W.

1. An extension of the hypergeometric function, *Quarterly Journal of Math.*, Oxford series, 14, 1943, pp. 55–78.
2. Some hypergeometric identities, *Journal London Math. Soc.*, 26, 1951, pp. 42–44.

Churchill, R. V.

1. *Operational Mathematics*, 2d ed. New York: McGraw-Hill Book Co., 1958.
2. Fourier Series and Boundary Value Problems. New York: McGraw-Hill Book Co., 1941.
3. *Introduction to Complex Variables and Applications*. New York: McGraw-Hill Book Co., 1948.

Copson, E. T.

1. *Theory of Functions of a Complex Variable*. London: Oxford Univ. Press, 1935.

Curzon, H. E. J.

1. On a connexion between the functions of Hermite and the functions of Legendre, *Proc. London Math. Soc.*, series 2, 12, 1913, pp. 236–259.

Dickinson, D. J., Pollak, H. O., and Wannier, G. H.

1. On a class of polynomials orthogonal over a denumerable set, *Pacific Journal of Math.*, 6, 1956, pp. 239–247.

Dickinson, D. J.

1. On Bessel and Lommel polynomials, Michigan thesis, 1953. (See also 2 below).
2. On Lommel and Bessel polynomials, *Proc. Amer. Math. Soc.*, 5, 1954, pp. 946–956. This is partly a digest of Dickinson [1].
3. On certain polynomials associated with orthogonal polynomials, *Boll. Unione Mat. Ital.*, (3), 13, 1958, pp. 116-124.

Dixon, A. C.

1. Summation of a certain series, *Proc. Lond. Math. Soc.*, series 1, 35, 1903, pp. 285–289.

Erdélyi, A.

1. With Magnus, W., Oberhettinger, F., Tricomi, F. G., et al. *Higher Transcendental Functions.* New York: McGraw-Hill Book Co., 1953, volume 1.
2. ———, vol. 2.
3. ———, vol. 3.
4. Expansions of Lamé functions into series of Legendre functions, *Proc. Royal Soc. of Edinburgh*, 62, 1948, pp. 247–267.
5. Hypergeometric functions of two variables, *Acta Math.*, 83, 1950, pp. 131–164.

Fasenmyer, Sister M. Celine

1. Some generalized hypergeometric polynomials, *Bull. Amer. Math. Soc.*, 53, 1947, pp. 806–812.
2. A note on pure recurrence relations, *Amer. Math. Monthly*, 56, 1949, pp. 14–17.

Favard, J.

1. Sur les polynomes de Tchebicheff, *Comptes Rendus Acad. Sci. Paris*, 200, 1935, pp. 2052–2053.

Gottlieb, M. J.

1. Concerning some polynomials orthogonal on a finite or enumerable set of points, *Amer. Journal of Math.*, 60, 1938, pp. 453–458.

Grosswald, Emil

1. On some algebraic properties of Bessel polynomials, *Trans. Amer. Math. Soc.*, 71, 1951, pp. 197–210.

Hall, N. A.

1. A formal expansion theory for functions of one or more variables, *Bull. Amer. Math. Soc.*, 46, 1940, pp. 824–832.

Hobson, E. W.

1. *The Theory of Spherical and Ellipsoidal Harmonics.* Cambridge: Cambridge Univ. Press, 1931; New York: Chelsea, 1965.

Huff, W. N.

1. The type of the polynomials generated by $f(xt)\ \varphi(t)$, *Duke Math. Journal*, 14, 1947, pp. 1091–1104.

Huff, W. N. and Rainville, E. D.

1. On the Sheffer A-type of polynomials generated by $\varphi(t)f(xt)$, *Proc. Amer. Math. Soc.*, 3, 1952, pp. 296–299.

Jackson, D.

1. *Fourier Series and Orthogonal Polynomials*, Carus Monograph No. 6., Math. Assoc. of Amer., Menasha, 1941.

Kampé de Fériet, J.

1. *La Fonction Hypergéométrique*. Paris: Gauthier-Villars, 1937.

Kazarinoff, N. D.

1. Asymptotic expansions for the Whittaker functions of large complex order m, *Trans. Amer. Math. Soc.*, 78, 1955, pp. 305–328.
2. Asymptotic forms for the Whittaker functions with both parameters large, *Journal of Math. and Mechanics*, 6, 1957, pp. 341–360.

Krall, H. L., and Frink, Orrin

1. A new class of orthogonal polynomials: the Bessel polynomials, *Trans. Amer. Math. Soc.*, 65, 1949, pp. 100–115.

Lagrange, Rene

1. *Polynomes et Fonctions de Legendre*. Paris: Gauthier-Villars, 1939.

Langer, R. E.

1. The solutions of the differential equation $v''' + \lambda^2 zv' + 3\mu\lambda^2 v = 0$, *Duke Math. Journal*, 22, 1955, pp. 525-541.

MacRobert, T. M.

1. *Spherical Harmonics*. New York: Dover, 1948.

Magnus, Wilhelm, and Oberhettinger, Fritz

1. *Formulas and Theorems for the Special Functions of Mathematical Physics*. Berlin: Springer, 1969.
2. *Anwendung der Elliptischen Funktionen in Physik und Technik*. Berlin: Springer, 1949.

Mehlenbacher, Lyle E.

1. The interrelations of the fundamental solutions of the hypergeometric equation; logarithmic case, *Amer. Journal of Math.*, 60, 1938, pp. 120–128.

Orr, W. M.

1. Theorems relating to the product of two hypergeometric series, *Trans. Camb. Phil. Soc.*, 17, 1899, pp. 1–15.

Palas, F. J.

1. The polynomials generated by $f(t) \exp[p(x)u(t)]$, Oklahoma thesis, 1955.

Pasternack, Simon

1. A generalization of the polynomial $F_n(x)$, *Phil. Mag.*, series 7, 28, 1939, pp. 209–226.

Pennell, W. O.

1. Recurrence formulae involving Bessel functions of the first kind, *Amer. Math. Monthly*, 36, 1929, pp. 385–386.

Preece, C. T.

1. The product of two generalized hypergeometric functions, *Proc. London Math. Soc.*, series 2, 22, 1924, pp. 370–380.

Rainville, E. D.

1. *Elementary Differential Equations*, 2d ed. New York: The Macmillan Co., 1958.
2. *Intermediate Differential Equations*. New York: Chelsea, 1972.
3. The contiguous function relations for ${}_pF_q$ with applications to Bateman's $J_n^{u,v}$ and Rice's $H_n(\zeta, p, v)$, *Bull. Amer. Math. Soc.*, 51, 1945, pp. 714–723.
4. Certain generating functions and associated polynomials, *Amer. Math. Monthly*, 52, 1945, pp. 239–250.
5. Notes on Legendre polynomials, *Bull. Amer. Math. Soc.*, 51, 1945, pp. 268–271.
6. Generating functions for Bessel and related polynomials, *Canadian Journal of Math.*, 5, 1953, pp. 104–106.
7. Symbolic relations among classical polynomials, *Amer. Math. Monthly*, 53, 1946, pp. 299–305.
8. A relation between Jacobi and Laguerre polynomials, *Bull. Amer. Math. Soc.*, 51, 1945, pp. 266–267.

Rice, S. O.

1. Some properties of ${}_3F_2(-n, n+1, \zeta; 1, p; v)$, *Duke Math. Journal*, 6, 1940, pp. 108–119.
2. Sums of series of the form $\sum_0^\infty a_n J_{n+\alpha}(z) J_{n+\beta}(z)$, *Phil. Mag.*, series 7, 35, 1944, pp. 686–693.

Saalschütz, L.

1. Eine Summationsformel, *Zeitschrift für Math. und Phys.*, 35, 1890, pp. 186–188.

2. Über einen Spezialfall der hypergeometrischen Reihe dritter Ordung, *Zeitschrift für Math. und Phys.*, 36, 1891, pp. 278–295, 321–327.

Sheffer, I. M.

1. Some properties of polynomial sets of type zero, *Duke Math. Journal*, 5, 1939, pp. 590–622.
2. Some applications of certain polynomial classes, *Bull. Amer. Math. Soc.*, 47, 1941, pp. 885–898.
3. Note on Appell polynomials, *Bull. Amer. Math. Soc.*, 51, 1945, pp. 739–744.
4. A differential equation for Appell polynomials, *Bull. Amer. Math. Soc.*, 41, 1935, pp. 914–923.
5. Concerning Appell sets and associated linear functional equations, *Duke Math. Journal*, 3, 1937, pp. 593–609.
6. Expansions in generalized Appell polynomials, and a class of related linear functional equations, *Trans. Amer. Math. Soc.*, 31, 1929, pp. 261–280.
7. On the properties of polynomials satisfying a linear differential equation: Part I., *Trans. Amer. Math. Soc.*, 35, 1933, pp. 184–214.

Shively, R. L.

1. On pseudo Laguerre polynomials, Michigan thesis, 1953.

Shohat, J.

1. *Théorie Générale des Polynomes Orthogonaux de Tchebichef.* Paris: Gauthier-Villars, 1935.
2. The relation of the classical orthogonal polynomials to the polynomials of Appell, *Amer. Journal of Math.*, 58, 1936, pp. 453–464.

Smith, F. C.

1. On the logarithmic solutions of the generalized hypergeometric equation when $p = q + 1$, *Bull. Amer. Math. Soc.*, 45, 1939, pp. 629–636.
2. Relations among the fundamental solutions of the generalized hypergeometric equation when $p = q + 1$. Nonlogarithmic cases, *Bull. Amer. Math. Soc.*, 44, 1938, pp. 429–433.
3. See [2] above. Part II, Logarithmic cases, *Bull. Amer. Math. Soc.*, 45, 1939, pp. 927–935.

Smith, R. T. C.

1. Generating functions of Appell form for the classical orthogonal polynomials, *Proc. Amer. Math. Soc.*, 7, 1956, pp. 636–641.

Szegö, G.

1. *Orthogonal Polynomials.* New York: Amer. Math. Soc. Colloquium Publ., 1939, vol. 23.
2. On an inequality of P. Turan concerning Legendre polynomials, *Bull. Amer. Math. Soc.*, 54, 1948, pp. 401–405.

Toscano, Letterio

1. Funzioni generatrici di particolari polinomi di Laguerre e de altri da essi dipendenti, *Boll. Unione Mat. Ital.*, series 3, 7, 1952, pp. 160–167.

Touchard, Jacques

1. Nombres exponentiels et nombres de Bernoulli, *Canadian Journal of Math.*, 8, 1956, pp. 305–320.

Truesdell, C.

1. On the functional equation $\partial F\ (z,\alpha)/\partial z = F(z,\ \alpha+1)$, *Proc. Nat. Acad. Sci.*, 33, 1947, p. 88.

2. *Special Functions*. Princeton: Princeton Univ. Press, *Annals of Math. Studies No. 18*. 1948.

Watson, G. N.

1. *A Treatise on the Theory of Bessel Functions*, 2d ed. Cambridge: Cambridge Univ. Press, 1944.

2. Dixon's theorem on generalized hypergeometric functions, *Proc. Lond. Math. Soc.*, series 2, 20, 1922, pp. xxxii–xxxiii.

3. Notes on generating functions of polynomials: (4) Jacobi polynomials, *Journal London Math. Soc.*, 9, 1934, pp. 22–28.

4. The product of two hypergeometric functions, *Proc. London Math. Soc.*, series 2, 20, 1922, pp. 189–195.

Weber, Maria, and Erdélyi, Arthur

1. On the finite difference analogue of Rodrigues' formula, *Amer. Math. Monthly*, 59, 1952, pp. 163–168.

Webster, M. S.

1. On the zeros of Jacobi polynomials, with applications, *Duke Math. Journal*, 3, 1937, pp. 426–442.

Weisner, Louis

1. Group-theoretic origins of certain generating functions, *Pacific Journal of Math.*, 5, 1955, pp. 1033–1039.

Whipple, F. J. W.

1. Some transformations of generalized hypergeometric series, *Proc. Lond. Math. Soc.*, series 2, 26, 1927, pp. 257–272.

Whittaker, E. T.

1. An expression of certain known functions as generalized hypergeometric functions, *Bull. Amer. Math. Soc.*, 10, 1904, pp. 125–134.

Whittaker, E. T., and Watson, G. N.

1. *Modern Analysis*, 4th ed. Cambridge: Cambridge Univ. Press, 1927.

Wilkins, J. E.

1. Neumann series of Bessel functions, *Trans. Amer. Math. Soc.*, 64, 1948, pp. 359–385.

Wyman, Max, and Moser, Leo

1. On some polynomials of Touchard, *Canadian Journal of Math.*, 8, 1956, pp. 321–322.

Index

LECTURES ON THE CALCULUS OF VARIATIONS

By O. BOLZA

A standard text by a major contributor to the theory. Suitable for individual study by anyone with a good background in the Calculus and the elements of Real Variables.

—2nd (c.) ed. 1961-71. 280 pp. 5x8. 8284-0145-4. Cl.
8284-0152-7. Pa.

VORLESUNGEN UEBER VARIATIONSRECHNUNG

By O. BOLZA

A standard text and reference work, by one of the major contributors to the theory.

—1909-63. ix + 715 pp. 5⅜x8. 8284-0160-8.

THEORIE DER KONVEXER KOERPER

By T. BONNESEN and W. FENCHEL

"Remarkable monograph."

—*J. D. Tamarkin, Bulletin of the A. M. S.*

—1954-71. 171 pp. 5½x8½. 8284-0054-7.

THE CALCULUS OF FINITE DIFFERENCES

By G. BOOLE

A standard work on the subject of finite differences and difference equations by one of the seminal minds in the field of finite mathematics.

Numerous exercises with answers.

—5th ed. 1970. 341 pp. 5⅜x8. 8284-1121-2. Cloth
—4th ed. 1958. 336 pp. 5⅜x8. 8284-0148-9. Paper

A TREATISE ON DIFFERENTIAL EQUATIONS

By G. BOOLE

Including the Supplementary Volume.

—5th ed. 1959. xxiv + 735 pp. 5⅜x8. 8284-0128-4.

HISTORY OF SLIDE RULE, *By F. CAJORI. See* BALL

INTRODUCTORY TREATISE ON LIE'S THEORY OF FINITE CONTINUOUS TRANSFORMATION GROUPS

By J. E. CAMPBELL

Partial Contents: CHAP. I. Definitions and Simple Examples of Groups. II. Elementary Illustrations of Principle of Extended Point Transformations. III. Generation of Group from Its Infinitesimal Transformations. V. Structure Constants. VI. Complete Systems of Differential Equations. VII. Diff. Eqs. Admitting Known Transf. Groups. VIII. Invariant Theory of Groups . . . XIV. Pfaff's Equation . . . XXI-XXV. Certain Linear Groups.

—1903-66. xx + 416 pp. 5⅜x8. 8284-0183-7.

INTEGRALGEOMETRIE

By W. BLASCHKE and E. KÄHLER

THREE VOLUMES IN ONE.

VORLESUNGEN UEBER INTEGRALGEOMETRIE, Vols. I and II, by *W. Blaschke*.

EINFUEHRUNG IN DIE THEORIE DER SYSTEME VON DIFFERENTIALGLEICHUNGEN, by *E. Kähler*.

—1936/37/34-49. 222 pp. 5½x8½. 8284-0064-4. Three vols. in one.

FUNDAMENTAL EXISTENCE THEOREMS,

by G. A. BLISS. See EVANS

THEORY AND APPLICATIONS OF DISTANCE GEOMETRY

By L. M. BLUMENTHAL

"Clearly written and self-contained. The reader is well provided with exercises of various degrees of difficulty."—*Bulletin of A.M.S.*

—2nd (c.) ed. 1953-71. 359 pp. 5⅜x8. 8284-0242-6.

A HISTORY OF FORMAL LOGIC

By I. M. BOCHEŃSKI

Translated and edited by PROFESSOR IVO THOMAS.

A history and source book, by one of the world's leading authorities. Generous selections, from the Greeks to Peano, Russell, Frege, and Gödel, threaded together by explanatory comment. Within schools, the arrangement is by subject.

"Covers the whole period from pre-Socrates to Gödel . . . It sets such a high standard of excellence that one may doubt whether it will have any serious rivals for a long time to come."—*The Journal of Symbolic Logic.*

—2nd (c.) ed. 1970. xxii + 567 pp. 5⅜x8. 8284-0238-8.

VORLESUNGEN UEBER FOURIERSCHE INTEGRALE

By S. BOCHNER

—1932-48. vi + 229 pp. 5½x8½. 8284-0042-3.

ALMOST PERIODIC FUNCTIONS

By H. BOHR

Translated by H. COHN. From the famous series *Ergebnisse der Mathematik und ihrer Grenzgebiete,* a beautiful exposition of the theory of Almost Periodic Functions written by the creator of that theory.

—1951-66. 120 pp. 6x9. Lithotyped. 8284-0027-X.

WISSENSCHAFTLICHE ABHANDLUNGEN

By L. BOLTZMANN

—1909-68. 1,976 pp. 5⅜x8. 8284-0215-9.
Three volume set.

LECTURES ON ERGODIC THEORY

By P. R. HALMOS

CONTENTS: Introduction. Recurrence. Mean Convergence. Pointwise Convergence. Ergodicity. Mixing. Measure Algebras. Discrete Spectrum. Automorphisms of Compact Groups. Generalized Proper Values. Weak Topology. Weak Approximation. Uniform Topology. Uniform Approximation. Category. Invariant Measures. Generalized Ergodic Theorems. Unsolved Problems.

"Written in the pleasant, relaxed, and clear style usually associated with the author. The material is organized very well and painlessly presented."

—*Bulletin of the A.M.S.*

—1956-60. viii + 101 pp. 5⅜x8. 8284-0142-X.

ELEMENTS OF QUATERNIONS

By W. R. HAMILTON

Sir William Rowan Hamilton's last major work, and the second of his two treatises on quaternions.

—3rd ed. 1899/1901-68. 1,185 pp. 6x9. 8284-0219-1. Two vol. set

RAMANUJAN: Twelve Lectures on His Life and Works

By G. H. HARDY

The book is somewhat more than an account of the mathematical work and personality of Ramanujan; it is one of the very few full-length books of "shop talk" by an important mathematician.

—1940-68. viii + 236 pp. 6x9. 8284-0136-5.

GRUNDZUEGE DER MENGENLEHRE

By F. HAUSDORFF

The original, 1914 edition of this famous work contains many topics that had to be omitted from later editions, notably, the theories of content, measure, and discussion of the Lebesgue integral. Also, general topological spaces, Euclidean spaces, special methods applicable in the Euclidean plane, the algebra of sets, partially ordered sets, etc.

—1914-65. 484 pp. 5⅜x8. 8284-0061-X.

SET THEORY

By F. HAUSDORFF

Hausdorff's classic text-book is an inspiration and a delight. The translation is from the Third (latest) German edition.

"We wish to state without qualification that this is an indispensable book for all those interested in the theory of sets and the allied branches of real variable theory."—*Bulletin of A. M. S.*

—2nd ed. 1962-67. 352 pp. 6x9. 8284-0119-5.

FOUNDATIONS OF ANALYSIS

By E. LANDAU

"Certainly no clearer treatment of the foundations of the number system can be offered. . . . One can only be thankful to the author for this fundamental piece of exposition, which is alive with his vitality and genius."—*J. F. Ritt, Amer. Math. Monthly.*

—2nd ed. 1960-66. xiv + 136 pp. 8284-0079-2.

ELEMENTARE ZAHLENTHEORIE

By E. LANDAU

"Interest is enlisted at once and sustained by the accuracy, skill, and enthusiasm with which Landau marshals . . . facts and simplifies . . . details."

—*G. D. Birkhoff, Bulletin of the A. M. S.*

—1927-50. vii + 180 + iv pp. 5½x8½. 8284-0026-1.

ELEMENTARY NUMBER THEORY

By E. LANDAU

The present work is a translation of Prof. Landau's famous *Elementare Zahlentheorie*, with added exercises by Prof. Paul T. Bateman.

—2nd ed. 1966. 256 pp. 6x9. 8284-0125-X.

Einführung in die Elementare und Analytische Theorie der ALGEBRAISCHE ZAHLEN

By E. LANDAU

—2nd ed. 1927-49. vii + 147 pp. 5⅜x8. 8284-0062-8.

NEUERE FUNKTIONENTHEORIE, *by E. LANDAU.*
See WEYL

Mémoires sur la Théorie des SYSTÈMES DES ÉQUATIONS DIFFÉRENTIELLES LINÉAIRES, Vols. I, II, III

By J. A. LAPPO-DANILEVSKIĬ

THREE VOLUMES IN ONE.

A reprint, in one volume, of Volumes 6, 7, and 8 of the monographs of the Steklov Institute of Mathematics in Moscow.

"The theory of [systems of linear differential equations] is treated with elegance and generality by the author, and his contributions constitute an important addition to the field of differential equations."—*Applied Mechanics Reviews.*

—1934/5/6-53. 689 pp. 5⅜x8. 8284-0094-6.
Three vols. in one

MATHEMATICAL AND PHYSICAL PAPERS

By J. LARMOR

—2nd (c.) ed. 1929-72. 1,554 pp. 6x9. 8284-0249-3.
Two vol. set. **In prep.**

VORLESUNGEN UEBER DIFFERENTIAL-GLEICHUNGEN MIT BEKANNTEN INFINITESIMALEN TRANSFORMATIONEN

By S. LIE

Edited by G. Scheffers. A textbook on the integration of ordinary and partial differential equations in which the Lie theory for solving such equations is expounded.

—1891-1967. xiv + 568 pp. 6x9. 8284-0206-X.

THEORIE DER TRANSFORMATIONSGRUPPEN

By S. LIE

The Bible of transformation groups, a work of vast range and great originality. It is written with especial simplicity of language. There are frequent *résumés* throughout the work and an abundance of illustrations.

—2nd (corr.) ed. 1888/93-70. 2,090 pp. 6x9. 8284-0232-9. Three vol. set.

VORLESUNGEN UEBER CONTINUIERLICHE GRUPPEN MIT GEOMETRISCHEN UND ANDEREN ANWENDUNGEN

By S. LIE

An expository work by the originator of the Lie theory of continuous groups.

—1893-71. xii + 810 pp. 6x9. 8284-0199-3.

LES ENSEMBLES ANALYTIQUES

By N. LUSIN

Leçons sur les Ensembles Analytiques et leurs Applications, with a preface by H. Lebesgue and a note by W. Sierpinski.

—1930-71. xv + 328 pp. 5⅜x8. 8284-0250-7.

COMBINATORY ANALYSIS, Vols. I and II

By P. A. MACMAHON

TWO VOLUMES IN ONE.

A broad and extensive treatise on an important branch of mathematics.

—1915/16-64. 660 pp. 5⅜x8. 8284-0137-3. 2 v. in 1.

INTRODUCTION TO COMBINATORY ANALYSIS, *by P. A. MACMAHON. See* KLEIN

THEORY OF NUMBERS

By G. B. MATHEWS

CHAPTER HEADINGS: I. Elementary Theory of Congruences. II. Quadratic Congruences. III. Binary Quadratic Forms; Analytical Theory. IV. Binary Quadratic Forms; Geometrical Theory. V. Generic Characters of Binary Quadratics. VI. Composition of Forms. VII. Cyclotomy. VIII. Determination of Number of Improperly Primitive Classes for a Given Determinant. IX. Applications of the Theory of Quadratic Forms. X. The Distribution of Primes.

—2nd ed. 1892-62. xii + 323 pp. 5⅜x8. 8284-0156-X.

L'APPROXIMATION, *by VALLÉE POUSSIN.*
See BERNSTEIN

GRUPPEN VON LINEAREN TRANSFORMATIONEN

By B. L. VAN DER WAERDEN

—1935-48. 94 pp. $5\frac{1}{2}$x$8\frac{1}{2}$. 8284-0045-8.

SYMBOLIC LOGIC

By J. VENN

A classic.

—2nd ed. 1894-1971. xviii + 540 pp. $5\frac{3}{8}$x8. 8284-0251-5.

THE LOGIC OF CHANCE

By J. VENN

One of the classics of the theory of probability. Venn's book remains unsurpassed for clarity, readability, and sheer charm of exposition. No mathematics is required.

CONTENTS: PART ONE: Physical Foundations of the Science of Probability. CHAP. I. The Series of Probability. II. Formation of the Series, III. Origin, or Causation, of the Series. IV. How to Discover and Prove the Series. V. The Concept of Randomness. PART TWO: Logical Superstructure on the Above Physical Foundations. VI. Gradations of Belief. VII. The Rules of Inference in Probability. VIII. The Rule of Succession. IX. Induction. X. Causation and Design. XI. Material and Formal Logic . . . XIV. Fallacies. PART THREE: Applications. XV. Insurance and Gambling. XVI. Application to Testimony. XVII. Credibility of Extraordinary Stories. XVIII. The Nature and Use of an Average as a Means of Approximation to the Truth.

—1962. 4th ed. (Repr. of 3rd ed.) xxix + 508 pp. $5\frac{3}{8}$x8.
8284-0173-X. Cloth
8284-0169-1. Paper

DIE GESCHICHTE DER TRIGONOMETRIE

By A. VON BRAUNMUEHL

A scholarly history of Trigonometry.

—1900/03. 535 pp. $5\frac{3}{8}$x8. Two vols. in one. **In prep.**

STATISTICAL DECISION FUNCTIONS

By A. WALD

"A remarkable application to statistical theory of the methods and spirit of modern mathematics . . . makes effective use of the modern theory of measure and integration, and operates at a high level of rigor and abstraction . . . Its ultimate liberating effect on statistical theory will be great. It is to be hoped that so rich and stimulating a book as this will reach an audience among mathematicians."—*Bulletin of A.M.S.*

—1950-71. ix + 179 pp. $5\frac{3}{8}$x8. 8284-0243-4.